D0899958

Historical Geography of Crop Plants

A SELECT ROSTER

Jonathan D. Sauer

CRC Press
Boca Raton Ann Arbor London Tokyo

Library of Congress Cataloging-in-Publication Data

Sauer, Jonathan D.
Historical geography of crop plants: a select roster / Jonathan D. Sauer.
p. cm.
Includes bibliographical references and index.
ISBN 0-8493-8901-1
1. Crops—History. 2. Crops—Geographical distribution. I. Title.
SB71.S38 1993
630'.9—dc20 92-45590
CIP

This book represents information obtained from authentic and highly regarded sources. Reprinted material is quoted with permission, and sources are indicated. A wide variety of references are listed. Every reasonable effort has been made to give reliable data and information, but the author and the publisher cannot assume responsibility for the validity of all materials or for the consequences of their use.

Direct all inquiries to CRC Press, Inc., 2000 Corporate Blvd., N.W., Boca Raton, Florida 33431.

PRINTED IN THE UNITED STATES OF AMERICA
1 2 3 4 5 6 7 8 9 0
Printed on acid-free paper

In memory of Carl O. Sauer,
my first and longest tenured geography teacher

Acknowledgments

For helpful criticism of portions of the manuscript, I am indebted to Charles F. Bennett, Paul A. Fryxell, Walton C. Galinat, Paul Grun, Jack Harlan, Charles B. Heiser, Jr., Lawrence Kaplan, Steven D. Patterson, C. Arthur Schroeder, Norman W. Simmonds, and Thomas W. Whitaker. I am indebted to Frederick J. Simoons for sage advice on sources and presentation. For research assistance and for patiently converting a very messy manuscript into edited form, I am deeply indebted to Charmion Burns.

Contents

INTRODUCTION

I believe the time is ripe for a systematic survey of the historical geography of major crops. The late 20th century has seen drastic changes in the crops themselves and in scientific knowledge of them. Their prehistories have been much clarified by archaeological research with improved methods of recovery, dating, and identification of plant remains. Their taxonomy has been greatly clarified partly by modern cytogenetics and comparative biochemistry, including isozyme analysis. Origins have been clarified by field study of geography and ecology of wild relatives.

Good, up-to-date syntheses, which will be cited, are available for certain crops. However, much relevant information is scattered through technical literature and embedded in detail of interest only to specialists. Also, recent research reports commonly reach conclusions only after an exhaustive review of older literature and outdated hypotheses. I have tried to spare readers from controversies that have been resolved.

The more clearly we can see individual crop histories, the grander and more complex the overall story has become. Conceptual models of domestication and spread that appeared plausible in the 1950s now appear simplistic and, at best, partial truths. In the 1950s, plant domestication was commonly visualized as simply a discovery or invention, the result of human intelligence at a particular time and place. Agriculture spread from a few original centers simply because it gave greater yields than gathering wild foods. Archaeologists were still hoping to find "the first farming village" in their regions.

It now appears that crop domestication has often been a process, diffuse in space and time, rather than an event. Even within a single crop species, wide-ranging wild progenitors have been repeatedly domesticated at different times and in different regions. Plants have been cultivated for multiple and changing reasons. There have been cases of spread of crops as an association, but they are not the rule.

Case histories of crops will be presented individually, not as groups or complexes. Each will begin with the wild progenitor and follow changes in geography and heredity as a straightforward narrative. Histories are considered as interesting in their own right, not as supporting or refuting some general theory. Comparisons and generalizations, which will be limited, are postponed to the conclusions.

Roster

The primary criterion for inclusion is availability of a substantial body of relevant information, including recent research. This permits inclusion of almost all major crops. There are some exceptions, namely crops that have been static or declining in importance, e.g., indigo, flax, East Indian spices, tea. These are omitted because I can say little about them that is not old, common knowledge. On the other hand, a few crops are included that are of secondary importance because good, new information is available, e.g., sugi, North American wildrice, triticale, kiwifruit. Crops grown not only for food but for a wide variety of other products are included. However, two large categories of cultivated plants are excluded: (1) those planted as ornamentals or for scent and (2) those used for medicine or drugs. They would require books of their own based on very specialized information.

Organization

Crop histories are organized under a simple taxonomic hierarchy, as is evident from the Table of Contents. The conventional economic botany classification by commodity types is unsuitable for a historical geography. Among other reasons are problems with crops that have multiple uses, the main use commonly changing during the crop history. Moreover, brief attention to the family and genus to which a crop belongs helps in understanding crop traits inherited from wild progenitors. Even fairly closely related families, such as the Rosaceae and Fabaceae, have been following independent evolutionary lines over long geologic time spans, and their basic biological differences are evident in their crop derivatives.

An alphabetical sequence of families is used for ready reference. Although unattractively arbitrary, this avoids the dilemma of choosing an appropriate phylogenetic system—the most familiar systems, such as the Englerian, are obsolescent; the more modern and natural systems, such as the Conquist, are not yet generally familiar. The alphabetical sequence is not pushed below the family level. In families with multiple crops, genera and species are arranged according to source regions and chronology of domestication.

Species vs. Subspecies

As in wild species, the choice between lumping and splitting, i.e., the choice of levels in the taxonomic hierarchy, comes up constantly in crops. Nontaxonomists may take this too seriously, not recognizing that it is often a matter of taste and judgement. As in their wild ancestors, speciation in crops can be a gradual process, and borderline cases are judgement calls. I have tried to follow prevailing usage in each family, citing former or less widely used names in parentheses as synonyms. However, in some crops particularly the cereal grasses and grain legumes, there is no general consensus. In cases where the cultigen is fully domesticated and clearly distinct from the wild progenitor, some experts separate them on the species level and some as subspecies. I have treated them as a species for the following three reasons:

- In general usage, subspecies are geographically segregated but weakly differentiated populations. These crops are usually sympatric with but sharply differentiated from their wild progenitors.
- Secondly, the crops were usually named before the wild relatives were discovered. Thus the crop binomial has priority, and the wild population becomes a subset of its progeny. This is like calling the wolf a subspecies of the dog.
- Finally, and more importantly, multinomials are too clumsy to be useful. Even the people who coin the names soon tire of them and revert to an informal binomial or drop the Latin entirely. For example, those who classify certain wheats as *Triticum turgidum* subsp. *durum* and *T. turgidum* subsp. *dicoccoides* end up calling them *durum* wheat and wild emmer, respectively. If a cultigen is distinct enough to rate a formal scientific name at all, it seems better to use a binomial.

Angiosperms — Flowering Plants

The two subdivisions of the flowering plants, monocots and dicots, both arose early in the Cretaceous period perhaps from separate gymnosperm ancestors, but almost certainly not from the conifers. Although the angiosperms have existed for less than half as long as the gymnosperms, they have diversified into hundreds of thousands of species as compared to the hundreds of living gymnosperm species. One current hypothesis is that angiosperms originated as ecological pioneers in coastal, lakeshore, and streambank habitats, occupying the margins of extensive gymnosperm forests with fern understories. In such linear and disjunct habitats, flowers attractive to insects and other animals would have allowed more effective pollination than the wind pollination characteristic of gymnosperms. Aside from attractiveness to animals, a key innovation of the angiosperm flower was the enclosure of the ovules or young seeds in an ovary. Instead of being deposited directly on the ovule, the angiosperm pollen has to grow a long pollen tube to deliver the sperm to the enclosed ovules. The angiosperm flower evolved specialized stigmatic structures to receive the animal borne pollen. Later on, as some angiosperms became more abundant and dominant, e.g., grasses or oaks, they reverted to wind pollination and dispensed with attractive flowers.

The ovary may have initially served merely to protect the young ovules, but it eventually evolved in a variety of ways that aided seed dispersal, such as a buoyant husk for sea dispersal, a fleshy fruit for internal animal dispersal, plumose appendages for wind dispersal, burrs for external animal dispersal. These specializations aided in diversification and occupation of scattered niches. The result has been a greatly enriched flora available for human exploitation and domestication.

DICOTS

Actinidiaceae — *Actinidia* Family

Four genera and several hundred species of trees, shrubs, and woody vines, most numerous in the Asian and American tropics, belong to the family. Only *Actinidia* is of economic significance.

***ACTINIDIA* — KIWIFRUIT** (Beutel et al. 1976; Corbet 1987; Darrow 1975; Schroeder and Fletcher 1967; Simoons 1990)

Actinidia includes about 25 species of woody vines native to Southeast and East Asia from Java and the Himalayas to eastern Siberia and Sakhalin Island.

Several species have been widely planted in gardens for their attractive flowers, foliage, and edible fruits. The fruits are berries with juicy pulp and numerous tiny seeds. Presumably the seeds are dispersed by monkeys and other climbing mammals attracted by the scent. The fruits are brown and inconspicuous when ripe, showing no characteristics of species attractive to birds.

Fruits of the wild species were of trivial importance in traditional Asian diets. The one species that is a commercial crop was domesticated outside of Asia.

***Actinidia chinensis* (incl. *A. deliciosa*) — Kiwifruit, Chinese Gooseberry**

The name kiwifruit, coined as a marketing ploy, has replaced the former name of Chinese gooseberry, which is just as well. The species is not related to a gooseberry, and the fruit is quite different. It is more like a strawberry in flavor and texture and is used in all the ways that a strawberry is used. It is larger, however, about the size of a chicken egg, covered with brown fuzz, with a bright emerald or red flesh. It needs no sugar when ripe. The vitamin C content is over twice that of any citrus. Kiwifruit also contains a proteolytic enzyme and can be used as a meat tenderizer.

After reaching mature size, the fruits ripen surprisingly slowly. At ordinary temperatures, they ripen in about 2 months; the process can be accelerated by keeping them enclosed with bananas or apples as sources of ethylene gas. In cold storage, the fruits are regularly kept for 4 to 6 months and can remain good for a year. This keeping quality has been crucial in permitting the crop to be grown far from its markets.

The species is native to southeastern China and adapted to a subtropical climate. It is deciduous and can tolerate some winter frost, but is very cold sensitive after coming into leaf in the spring.

The first record of planting was about 1900 by foreign residents in the Yangtze Valley. Seeds from China were sent in the next decade to England, the U.S., and New Zealand. By 1910, some vines were fruiting in the North Island of New Zealand and California.

Experimental planting was carried on in research stations and private gardens, but it took time for any marketing to develop. New Zealand pioneered commercial planting. By the 1930s, New Zealand growers had developed named cultivars, selected from the highly heterogeneous seedlings. These were propagated as clones, either as cuttings or by budding and other grafting on seedling rootstocks. The vines are dioecious, and grafting was also used to add a few male branches for pollination onto female vines.

Commercial planting began in earnest in New Zealand after the end of World War II, mostly around the Bay of Plenty and other widely scattered areas on the North Island. Much of it was in combination with established citrus orchards. The crop is labor-intensive, requiring training on heavy trellises, shelter from wind, careful pruning, and artificial pollination. The New Zealand *Actinidia* vineyards are mostly less than 4 ha in size. Their combined production grew to thousands of tons a year and for decades monopolized the world export market, but no longer.

In California, commercial planting of kiwifruit developed only after the fruit had become a hallmark of expensive, modish cuisine during the late 1960s. Improved named cultivars had been introduced from New Zealand in the 1930s and were available from state experiment stations and commercial nurseries. A boom in planting developed during the 1970s, especially among avocado growers in the south and peach growers in the Central Valley. Here also, plantings were usually of a few hectares and labor-intensive. California growers had an extra expense of irrigation; kiwifruit needs lots of water during the long summer drought. However, the sale price during the 1980s was high enough that gross receipts from kiwifruit per hectare were commonly over twice that from any other crop. During the 1980s, annual California production rose to about 4000 tons compared to 6000 tons in New Zealand. California, like New Zealand, was able to ship in quantity to distant markets, including Japan and Europe. The New Zealand and California harvests were neatly timed to provide year-around supply.

This joint monopoly was too good to last. Southern hemisphere planting has spread to Australia and Chile. Currently commercial planting is booming in southern Europe, especially Italy. With government encouragement, surplus Italian vineyards are being replanted to kiwifruit, and Italy is expected to become the world's leading *Actinidia* exporter in the early 1990s.

Amaranthaceae — Amaranth Family

This is a cosmopolitan family of over 50 genera and nearly 1000 species. The most abundant species are colonizing herbs of shorelines and other open habitats and were evidently preadapted to colonize artificial habitats as weeds. A few genera have been domesticated as ornamentals, e.g., *Celosia* (cockscomb), *Iresine* (bloodleaf), and *Gomphrena* (globe amaranth). Only *Amaranthus* has become a crop.

AMARANTHUS— GRAIN AMARANTH (Early references: J. Sauer 1967b, 1976. Later references: Bye 1972; Coons 1982; Fritz 1984; Kauffman 1992; Kulakow et al. 1985; Nabhan 1980)

Amaranthus includes 75 or so wild and weedy species native to tropical and temperate regions of the whole world but is most diverse in the Americas. Natural habitats include desert washes, streambanks, lakeshores, brackish marshes, and ocean beaches. Natural dispersal in these patchy habitats is evidently largely by birds, many kinds of migratory birds regularly feeding on the seeds and excreting a fraction of them undigested. Since Columbus, various species have spread between the Old and New Worlds as weeds.

Amaranths have C_4 type photosynthesis, like various tropical grasses including maize and sugar cane, which means they are adapted to bright sunshine and can grow rapidly with limited loss of water by transpiration. They are quick-growing annuals, completing their life cycles and producing astronomical quantities of seed in a single growing season. They evidently expend little food on defense against herbivory and are highly palatable to many animals. In many regions, the wild species are gathered when young and cooked as potherbs.

Prehistoric domestications of quite different native amaranths took place independently in Asia and the Americas. In Asia, the cultigens were selected primarily as potherbs, quick growing and tender, without development of large inflorescences or heavy seed yield. In the Americas, although thinnings of the crop were used as potherbs, selection was mainly for increased grain production. The New World wild species included some fairly tall plants with relatively large, compound inflorescences, and the cultigens derived from them are much more robust, often taller than a person and with enormous inflorescences. Selection of the cultigens has been for quantity of seeds produced, not for larger seed as in true cereals. The cultigens, like wild amaranths, produce seed approximately 1 mm in diameter. Also, amaranths are peculiar among grain crops in that the cultigens retain the dehiscent fruits of the wild progenitors. The lid of the single-seeded ovary pops off. Some of the seeds of the crop escape harvest, but the bulk remains caught in the densely packed inflorescences until the plants are cut and dried. In Native American agriculture, the dried inflorescences are usually rubbed between the hands into fragments, and the chaff is removed from the amaranth grain by winnowing. In both Asian and American cultigens, mutations for bright red, blood red, and other striking pigmentation of the plants were selected.

Amaranthus tricolor and *A. lividus* — Potherb Amaranths

The Asian cultigens include a variety of races assigned to *A. tricolor* and *A. lividus* (both s.l.). They are widely grown garden vegetables in India, the East Indies, Southeast Asia, and the Far East and have been introduced in modern times elsewhere under such names as Chinese spinach, Malabar spinach, and tampala. Ornamental races include Joseph's coat.

Amaranthus cruentus, A. hypochondriacus, and *A. caudatus* — Grain Amaranths

The rest of this chapter deals with the New World grain amaranths, which are assigned to three species: *A. cruentus*, *A. hypochondriacus*, and *A. caudatus* (all s.l.). The three species are quite distinct in technical flower characters, mainly inherited from wild ancestors. Each species, however, includes much genetic diversity for plant pigmentation and seed color controlled by simple Mendelian recessives favored by human selection. All three species have both dark- and light-colored seeds. The recessive pale seed character, which is unknown in wild and weed amaranths, is associated with loss of dormancy, so the seed germinates promptly when sown. The dark grain is quite edible and was regularly harvested by hunting-gathering tribes of North America. However, the light-colored grain pops better, tastes better, and is generally selected for sowing. Most grain amaranth crops, however, include a few dark-seeded plants that are difficult to completely eliminate because of self-sown volunteers. If selection for pale seed is relaxed, as happens when the species are grown as ornamentals or dye-plants, the dark seed form gradually becomes predominant.

Careful attention to seed color and other mutations was undoubtedly crucial in helping prehistoric farmers to isolate the crop genetically from wild and weed amaranths. In the case of *A. caudatus*, xenia was also involved; as in maize, certain genetic characteristics of the seed are visible through the seed coat; common *A. caudatus* cultivars have gaudy pink cotyledons. Other mutations for plant pigmentation would have also allowed an observant cultivator to pull out unwanted hybrids before they flowered and contributed their genes to the next harvest.

Native Americans used amaranth grain in the same ways as maize: popped, parched, milled for tortillas and tamales, as gruel, and so on. It yields about as much per plant as maize and can be grown by similar techniques. There is an important difference in the nutritional value, however. Amaranth grain is exceptionally high in an essential amino acid, lysine, for which maize is a relatively poor source. Grain amaranth cultivation by native Americans was geographically much more restricted than maize cultivation. Grain amaranths were a crop of semi-arid regions, with their greatest importance in cool, tropical highlands. The historical record is best for the Aztec Empire, where the tribute levied on each province by the Emperor Moctezuma at the time of the Spanish Conquest was recorded. This tribute included at least 10,000 bushels a year from each of 17 provinces of a grain the Aztecs called *huauhtli*, probably a generic term for both amaranth and chenopod seeds. Most of this was probably harvested from cultivated pale-seeded *A. hypochondriacus*. The total *huauhtli* tribute was about 200,000 bushels, compared to 230,000 of beans and 280,000 of maize. This does not mean that amaranths were nearly equal to maize and beans in the daily diet, however, because much of the *huauhtli* tribute to Tenochtitlan was taken by the priesthood for ceremonial purposes. A paste of *huauhtli* grain was used for innumerable rituals, small and large. For example, *huauhtli* dough was used to make a huge idol of the Aztec war god Huitzlipochtli. It was paraded through the city, accompanied by ceremonial dances, singing, and band music. It was returned to the great temple pyramid, and after more human sacrifices than on any other day, the idol was broken up and distributed to the populace to be eaten with great reverence as the bones and flesh of the god. To Spanish Catholics, this was obviously a Satanic parody of Holy Communion. After the Conquest, the Spanish clergy attempted to suppress use of amaranths in heathen ritual, but stories of forcible Spanish destruction of the crop are a fabrication. Use of amaranth dough for ceremonial figures, cakes, and bread was not peculiar to the Aztec Empire, but extended over much of Mexico and far beyond. The bright coloration, often blood red, of the cultivars probably was involved.

The aboriginal distribution of the three grain species can be roughly sketched from archaeologic remains and historical ethnographic records. *A. cruentus* (incl. *A. paniculatus*) may have been the original domesticate. Pale-seeded forms of it from caves at Tehuacan, Puebla, date from about 4000 B.C. As in maize and other crops that enter history in these caves, it is unlikely that *A. cruentus* was a local domesticate. It may have had long prior evolution and migration as a crop. Its wild progenitor is almost certainly *A. hybridus*. *A. hybridus* (incl. *A. quitensis*) grows today and probably grew then over an enormous range in the deciduous forest and prairie regions of eastern North America; tropical highlands of Mexico, Central, and South America; and plains of temperate South America. Ethnographic records documented by plant collections show survival of *A. cruentus* as a pale-seeded grain crop in widely scattered regions of Mexico and Central America: in Puebla, in Chihuahua among the Pima Bajo, among the Mixtec of Oaxaca, and among the highland Maya of Guatemala. An extremely dark red variety was grown by Pueblo peoples of Arizona and New Mexico to color ceremonial maize bread. Since at least the mid-19th century, dark-seeded, deep red *A. cruentus* has been a common ornamental in Central and South America among both Indian and non-Indian people.

A. hypochondriacus (incl. *A. leucocarpus*) joined *A. cruentus* in the archaeologic record at Tehuacan about 500 A.D., again as a pale-seeded cultivar probably domesticated long before and some distance away. *A.*

hypochondriacus is closely related to both *A. cruentus* and a wild species *A. powellii*, and it seems likely that it is a hybrid between the two. *A. powellii* probably contributed genes for broader ecologic tolerances, especially for less moisture. *A. powellii* is native to canyons and desert washes of the western Sierra Madre. It invades irrigated fields as a weed. *A. hypochondriacus* remains the common grain amaranth of the highlands of central Mexico today, often grown in substantial plantings. Pale-seeded *A. hypochondriacus* has continued to be grown into the 20th century in the northwestern Sierra Madre by some of the least acculturated Indian peoples of Mexico, including the Tarahumar, Mayo, Warihio, Huichol, and Yaqui. The crop had spread prehistorically from northwestern Mexico into what is now the southwestern U.S. Well preserved bundles of *A. hypochondriacus* have been found in cliff dwellings of that region. Caches of *A. hypochondriacus* with pale grain have also been found in Ozark rock shelters and dated at about 1100 A.D., but there is no historical record of the crop among Indian groups in the eastern U.S. In his explorations of the Colorado River region of Arizona and Utah in 1872 to 1873, John Wesley Powell obtained amaranth seed by trade with the Southern Paiute. The harvest included both the pale-seeded cultigen *A. hypochondriacus* and dark seeds from wild species growing in the fields as volunteers. Similar mixed harvests of the grain were noted by Isabel Kelly in the 1930s among several bands of Southern Paiute in their irrigated gardens. There is no subsequent record of grain amaranths among Indian peoples north of Mexico.

The origin of *A. caudatus* is problematical. The simplest hypothesis would be that it arose in antiquity by introduction to South America of cultivated *A. cruentus*, followed by hybridization with a local wild species. One problem is that there is a wide region in southern Central America and northern South America where *A. cruentus* is grown only as a dark-seeded ornamental and ceremonial dye plant. Possibly the grain crop passed through this region, but was not maintained there. In any case, pale-seeded *A. caudatus* was an aboriginal grain crop in the temperate Andean valleys of Peru, Bolivia, and northwestern Argentina.

After the Spanish Conquest of their homelands, the three cultigen species were introduced to the Old World, probably originally as curiosities and ornamentals rather than grain crops. No Europeans, other than Spaniards, would have been likely to encounter them during the early colonial period. The first available European records, however, are not from Spain but in the Renaissance herbals of central and northern Europe. Unfortunately, the three species cannot usually be sorted out in the early European sources. By the 18th century, when herbarium specimens were being preserved that allowed accurate identification, all three were widely distributed.

During the 19th century, *A. cruentus* became a cosmopolitan ornamental of tropical and warm temperate regions. The common form is remarkably homogeneous with dark red plant color and dark seeds. This variety outran European exploration of Africa and was already present in gardens in some of the remotest interior regions when explorers such as Livingstone, Speke and

Grant, Burton, and Schweinfurth arrived. At about the same time, the species was spread through gardens in remote parts of Asia and the East Indies. Occasionally, the species acquired ritual and dye uses. Commonly it is grown as a potherb. A remarkably intensive pattern of cultivation as a potherb has been developed by market-gardeners in West Africa, where it is one of the most important green vegetables. There is no record that the pale-seeded form of *A. cruentus* was ever introduced to the Old World, nor is there any record of the species as a grain crop there.

A. hypochondriacus, by contrast, was introduced to the Old World in both dark and pale seed forms and in a wide variety of plant color forms. In Europe, the species was not used for grain, and dark-seeded forms gradually replaced the pale-seeded forms; the latter are present in European grown specimens from the 16th until the mid-19th century. By then pale seed forms had been taken on to Asia to become a grain crop. The earliest records of the crop there are from the 18th century in Ceylon and South India, where it became a staple of the hill tribes. During the 19th century, it became widespread in India, most importantly in the Himalaya, and across the interior of China to eastern Siberia. The grain is used much as in Mexico, parched, popped, milled for tortilla-like chapatis. It is also used in peculiar Asian recipes. In India especially, it has acquired ceremonial importance for certain festivals. An outpost of the crop was established in East Africa during World War II to supply the local Indian communities when supplies of the grain from India were interrupted. Several hundred hectares a year were grown in Kenya in the 1980s.

A. caudatus, like *A. cruentus*, reached the Old World as a limited sampling of its New World diversity. The species is clearly recognizable in Flemish, Swiss, and English botanical literature of the early 17th century, attributed both to Peru and Asia. The common ornamental form, which is now cosmopolitan, has long, soft, unbranched, pendant inflorescences with bright pink flowers and pale seeds. Similar forms occur in the Andes but together with a great variety of other forms. The species is grown for grain at a few places in India and China, usually together with *A. hypochondriacus*. The species has been present in Ethiopia since the early 19th century, planted both as an ornamental and a minor grain crop.

By the mid-20th century, grain amaranths had declined to a minor relic in most of their ancient homelands and only in India was their cultivation actively expanding. Since the 1970s, widespread interest in expanding the crop has developed, thanks largely to publications by the Rodale Press of Emmaus, Pennsylvania, seconded by the Board on Science and Technology for International Development of the National Academy of Sciences. The excellent protein content of the grain, the hardiness of the plants, and their adaptability to cultivation with hand tools seemed to give the crop great promise, particularly as a noncommercial, home-grown cereal for overpopulated and undernourished regions. Experimental programs are currently evaluating the crop in many countries. Ironically, the most notable accomplishment so far has been in the U.S., particularly in the wheat belt in the High Plains, where the crop is grown with heavy machinery for sale through health food stores to an overfed population. The commercial crop in the U.S. includes both *A. cruentus* and *A. hypochondriacus* plus hybrids. From a few hundred hectares in 1983, it expanded to about 1000 ha in 1988. With true serendipity, high latitude farmers growing this once tropical crop have turned cold, autumn weather to their advantage. Frost-killed stands dry out enough to be harvested with an ordinary wheat combine, greatly facilitating production of clean grain. A cultivar distributed by the University of Nebraska under the name of Plainsman is currently being planted by High Plains farmers. Plainsman is derived from a hybrid made in 1977 at the Rodale Research Center in Pennsylvania. A common Mexican landrace of *A. hypochondriacus* was crossed with its presumed primary wild progenitor *A. hybridus*. Subsequent selection led to a cultivar similar to *A. hypochondricus* except shorter in stature and better suited for combine harvesting.

Anacardiaceae — Sumac Family

The Anacardiaceae include about 60 genera and 400 species of trees, shrubs, and lianas, with members throughout the tropics and in many temperate regions. Some genera are exploited as sources of tannins and lacquers. Some are widely planted as ornamentals, e.g., *Schinus* (pepper trees), others for edible seeds, e.g., *Pistacia* (pistachio), and for their fruits, e.g., *Spondias* (mombin or Otaheite apple). The most important crops are the cashew and mango.

ANACARDIUM — CASHEW (Burkill 1935; Johnson 1972; Patiño 1963; Purseglove 1968)

The Anacardium is a small tropical South American genus of about 10 species of which only one is an incipient domesticate.

Anacardium occidentale — Cashew

The cashew is a low, wide-spreading evergreen tree native to Brazil. Johnson (1972) found that the species grows wild in coastal sands of Northeast Brazil, a region of long dry seasons. The cashew trees grow on beach ridges and active dunes as members of the *restinga* vegetation, an open woodland. The cashews are exposed to the onshore Trade Winds and salt spray. Their deep tap root and extensive lateral roots are well adapted to the habitat. The leathery leaves are characteristic of coastal thicket plants.

In the published literature, *A. occidentale* is reported to be also native to vast areas of interior Brazil, including all the states and territories of Amazonia and the plateau region of southern Brazil. Tropical coastal thicket trees in general commonly are shared with open inland habitats; they do not need salt spray, but benefit from its exclusion of competitors. The cashew may have colonized scattered, naturally open, inland sites before human intervention in the ecology of Brazil. At present, however, most of the inland occurrences of the species are in artificially disturbed habitats. The species is weedy, and its seedlings colonize such habitats, whether dispersal is natural or initiated by human planting.

The cashew has an extraordinary adaptation for seed dispersal. The true fruit contains a single seed — the cashew nut, which is protected against predators by a tough shell loaded with an acrid, poisonous oil. The pedicel on which the fruit is borne becomes greatly enlarged into an attractive, juicy false fruit — the cashew apple, which actually looks more like a pear. At maturity, the apple is red or yellow, which would suggest attractiveness to birds, but most observations of animal dispersal have been of fruit bats. The apple, with nut attached, can also float, but probably not far. A successful seedling usually begins producing seed of its own within 3 years.

Brazilian Indian use of cashew nuts and apples is well documented in French, Portuguese, and Dutch accounts between 1550 and 1650. The Tupi name *acajú* became *caju* in Portuguese and *cashew* in English. The juice of the cashew apple has been and still is fermented to make wine. In 1558, Thevet published a drawing of Indians harvesting what were unmistakeably cashew fruits and squeezing the juice from the apples. The cashew was probably spread by the Indians as a dooryard garden tree, but there is no record of systematic planting. It may have been spread by prehistoric Indians into the Guianas and eastern Venezuela. It was probably a dooryard garden plant of the Caribs in the Lesser Antilles. It was not among the plants of the Arawaks of the Greater Antilles, nor was it in Colombia, Central America, or Mexico at the time of the Spanish Conquest. First Spanish accounts were from Venezuela in the mid-16th century.

The Indians roasted cashew nuts in open fires, burning off the caustic shell oil. The Portuguese were quick to adopt the simple Indian techniques of roasting the nuts and making wine from the cashew apples. They occasionally sent some nuts to Lisbon as early as the mid-17th century.

By 1750, cashews were widely planted throughout tropical America, not just for the nuts but as a multiple-purpose garden tree. It made a fine dooryard shade tree, provided the lower branches were pruned. It was evergreen and pest-free. The sap of the trunk could be tapped for an insect-repellent, protective varnish. The cashew apple yielded tasty, fresh juice and could be made into preserves. The wine could be distilled for brandy. Excess volunteer trees were cut for firewood and charcoal. Commercial cashew plantations in tropical America were not begun until the 20th century.

Meanwhile, the species had become pantropical. The Portuguese introduced it to India in the 1560s, perhaps more as a source of wine and brandy than for the nuts. Cashew trees were reported in gardens of Cochin on the Malabar Coast and Goa in the 1570s and 1580s. Four hundred years later, India remained the world's main producer of cashew liquor at a rate of about 250,000 gallons a year. From India, the species spread to other parts of tropical Asia, still bearing the name of *caju* or variants in different languages. It seems likely that the Portuguese also introduced the cashew early to both East and West Africa, but the earliest report of the species in Africa is from the late 18th century. By then, an independent French introduction had taken place from Cayenne to Ile de France, now Mauritius, in the Indian Ocean.

During the 19th century, cashew trees became widespread in the Indian Ocean region, initially by planting in gardens but soon escaping as weeds into artificially disturbed sites and as fully naturalized members of seashore woodlands. Local dispersal was sometimes by fruit bats, monkeys, elephants, and perhaps by floating. By the mid-20th century, the Indian Ocean region had over a million hectares under cashews, by very rough estimates. These were generally casually planted and self-sown trees mixed with other species, not as plantations. The largest stands were in Mozambique, Tanganyika, Madagascar, and India.

The first record of international trade in cashews was in 1907 when India exported 430 tons of kernels to Britain. That same year, India imported about 1/2 ton of unshelled nuts from East Africa. Increasing quantities of African cashews were soon being processed in India for reexport. World demand for cashews began growing rapidly in the 1930s. Until 1960, India managed to keep up with this demand without much change in technology. Unmechanized factories used hand labor to roast and shell the nuts. The work was dangerous because of the poisonous oil, which was sold as a valuable by-product for insect repellent and protective varnishes, for brake linings and lubricants, and other uses.

Drastic change began in the industry about 1960. Regular plantations of cashews in pure stands expanded rapidly in India, Madagascar, East Africa, and especially Mozambique. Tentative planting began in several West African countries. Indian research stations began experimenting with vegetative propagation and grafting of superior cashew trees. Fresh germ plasm was introduced by seed imported from East Africa and Brazil. Artificial hybrids were made by Indian breeders to combine desirable traits found in different parents, e.g., large, uniform nuts and high yield.

Research projects in Africa sought to insert cashew trees into the shifting cultivation cycle. It was hoped that cashews planted at the close of the cropping phase would survive and yield during the 30- or 40-year fallow period while the forest regrowth took place.

By 1970, Madagascar and Africa had established cashew processing factories to break their long dependence on India. Africa had nine factories, six of which were in Mozambique. The most successful of these were using Italian-engineered machinery and had a capacity of over 15,000 tons of nuts a year.

Meanwhile in Brazil, processing of cashew apple, and nuts by traditional methods for local use continued with little change until the mid-20th century.

The first factory to make products in quantity for long-distance shipment opened in Ceará in the 1930s. For decades, it remained unique and processed nuts trucked in from several northeastern states. Dramatic changes began about 1960 under a program initiated by the state government of Ceará to develop large-scale cashew plantations integrated with processing factories. In the 1960s, cashew plantations were started along a 2000-km stretch of the Northeast Brazil coastal plain, usually cleared from forest on sandy land not preempted by coconuts or sugar cane. By the time of Johnson's field survey in 1971, Brazilian cashew production had increased tenfold over the flat pre-1960s level to about 50,000 tons a year. Eighteen factories were in operation, mostly in Ceará. Brazilian processing methods were original developments involving locally made machinery and much hand labor. Much of the production, both of kernels and by-products, was for domestic sale, but by 1970 about 6500 tons of kernels, mainly still raw, were being exported. Brazilian experiment stations were initiating research on cashew planting and genetic selection.

A. occidentale is now a significant plantation crop in international trade, but is still essentially wild genetically and only an incipient domesticate.

***MANGIFERA* — MANGO** (Burkill 1935; Patiño 1969; Popenoe 1950; Purseglove 1968; Singh 1976)

Mangifera includes about 40 species of large evergreen trees native to forests of Southeast Asia and the East Indies. The fruits have a fleshy, outer tissue attractive to birds and bats with a woody, fibrous inner pit protecting the seed. Before the seed is ripe, the fleshy tissue contains an irritant latex. This is absent in ripe fruits, but most wild mango fruits are too resinous and fibrous to be very attractive to humankind. A few species are casually planted in their native regions, but only one is a significant fruit crop.

***Mangifera indica* — Mango**

M. indica was evidently domesticated in the northeastern India-Burma region, where it still grows wild in the hills of Assam and adjacent areas. The wild trees, like their cultivated derivates, can grow to over 40 m in height and live for over a century. They are adapted to the tropical monsoon climate with a long season of heavy rain and a short very dry season. Flowering and fruiting are inhibited by rain. Although the species can be grown in equatorial rain forest climates, fruit yields there are poor.

The mango was probably being planted in India by 2000 B.C. and is prominently recorded in ancient Sanskrit writings. Human selection has been concentrated on the fruit and has produced cultivars with much larger, better flavored fruits with less resin and fiber and smaller pits. This was a rather remarkable accomplishment because of the long generation time and uncontrollable cross-pollination. The trees are always highly heterozygous and seedlings consequently highly variable. In a few ancient cultivars of southern India, this difficulty was overcome by selection of lines with apomictic seeds genetically identical to the mother plant, thus allowing propagation of superior clones by seedlings. Eventually, but probably not in antiquity, methods of approach grafting or inarching were developed, allowing vegetative propagation of selected clones. Approach grafting is slow and difficult because the graft has to be completed before the scion is severed from its parent tree. The fruit of a great majority of mango trees is still of low quality. Commonly trees come up as volunteers from discarded pits around settlements along roadsides or field margins.

Spread of *M. indica* from northeastern India, whether as useful volunteers or deliberate plantings, had to be gradual. A ripe mango could not be carried far before spoiling, and the seed in the dry pit remains viable for only about a month. Also, mangoes have never been essential in the diet, although they do provide some sugar and vitamins and have been popularly considered to have medicinal value. As improved cultigens were developed, the mango eventually became widespread in India and Ceylon. It became highly esteemed not only as a fresh fruit, but also dried, pickled in chutney, preserved in sugar, salt or wine, and used in many complex recipes. The mango came to figure symbolically in Hindu and Buddhist mythology and ceremonies. By the time of Alexander the Great, mangoes were important over much of India. Eighteen centuries later the Moghul Emperor Akbar ordered planting of thousands of mango trees in huge orchards.

Dispersal of *M. indica* eastward to tropical Asia through Indo-China and down the Malay Peninsula had probably begun by about 400 B.C. Later on, Buddhist monks may have aided the spread. However, the spread was very slow and erratic, hindered by low quality of most seedling trees and by low yields in wetter climates. It has been suggested that the mango was not introduced to the Philippines until Islamic expansion in the 15th century.

By the time of arrival of Europeans in Southeast Asia, the species was still far short of filling its climatically suitable range. It was not yet recorded in Chinese sources except as a foreign fruit known from travelers' accounts. Mangoes were not widespread in the East Indies until after the European incursion. The first records in several regions, e.g., at Macao, were associated with Portuguese colonies. Perhaps successful introduction came with arrival of superior cultivars brought directly from Goa and other Indian ports on Portuguese ships. During the 16th century, techniques of budding, quicker and easier than approach grafting, were introduced from Europe, greatly aiding the expansion of the best clones as scions grafted on seedling rootstocks. However, propagation of mangoes by seed continued in general folk use.

During the 17th century, unusual mango cultivars were introduced to Manila, the northern Philippines in general, and to Guam where no mangoes were growing when the Spaniards arrived. These regions now have good quality mangoes propagated by apomictic seeds like those of southern India. Perhaps the Spaniards obtained them via the Portuguese colony at Macao.

It has been suggested that the westward expansion of the mango from India began in medieval times with Persian and Arab trade to East Africa. If the mango was grown in East Africa at the time of the Portuguese arrival, it was unimportant. During the 17th century, the Portuguese planted mangoes in coastal areas of both East and West Africa; but acceptance by the Africans was slow, and spread into the interior was erratic. Mango trees were present in a few interior market towns in West Africa, e.g., Fouta Djallon, when European explorers arrived in the late 19th century, but most of the spread came later.

The earliest known successful introduction of *M. indica* to the New World was to Bahia in Brazil about 1700 with plantings elsewhere along the Brazilian coast soon after. Perhaps seed was brought from West Africa; the brief viability of mango seed would have been enough to withstand fast Trade Wind passage. If the introduction was from the Indian Ocean, young growing trees must have been transported. Such transport is known to have been attempted before 1700 by British East India ships returning home; in the 1690s young mango trees were deposited by them in Barbados and also taken to England for hothouse culture. Nothing came of this particular effort, but in 1742, the mango was successfully introduced to Barbados from Rio de Janeiro. Soon other direct introductions to the West Indies were made from the Indian Ocean.

In the latter half of the 18th century, British and French colonial botanical gardens played a leading role in mango introduction. Mango trees, along with East Indian spice trees, were planted in the St. Vincent botanical garden when it was started in 1766. The famous French botanical gardens on the Indian Ocean islands of Mauritius (then Ile de France) and Reunion (then Ile Bourbon) relayed Indian mango cultivars to Martinique and other West Indian colonies from 1760. In 1782, a French ship bound for St. Domingue with mango seedlings from the Indian Ocean was captured by the British, and the trees diverted to their botanical gardens in St. Vincent and Jamaica. Captain Bligh, on his second voyage in 1793, delivered mango trees from Timor to the St. Vincent and Jamaica gardens, along with the famous breadfruit. Mango trees were soon growing on sugar plantations and around villages on many other West Indian islands.

Meanwhile, a quite separate introduction had taken place across the Pacific from the Philippines to Mexico. The date is uncertain, but it was not very early. Mangoes were not grown around Manila until over a century after the trade with Mexico began, nor were they grown in Mexico until late in the 18th century when Mexico acquired, under the name of mango of Manila, the unusual apomictic variety grown in the Philippines and Guam. This variety is still common in Mexico and yields uniform, fairly good quality fruit from seedling trees. It was taken on to Cuba at an uncertain date, probably around 1800. The great expansion of mango growing over tropical Central and South America took place between 1800 and 1850, mainly by casually planted and volunteer seedling trees often yielding poor quality fruit.

Geographic spread of *Mangifera* was essentially completed in the last half of the 19th century with its introduction to such far-flung places as Florida, Hawaii, Fiji, Queensland, and Natal.

In the late 19th and the 20th century, there have been innumerable exchanges of mango cultivars in a worldwide network of botanical gardens, experiment stations, and commercial growers. There have been sporadic efforts at scientific mango breeding by hybridization and selection, but they have not been notably successful. The best commercial varieties all originated as chance seedlings and have been propagated clonally. For example, the Mulgoba variety was introduced from Bombay to Florida, where one of its seedlings was named the Haden and propagated clonally. The Haden has become one of the main grafted commercial varieties of Central America.

Asteraceae (Compositae) — Daisy Family

Although the family arose late in geological time, it has more members, about 20,000 species in 1000 genera, than any other dicot family. Composites are abundant worldwide, especially in open habitats such as seashores, riverbanks, and in grasslands and shrublands subject to frequent wildfire. Many members were preadapted to artificially disturbed habitats and have become common weeds.

Composites are strongly represented among garden ornamentals, including *Chrysanthemum*, *Dahlia*, *Zinnia*, *Coreopsis*, and a hundred other genera. Although the individual flowers are small, the characteristic clustering of many flowers into compact heads produces an inflorescence that is attractive to insects and people. Considering its diversity and ecology, the family is poorly represented among crop plants. *Chrysanthemum* is cultivated as a source of the insecticide pyrethrum, *Parthenium* for guayule rubber, and *Carthamus* for safflower oil. There are a few minor garden vegetables: *Cichorium* (endive and chicory), *Tragopogon* (salsify), *Cynara* (artichoke), and *Polymnia* (yacon). Only two genera have contributed major crops.

LACTUCA **— LETTUCE** (Hedrick 1919; Hutchison 1946; Ryder 1986; Ryder and Whitaker 1976, 1980; Whitaker 1969)

Lactuca includes a hundred or so species of annual and perennial herbs widely distributed in the temperate regions of the northern hemisphere. The wild species do not form heads, but generally have an erect, branched stem with bitter, often prickly leaves and a milky latex. The achenes are wind-borne like those of the closely related dandelions, an adaptation for colonizing scattered open habitats, such as coastal dunes, stream alluvium, and rocky slopes. *Lactuca* seeds have a dormancy mechanism involving phytochrome pigment. After burial in moist soil and incubation in the dark, germination is triggered by even momentary exposure to dim light, as happens when there is some kind of soil disturbance. Germination would thus tend to be postponed until an open niche became available. Like other pioneers, *Lactuca* species were preadapted to become weeds where people disturbed competing vegetation by digging ditches or cultivation. Only one species has been domesticated.

Lactuca sativa (incl. *L. scariola*) — Lettuce

L. sativa (incl. *L. scariola*) includes all the lettuce cultivars: cos or romaine, leaf lettuce, butterhead, iceberg or crisphead, and others. The apparent wild progenitor is *L. serriola* (not the same as *L. scariola*) a native of the Mediterranean region, where it grows on sand dunes and rocky slopes. It has become far more abundant and widespread as a weed of field edges, roadsides, walls, and ditches.

The first evidence that the cultigen *L. sativa* had been developed under domestication comes from paintings in Egyptian tombs, beginning about 4500 B.C. and continuing through the Old and Middle Kingdoms. Egyptologists interpret these as representing the primitive romaine or cos cultigen. It is possible that cultivation was begun for the sake of the seeds, which contain an excellent edible oil. At any rate, the cultigen has larger achenes than the wild progenitor. It also has a recessive mutation causing the flower to retain the ripe achenes until harvest. This change would have been favored, of course, regardless of whether the crop was planted for seed or foliage.

Clear historical records of cultivated lettuce begin in Greece about 450 B.C. and continue through later Greek sources, including Aristotle. In Rome, various cultivars were recorded by the 1st century A.D., and the crop was evidently already common, as it has remained in the Mediterranean region in general.

The spread of *L. sativa* eastward is obscure. Herodotus mentions it as fare for Persian royalty about 660 B.C., and the species has long been grown in northern India under a Persian name. Different authorities place its introduction to China in the 5th and 7th centuries A.D. A distinctive cultivar, called stem lettuce, was selected in China.

The Romans presumably introduced cultivation to northern Europe, where it was grown in summer rather than as a winter annual. It survived the Middle Ages in obscurity. Chaucer mentioned it in England, and it is well documented in the 16th and 17th century herbals in various European countries, the varieties including cos, leaf, and head lettuce. During the 17th and 18th centuries, selection of mutants in European gardens resulted in great proliferation of the named varieties, especially in France, Holland, and England.

Meanwhile, *L. sativa* had been widely introduced to the New World tropics. The Spaniards were growing it in Hispaniola by 1494 and soon after elsewhere in the West Indies and on the mainland. Portuguese, French, and other Europeans introduced it to their tropical American colonies early in the colonial period. The crop was easily grown and usually produced seed in both lowland and highland tropical climates. However, lettuce does not seem to have been taken seriously by most American Indian groups. Some of them in the Andes grew it mainly as fodder for their Guinea pigs.

In North America, lettuce was a very minor crop until the 19th century when it became widespread in home gardens and in truck farming for local markets.

An assortment of quite diverse types of leaf lettuce and other cultivars were grown, a pattern that continues today.

A quite different pattern developed in the early 20th century in certain regions of California. John Steinbeck narrated the beginnings of the story in "East of Eden". California entrepreneurs discovered that the tight-headed, chlorophyll deficient iceberg and similar cultivars could be shipped to distant markets under ice in carload lots. Such shipments began in a small way about 1915 and soon developed into an epitome of agribusiness. The Salinas Valley became North America's main source of mass market lettuce during most of the year with the Imperial Valley and other southern desert valleys joining in during the winter.

Since 1950, a series of new crisphead cultivars have succeeded the old iceberg cultivar as dominants in the mass market. These were developed by government experiment stations primarily to cope with problems of diseases and pests inherent in a large monoculture. Genes for resistance have been successfully transferred from the wild species *L. serriola*, *L. virosa*, and *L. saligna* to new cultivars, e.g., Valverde, Hope, Calmar, and Vanguard. Selective breeding has also improved durability of the heads under mechanical harvesting and shipping. The more delicate cultivars remain a legacy of folk selection.

***HELIANTHUS* — SUNFLOWER** (Beard 1981; Heiser 1965, 1978; Riesenberg and Seiler 1990; Weiss 1983)

Helianthus includes about 50 species in temperate North America and 20 species in the Andean region. It is mainly tall, coarse, annual or perennial herbs. The name sunflower is appropriate for three reasons: (1) the species are sunloving; (2) the developing flower heads track the sun during daylight; (3) the heads with their circles of yellow ray flowers are a solar image.

In prehuman times, the species were almost certainly rather rare colonists of desert washes, sand dunes, buffalo wallows, marshy forest openings, and other disjunct habitats. Unlike many composites, the genus lacks the dandelion-like pappus that functions in wind dispersal. *Helianthus* seeds, actually one-seeded fruits, are relatively large and heavy. They lack endosperm, but the embryo contains enough food, especially oil, to be fed upon and scattered by many kinds of birds and some rodents and large herbivores. Various wild sunflower species were gathered aboriginally by North American Indians. Indians in eastern North America also dug the tubers of *H. tuberosus*, the Jerusalem artichoke. Only one species was regularly planted and developed into a cultigen, namely *H. annuus*.

***Helianthus annuus* — Sunflower**

H. annuus in its native North America is a highly variable species, including wild, weedy, and cultivated forms. Truly wild forms (subsp. *jaegeri*) are known only in moist microhabitats in deserts of the southwestern U.S., including the Four Corners region. Weedy forms, dependent on human ecological disturbance (subsp. *lenticularis*) and domesticates (var. *macrocarpus*), became much more widespread during aboriginal times.

Heiser (1965) proposed an hypothesis that is still tenable. The species spread prehistorically over eastern North America as a weedy camp-follower. The process of domestication was gradual and geographically diffuse. Eventually Indian selection produced robust plants up to 2-m tall, unbranched, and with few heads or a single huge head.

Archaeological remains record Indian exploitation of H. annuus over a wide area of North America, usually in the form of achenes charred by roasting. The earliest remains of domesticated sunflowers are from Arizona about 1000 B.C. (Fritz 1990). By the beginning of the Christian era, primitive sunflower cultivars were present in Kentucky and Tennessee.

By the time European explorers arrived, *H. annuus* was a minor Indian garden crop in much of North America from the Great Lakes to Virginia and from the High Plains to northern Mexico. In addition to toasting and grinding the seeds, some Indian groups expressed the oil from them.

In Europe, *H. annuus* began appearing in the herbals in 1568. It was rapidly spread as a curiosity and ornamental. Its possible value as an oil crop was suggested in Britain in the early 18th century. Actual exploitation as an oilseed began in Russia late in the 18th century. In the late 19th century, sunflower breeding in Russia increased the oil content of the seeds from less than 30 to 50%. Head size and seed size were also increased. Russian breeders obtained genes for disease and pest resistance by crossing *H. annuus* cultivars with wild *Helianthus* species.

In the early 20th century, *H. annuus* became by far the most important oilseed crop in Europe being planted on millions of hectares each year. The Soviet Union remained the leading grower, but by 1930, extensive planting had spread to other eastern European countries, the Balkans, and Turkey. Sunflower seed oil became appreciated for its flavor and high content of polyunsaturated fatty acids. It is of top quality for salad oil, margarine, and cooking; it gives off little smoke in frying.

By 1950, sunflower oil was in demand, and the crop was planted extensively in China and in temperate southern Africa and South America, especially Argentina, Uruguay, and Chile. By 1970, Australia began planting sunflowers on a large scale. Since 1970, *H. annuus* has become a major crop in Spain.

It is remarkable that acceptance as a major crop was slowest in its homeland. In North America, in addition to the aboriginal cultivars, the Mammoth Russian had been introduced by immigrants in the late 19th century. The plants were admired in home gardens and were grown as a rather trivial source of birdseed and for roasting as snacks.

Significant planting began in the 1970s in the Red River Valley region of the Dakotas and Minnesota, when farmers who had been specializing in flax growing needed a new crop after water-based latex paints destroyed the market for linseed oil. The area planted to *H. annuus* in the U.S. rose from about 1000 ha in 1970 to 1,500,000 ha in 1980. Initially, the seed was exported to Europe, but domestic consumption soon skyrocketed. The crop has expanded to other regions of the U.S. and Canada, largely in high latitude regions with climates that severely restrict the choice of crops. Sunflower seeds are now produced by the millions of tons annually in the U.S., ranking second only to soybeans as a vegetable oil crop. The residue after oil extraction provides a very important high-protein feed for livestock. Sunflower oil has potential as a renewable fuel source; it can make up 25 to 50% of diesel fuel and would be economical with only a moderate increase in the average price of petroleum.

Modern breeders have reversed the millenia of selection for tall plants with huge heads and have developed dwarf cultivars with medium-sized heads better suited for mechanized harvesting. Most of the U.S. crop is now grown from F_1 hybrid seed, giving enhanced vigor and yield. Commercial seed companies produce the hybrid seed by interplanting selected inbred lines, just as with maize. The inbred line from which the commercial seed is obtained is prevented from self-pollination by a gene for male sterility. This gene was incorporated in an *H. annuus* cultivar in the 1970s. It was obtained from a common weedy species *H. petiolaris*.

Brassicaceae (Cruciferae) — Cabbage Family

The crucifers include several hundred genera and several thousand species of mostly annual or perennial herbs. Crucifers are more diverse and abundant in temperate and cold climates than in the tropics. Most are ecological pioneers of such places as desert sands, cliffs, and beaches. The small seeds are rich in fat and are commonly eaten and presumably widely dispersed by birds. The seeds and other parts of the plants characteristically contain glucosides that limit herbivory. When the plant tissues are crushed, the glucosides are enzymatically converted to various toxic, bitter, and irritant compounds, including a pungent volatile oil that makes mustard seed the world's favorite nontropical spice.

As ecological pioneers, many crucifers were preadapted to become weeds of roadsides, grain fields, and other human-modified habitats. People inevitably found uses for these conveniently available weeds. Many species are gathered for cooking as potherbs; the overly pungent ones are edible after repeated boiling in changes of water. As potherbs, the crucifers are of nutritional value more for vitamins and minerals than for energy. Some of the more mildly pungent kinds are eaten fresh as salad plants, e.g., *Raphanus sativus* (radish), *Lepidium sativum* (garden cress), *Rorippa nasturtium-aquaticum* (watercress). At the other extreme, *Armoracia rusticana* (horseradish) is cultivated for its high mustard oil content. The genus *Brassica*, which is discussed in detail below, is a multiple purpose domesticate.

Common garden ornamentals in the family include *Iberis* (candytuft), *Lobularia* (sweet alyssum), *Matthiola* (stocks), and *Erysimum* (wallflower).

***BRASSICA* — MUSTARD, RAPE, TURNIP, CABBAGE, BROCCOLI, CAULIFLOWER, KOHLRABI** (Burkill 1935; Bye 1979; Dickson and Wallace 1986; Gade 1972; Hedrick 1919; Hemingway 1976; Kitamura 1950; McNaughton 1976; Narain 1974; Patiño 1969; Simoons 1990; Thompson 1976; U 1935; Weiss 1983)

The genus *Brassica* consists of a dozen or so polymorphic species, broadly defined, or an indefinite number of more uniform groups. All the species are native to the winter rain Mediterranean region, usually growing in rocky, open habitats. Some became agricultural weeds in antiquity and expanded their ranges northward into Europe and eastward into Asia. As useful volunteers in fields and gardens, they were gradually modified by human selection, with a host of different cultivars eventually developed. The array of *Brassica* cultivars is so complex that their origins would probably have been undecipherable if they had not happened to be derived from ancestors with different basic chromosome sets. There are three basic diploid species with 8, 9, and 10 pairs of chromosomes, respectively. These three genomes are designated B, C, and A. Hybrids between these are, of course, normally sterile. Doubling of the chromosome sets in the sterile interspecific hybrids resulted in three new polyploid species, each sexually fertile but incapable of breeding with either parent species. The results are elegantly summarized in the so-called Triangle of U (U, 1935):

		B. nigra black mustard BB 2n=16		
	B. carinata Ethiopian mustard BBCC 16+18=34		*B. juncea* brown mustard AABB 16+20=36	
B. oleracea cabbage CC 2n=18		*B. napus* rutabaga AACC 18+20=38		*B. campestris* turnip AA 2n=20

Folk names in the genus do not correspond with the biological species. Rather they correspond to uses. For example, in English, cabbage is usually applied to crops producing large succulent leaves, rape to those producing oilseeds, mustard to those producing spice seeds, turnip to those producing swollen roots, regardless of which species they belong. Only if adjectives are added do the folk generic names have taxonomic meaning.

Brassica nigra (2n=16) — Black Mustard

B. nigra is native to the Mediterranean region, where it is reported as a fossil from prehuman Pleistocene time. The species probably joined Neolithic agriculture as a volunteer in wheat and barley fields, spreading with those crops through Eurasia and North Africa. Millenia later, *B. nigra* went with those crops overseas, its seeds often carried accidentally with the cereal grain. In California, for example, it was a weed during the Spanish period, and its remains are common in adobe bricks of the early missions.

Like other mustards, *B. nigra* is used as a green vegetable and as a source of mild tasting nonvolatile oil expressed from the seed. However, its main use has been as a spice. Use of mustard seed as a spice dates from the beginning of written history in Babylonia and India and is abundantly recorded in classical Greek, Roman, and biblical sources. Although other species of *Brassica* were also used, it seems likely that *B. nigra* was the main source in ancient history as it has been in modern history. Traditionally, mustard is prepared by mixing a small amount of seed of white mustard *Sinapis alba*, other spices, and a lot of *B. nigra* seed. In ancient times, volunteer plants were probably more than adequate without deliberate planting. Cultivation of *B. nigra* as a field crop came after Medieval times, along with commercial production in Düsseldorf, Dijon, and a few other European towns. Recently a major mustard growing region developed in western North America, particularly California. Smaller scale commercial crops are grown in Argentina and Australia.

Very little evolutionary change in *B. nigra* has taken place under cultivation except for selection of less branched and taller forms. Cultivated *B. nigra* retained the wild and weed type of dehiscent fruit, which scattered the seeds as they ripened. Much seed always escaped harvest to contribute volunteer plants to the next crop. The only way to obtain a good seed yield was repeated hand harvesting of ripening fruits. This is now past history in commercial mustard spice production. During the 1950s, growers suddenly abandoned the old *B. nigra* for indehiscent cultivars of *B. juncea*, which could be machine harvested, as noted below.

Brassica oleracea (incl. *B. cretica, B. insularis, B. rupestris*) (2n=18) — European Cabbage, Cole, Kale, and Derivatives

Wild populations grow on rocky coasts of Britain, the Bay of Biscay, and the Mediterranean from Spain to Greece. The species was probably taken into cultivation independently in different places. There are no genetic barriers to interbreeding within the 18 chromosome diploids whether wild or domesticated. The amazing polymorphism of *B. oleracea* cultivars seems to trace partly to diverse wild populations. Cauliflower, for example, seems to have inherited some of its peculiarities from *B. cretica*. Human, not natural selection is clearly responsible, however, for the nearly sterile white inflorescence of cauliflower, the solid heads of cabbages and brussels sprouts, and for survival of other bizarre mutants.

Use of both wild and cultivated *B. oleracea* is recorded from classical Greece and Rome. The only cultivar forms recorded from Greece seem to have been the primitive kales or coles, nonheading and perhaps differing from the wild mainly in having been selected for less bitter and more succulent shoots. Kales are still widely grown as potherbs and forage plants. Roman cultivation may have begun independently from other wild populations and achieved slightly more improvement. Rome had a loose-headed cabbage by the 1st century A.D., another kind grown for its flowering shoots (a precursor of broccoli?), and a marrow cabbage grown for its soft stem (a precursor of kohlrabi?). There ensues a long period that, as far as history of *Brassica* cultivation goes, was really a Dark Age in Europe.

About 1150 A.D., several varieties of cabbage were mentioned as cultivated in Moorish Spain and others, red and white, in Germany, but the varieties are unidentifiable. Only after 1600 are extant *Brassica* cultivars clearly recognizable in European botanical literature. During the 17th century, solid-headed cabbages, savoy cabbage, kohlrabi, sprouting broccoli, cauliflower, and possibly brussels sprouts were all being grown on the continent. In several cases, the herbals give the impression that the cultivars were recent arrivals in northern Europe from Italy or the Near East. Although Britain had some kind of cabbage by the 14th century, arrival of modern cultivars was much later. Brussels sprouts did not begin their conquest of English cuisine until the 19th century.

Spanish colonists began growing cabbages in the West Indies and on the tropical American mainland about 1500. The first introductions were evidently loose-headed savoy cabbage and kales, while solid-headed cabbage, broccoli, and cauliflower arrived later in the colonial period. In the tropical lowlands, the crop grew well enough, but did not produce seed and had to be propagated vegetatively or by fresh seed from Spain. In Spanish America, *B. oleracea* really came into its own in tropical highlands and in high southern latitudes.

Some kind of cabbage, probably kales and loose-headed types, were introduced to eastern North America early by Cartier to Canada in 1540 and by the English to Virginia by 1669. By the time of the American Revolution, cabbages had been taken into the gardens of the Indians in New York and Florida. Savoy cabbage, brussels sprouts, and other important modern cultivars did not appear in North American records until the 19th century.

B. oleracea also was taken by early Dutch and British colonists to the southern hemisphere. It was more important in those days than it is now because of its anti-scorbutic value. Pickled cabbage (sauerkraut) was a traditional folk defense against scurvy. Without understanding the bacteria responsible, empirical methods of anaerobic fermentation successfully preserved the vitamin C content of the cabbage. In regions with long winters without fresh vegetables and on long ocean voyages, sauerkraut was a godsend. Captain James Cook was a great believer in keeping his crews healthy by feeding them sauerkraut and recognized its role in his successful explorations. Cook planted cabbage seeds on Queen Charlotte Sound, New Zealand, on his first voyage in the hope of having a fresh supply of cabbage upon returning.

B. oleracea cultivars, particularly broccoli, have been adopted into Chinese and Japanese cuisine since the late 19th century, but they have to compete with traditional Oriental *Brassica* cultivars noted below.

Since the 1960s, modern scientific breeding programs, particularly in North America, Europe, and Japan, have changed the nature of head cabbage cultivars grown commercially. Breeding has improved disease and pest resistance and has led to more solid heads to withstand mechanical harvesting, bulk shipment, and storage. Also, yields have been boosted by using F_1 hybrid seed for planting the crop. As in maize, the hybrid seed is sold by commercial seed companies. Two inbred cabbage cultivars are interplanted with self-pollination blocked by self-incompatibility or male sterility.

Brassica campestris (2n=20) (incl. *B. rapa, B. oleifera, B. chinensis, B. pekinensis,* etc.) — Some Rapes, Mustards, Common Turnip, Oriental Cabbages

The wild type is a slender-rooted, branching annual native to temperate Eurasia. It was probably a minor member of the pioneer flora of naturally open habitats that became enormously more abundant as a Neolithic weed of wheat and other crops. As a useful weed, it was evidently domesticated independently in various regions, yielding very different but interfertile cultivars.

The most primitive cultivars appear to be the rapes, i.e., those grown primarily as oilseed crops. Sanskrit records of sarson, a variety of *B. campestris*, show that it has been an important oilseed crop in India since at least 1500 B.C. Sarson and other varieties of *B. campestris* are still very commonly grown as oilseeds in the subcontinent. Related varieties may have been equally ancient crops in southwestern Asia in preclassical times, but they did not invade the domain of the olive in the Mediterranean.

The Romans encountered *B. campestris* as a crop of the barbarians in Gaul, not as rapeseed but as the turnip. There are old Germanic, Slavic, and Celtic names for the turnip, and it evidently was selected as a root vegetable in northern Europe in antiquity. Rape as an oilseed crop finally arrived in northern Europe from the east in the late Middle Ages, but for centuries, it remained a very minor crop. (The sudden rise of *B. campestris* and *B. napus* as European oilseed crops in the mid-20th century is discussed below.) Turnips, for both table vegetables and livestock fodder, became increasingly important in northern Europe between the 15th and 18th centuries, largely because of development of new varieties. One of these is the stubble-turnip, selected for very rapid growth in the autumn after the grain crop, usually rye, has been harvested.

The turnip and other European cultivars of *B. campestris* have been introduced to the Americas on innumerable occasions. The first record of introduction of turnips to eastern North America was in 1540 to Canada by Cartier. Turnips were being grown in the Virginia colony by 1609 and were widely grown in other colonies by the 18th century. In Canada, *B. campestris* eventually became a huge oilseed crop.

Spaniards introduced turnips soon after 1500 to Hispaniola, other West Indian islands, and Panama. By about 1550, they were adopted into Indian gardens in Colombia.

Field mustard, the unimproved weed form of *B. campestris*, was independently introduced to the Americas and spread widely as a volunteer in wheat and barley fields. Indian farmers of the Andean valleys have accepted it as a useful weed. While they do not want it to dominate the grain fields, neither do they want to eliminate it. They pull the young mustard plants out of the grain to cook as a green vegetable and for livestock fodder. Enough of the weeds remain to produce another year's crop. A similar situation is found among the Tarahumar of northwestern Mexico. These Indians presumably obtained *B. campestris* as a weed of grain crops brought by early Spanish missionaries. The Indians not only gather the weed for greens, but also sow patches of it as an incipient domesticate, selecting seed for production of more greens and swollen roots.

B. campestris probably spread to China from India, both as a weed and as an oilseed crop, but its ancient history is obscure. The species became an extremely important garden crop in the Far East primarily as a leaf vegetable. A large number of highly improved cultivars were developed in China, Korea, and Japan for use as salad greens, cooked, and pickled. Two of the more important of these, Chinese celery cabbage or *pai-ts'ai* and Chinese white cabbage or *pak-choi*, are recognizable in writings of the 5th and 6th centuries A.D. Like *B. oleracea* in Europe, the crop became vital as a source of vitamins during the long winter period. Properly stored, the heads can be kept fresh a long time, and they can be preserved indefinitely by pickling. In Korea, *kimchi* became a staple, often eaten at every meal. It is somewhere between coleslaw and sauerkraut, made up mainly of *B. campestris*, with a lot of radish and spices added to the basic dish and endless other ingredients in fancy dishes. The crop is widespread and important in Japan with not only Chinese varieties but a great many additional locally developed cultivars. The brightly colored ornamental varieties of Japan provide winter color in a drab season. Some Oriental cabbage cultivars have been grown for the Paris market since the 1830s. Introduced widely elsewhere since then for overseas ethnic Asian markets, these have gradually moved into general markets.

Brassica napus (AACC=38) — Rutabaga, Some Rapes

This allopolyploid may have originated repeatedly where *B. oleracea* and *B. campestris* were cultivated together. The first recognizable record is of a white-rooted variety grown in Bohemia about 1620. By 1700, the rutabaga was being grown in France and northern Europe. Linnaeus found it in coastal sands in Sweden in the mid-18th century, probably escaped from cultivation. During the 18th century, it was introduced to Britain with the name Swede turnip. By 1806, it was in North America. Like many polyploid hybrids, the rutabaga is exceptionally vigorous and quick growing.

B. napus, like its parent species, is highly polymorphic with cultivars selected for different purposes. Some are grown not for the root but for oilseeds and as leafy vegetables and forage crops. These share names such as rape or kale with their diploid relatives. Not until early in the 19th century can their stories be untangled from the diploid crops. By then, they were widely grown as fodder and oilseed crops in northern and eastern Europe. Until very recently, European rapeseed oil from both *B. campestris* and *B. napus* was used mainly for illumination and lubricants. Seeds of the old European cultivars contained erucic acid, a long-chain fatty acid, and other complex organic compounds that made the oil unsuitable for food use and the high-protein meal unsuitable for livestock feed. Breeders finally eliminated these undesirable compounds in the so-called double zero cultivars. Since the 1930s, both *B. campestris* and *B. napus* have become major commercial oilseed crops in many European countries and Canada. The bright yellow flowers of the extensive rape fields have become a dominant landscape feature of these regions.

B. napus was introduced to Japan in 1887. Subsequently, breeders there have synthesized the allopolyploid anew with the Chinese cabbage, instead of the turnip, as the source of the CC genomes. The result is a heading cabbage-like *B. napus*. Like AACC hybrids synthesized by European plant breeders, all these new creations interbreed freely with the old *B. napus* cultivars.

Brassica juncea (AABB=36) — Indian or Brown Mustard

This species may have originated repeatedly in Southwest Asia and India where the two parent diploids *B. nigra* and *B. campestris* overlap. Like both parental species, *B. juncea* evidently spread widely as a weed, particularly in grain fields. It proved useful for greens, oilseeds, and spice. It is normally self-pollinated, and many pure-line cultivars were selected, especially in Southeast Asia, China, and Japan.

For commercial mustard manufacture, *B. juncea* was considered inferior to *B. nigra* until very recently. In the 1940s, a newly imported yellow-seeded Chinese variety of *B. juncea* began to be cultivated in the High Plains of North America. It and other *B. juncea* lines proved suitable for machine harvesting and rapidly supplanted *B. nigra*. The leading commercial mustard producing region became provinces of Canada and adjacent Dakotas and Montana. Subsequently, collecting of *B. juncea* germ plasm and rapid breeding of new spice cultivars have been very active in Canada, England, Germany, and France. Recent breeding in India, Pakistan, Japan, and Russia has been mainly for oil.

Brassica carinata (BBCC=34) — Ethiopian Mustard

The hybrid probably originated in northeastern Africa where weedy *B. nigra* and cultivated *B. oleracea* grew together. The species is not regularly cultivated elsewhere.

Chenopodiaceae — Chenopod Family

The chenopod family contains over 100 genera and 1500 species of herbs and shrubs native to nearly the whole world but especially prominent in arid regions and saline habitats, including salt marshes and the salt-spray zone on sea cliffs. Three genera have been domesticated as food crops.

BETA — CHARD, BEETS (Campbell 1976; Galloway 1989; Hedrick 1919)

Beta includes five wild species of small, often procumbent herbs native to seacoasts and other open habitats in Europe, the Near East, northwestern Africa, and the eastern Atlantic islands. They have occasionally been gathered and cooked as potherbs. Only one of these, *B. maritima*, has been taken into cultivation.

Beta maritima and *B. vulgaris* — Beets

B. maritima, the sea beet, is a biennial that grows wild along the seacoasts of the Mediterranean and the Atlantic coast of Europe as far north as the Baltic. Several flowers grow together to form a corky, multi-seeded fruit evidently adapted for drift dispersal. *B. maritima* differs from its cultivated derivatives, collectively called *B. vulgaris*, in being a smaller, often procumbent plant with ordinary looking leaves and a nonswollen root. Domesticated *B. vulgaris* differs by a series of simple gene mutations selected by gardeners, none of which would have been favored by natural selection. The crop, whether leafy or not, is always harvested when young, so seed for planting must be obtained from plantings set aside to mature as biennials.

The most primitive cultivar is the leaf beet or chard, cooked like spinach as a leafy vegetable or potherb. Starting with Aristotle about 350 B.C., chard was repeatedly noted by Greek writers. Color variants — red, yellow, dark green — had already been selected. Chard was the original *Beta* of the Romans and the standard garden beet of medieval Europe. It also was mentioned in Chinese literature from the 7th century A.D. on, probably having been introduced from Europe by caravan trade. A modern type of chard with erect, succulent leaves and broadly expanded petioles seems to have arrived in England from Italy or Sicily in the 17th century. The name Swiss chard was in use in England in the 18th century. Brightly colored chard cultivars are grown as ornamentals.

Beet roots were cooked and eaten from Roman times on, but the early accounts all refer to long, white, straight roots. The modern garden beet with a swollen red root appears clearly in the mid-16th century in the great herbals of Germany, England, and other European countries, sometimes attributed to recent introductions from Italy. Huge coarse white or yellow beet cultivars were developed for livestock feed in several European countries in the 18th century. Some of these had sucrose contents of 6%, double that of table beets.

By 1750, a process for extracting sucrose from red and white beets had been developed in Prussia. Manufacture of beet sugar was soon tried there and in France, but the crop was uneconomical until the Napoleonic wars when British blockades cut off supplies of cane sugar to the continent. Both the King of Prussia and Napoleon took an active personal role in developing the industry, providing substantial financial support for model factories and schools to train factory technicians and beet growers. With peace and resumption of cane sugar imports, beet sugar again became uneconomical, but the crop was saved by subsidies and by tariffs on imported sugar imposed by the continental beet-growing countries. European farms were able to insert an occasional sugar beet crop into their rotation in place of fallowing; the beet tops and sugar mill waste were fed to livestock, and the manure returned to the fields instead of losing a crop year for fallow. Selection gradually increased sucrose content of beets to about 20%, and the crop spread to many European countries. By 1900, European beet sugar production approached total world cane sugar production.

During the 19th century, *B. vulgaris* seed from Europe was widely distributed around the world. The garden beet was even planted in the tropics, mainly for consumption by Europeans. It does not flower in low latitudes, lacking the needed stimulus of short nights, and cultivation there depends on continued importation of seed.

In the 19th century, beets became well established in gardens of North America, Australia, and other temperate lands. The sugar beet followed the garden beet to North America in the mid-19th century, initially in the eastern and midwestern United States, but pioneer factories repeatedly failed. Successful beet sugar factories were established in the late 19th century, the first in California in 1870, later in other western and in upper midwestern states. A prominent pioneer in the industry was Claus Spreckels, who would later become a pioneer in Hawaiian sugar cane. Like cane, sugar beets have often been grown as large monocultures. In western North America, the beet sugar companies generally have large landholdings of their own, as well as contracting for beets from other growers. Like other monocultures, sugar beets have often been plagued by diseases and pests. An especially severe problem in western North America has been virus diseases. These have caused temporary abandonment of the crop in some areas until resistant varieties could be developed. Also, growers have cooperated to observe beet-free periods when no beets were left in the ground to harbor the viruses.

Much modern research has been devoted to sugar beets in both Europe and North America. In addition to disease resistance, success has been attained with improving yields and adapting the crop to machine planting of seed and machine digging of roots. Commercial seed is now generally obtained by interplanting selected parental lines, one male sterile, and allowing them to hybridize by wind pollination. The male sterile line is commonly a tetraploid producing diploid pollen, so triploid seed is produced to plant the crop.

SPINACIA — SPINACH (Hedrick 1919; Simoons 1990; Smith, P. 1976)

S. tetrandra, the possible wild ancestor of cultivated *S. oleracea*, is native to mountainous regions of southwestern Asia. The domesticate differs from the wild in having nonprickly fruits and smooth leaf margins, both characters presumably developed by human selection. The most primitive *S. oleracea* cultivars are grown in the Himalayas. The domesticate is most diverse in Afghanistan and adjacent regions of Central Asia, but the history of cultivation there is obscure.

The first mention of spinach as a garden vegetable was in China in the 7th or 8th century A.D. It evidently was introduced to Spain by 1100 A.D. via Islam. One Moorish writer in Spain called it the queen of vegetables. The Europeans adopted it promptly. Albertus Magnus described a primitive, prickly-seeded spinach in Germany about 1280. By 1550, a smooth-seeded cultivar was described in Germany. By then, spinach was already well known in French and English gardens.

CHENOPODIUM — QUINOA AND RELATIVES (Fritz 1990; Gade 1970; Hunziker 1943; Risi and Galway 1984; Simmonds 1976a; Smith, B. 1987; Wilson 1981, 1988, 1990)

Chenopodium is a cosmopolitan genus of over a hundred species, mostly annuals adapted to colonize scattered and temporary openings produced by natural or artificial disturbance. *Chenopodium* species commonly complete their life cycles within a few months. Under optimum conditions, an individual plant may produce over 100,000 seeds. The flowers are inconspicuous and wind-pollinated. Self-pollination predominates, each flower producing a single seed tightly enclosed within a papery fruit or pericarp. The fruits are the units of dispersal. *Chenopodium* seeds contain a relatively large embryo coiled around a central endosperm. They are important food of a great many different kinds of birds, and a fraction of them are evidently widely disseminated as a result.

Some species are used as medicinal and culinary herbs, e.g., *C. ambrosioides*, known in Mexico as *epazote*. Various species are commonly gathered as potherbs. Usually the wild and weed supply makes planting unnecessary. *C. bonus-henricus* or fat-hen was formerly casually planted as a potherb in Britain.

Actual domestication of *Chenopodium*, evidently involving at least three different wild progenitors, was accomplished only by prehistoric New World Indians. Two *Chenopodium* species were ancient grain crops in western South America: *C. pallidicaule*, *cañihua*, which is a diploid (2n=18), and *C. quinoa*, *quinoa*, which is a tetraploid (4x=36).

Cañihua is a low growing plant, up to a half meter tall, with small axillary inflorescences similar to those of wild *Chenopodium* species. Its wild progenitor is uncertain, but Hunziker found a closely related wild chenopod in the

Argentine Andes. *Cañihua* is extremely resistant to cold, drought, salinity, and pests. It is grown as a sort of insurance crop on the Altiplano of southern Peru and Bolivia at elevations of 4000 to 4500 m. It is often rotated with potatoes. The seed is broadcast in fields where potatoes have been dug and allowed to compete with grasses and other weeds until harvest. The plants are pulled or cut individually as they mature rather than harvested en masse. As the plants dry, the grain is removed by repeated threshing. The residue from threshing and winnowing is burned for ash, which is in great demand for chewing with coca leaves. The grain requires soaking and washing to remove the fruit coat and the bitter saponin. It is an excellent high-protein food, superior to true cereals, and also used for brewing.

Quinoa is also primarily a high Andean crop, but its upper limit is a few hundred meters lower than that of *cañihua*. *Quinoa* has also been much more strongly shaped by artificial selection, particularly for grain yield. Some local landraces have also been selected for reduced content of bitter saponin. *Quinoa* plants grow 1 to 2 m tall and produce massive compound inflorescences. Unlike true cereals, the cultigens have not been selected for nonshattering, but the massive crowded inflorescences trap the bulk of the loose grain.

Quinoa grain is outstanding nutritionally. It contains 15 to 20% protein, which has a high proportion of lysine and other essential amino acids, about 5% fat, which is made up of some essential fatty acids, and about 65% carbohydrate, plus some minerals. It was a significant component of the rich pre-Columbian crop complex of the Andean highlands.

The earliest archaeological report of *quinoa* is from before 1500 B.C. in the Azapa Valley in the northern Chilean desert (Rivera 1991); the grain may have been brought down from the highlands rather than locally grown. Dozens of landraces were developed prehistorically in the Andes. The center of diversity is in the central Andes of southern Peru and Bolivia. A few landraces extend into northern Peru and Ecuador, a few others extend down to medium elevations on the east slope of the Andes, and one, known as *quingua*, is grown at low elevations on the temperate coast of Chile. *Quingua* was reported as a crop of the Mapuche and Aruacanian Indians at the time of Spanish entry into southern Chile. The gross distribution of the crop in South America has not changed drastically since the arrival of the Spaniards except that it is no longer grown in southern Colombia, its original northern limit.

A crop such as *quinoa*, capable of yielding exceptionally nutritious grain in cold, arid, saline habitats, has inevitably been repeatedly introduced overseas. It reached the Himalaya in the 19th century, but without notable success. Currently, *quinoa* has been taken into the vast international network of germ plasm exchange and is being grown experimentally in innumerable countries, especially at high elevations and high latitudes. A possible initial obstacle to introduction outside the tropics may be daylength requirements, which might be most easily overcome by using seed of the Chilean *quingua* variety. A rapidly expanding market for the grain in health-food stores in the U. S. and

Europe has relied on Andean grain, milled to remove the outer bitter coat. Farm-scale cultivation in Colorado, Washington State, England, and Scandinavia is currently reported (U.S. National Research Council 1989).

The closest wild relative of *C. quinoa* is *C. hircinum*, another tetraploid, which is widespread in the plains of southeastern South America and at mid-elevations on the eastern slopes of the Andes. There are no diploid *Chenopodium* species in South America that are considered possible ancestors of *C. hircinum*. Its closest wild relative is a North American tetraploid *C. berlandieri*, which evidently originated as an allopolyploid hybrid between two North American diploids *C. neomexicana* and *C. watsonii*. Both diploid species have ranges centered in the southern Rocky Mountains and High Plains. The tetraploid *C. berlandieri* is a highly polymorphic species with an enormous range through western North America from Alaska to Guatemala and eastward to the Atlantic coast of the U.S. and the Gulf of Mexico. It seems very likely that *C. hircinum* evolved from a very ancient introduction of *C. berlandieri* to temperate South America by migratory birds.

The cultigen *C. quinoa* differs from the wild *C. hircinum* by recessive mutations that result in yellowish or reddish grain rather than the wild-type black grain. The grain of the domesticate lacks dormancy. It is also larger than the wild. *C. quinoa* and *C. hircinum* interbreed freely when in contact, and the hybrids are fully fertile. Hybrids, called *ajara* in folk taxonomy, have followed the crop northward up the Andes outside the range of pure *C. hircinum*. Although farmers try to eliminate *ajara* from their seed stocks, some gets by as a contaminant and more survives as a free-living weed that continues to interbreed with the crop.

An independent domestication of a chenopod took place in Mexico. The cultigen *C. nuttalliae* was evidently derived from the wild *C. berlandieri*. However, *C. nuttalliae* is not grown as a grain crop but as a green vegetable, known by the Nahaua name *cuauhzontli*. The tops of the plants are cut off with the immature inflorescences and cooked as a potherb, something like broccoli (J. Sauer 1950). Wilson (1990) suggests that *C. nuttalliae* was formerly a major grain crop in Mexico, known as *huauhtli* in Nahuatl, although during his field surveys his informants reported only use as a vegetable. Presumably Wilson was reasoning by analogy with *quinoa*, but the analogy is imperfect. *Quinoa* is an important crop only at elevations too high for maize or grain amaranths, which are better tasting and do not require long leaching. Mexico has very little area too high for maize or grain amaranths, and there would have been little point in growing *cuauhzontli* for its bitter grain.

There is one Mexican chenopod of uncertain taxonomic status that produces distinctive bright red fruits containing hard, nonbitter seeds (J. Sauer 1950; Wilson 1990). This is grown only in the Tarascan region of Michoacan and near Lake Texcoco. In both regions, it is used, along with amaranth and maize dough, to make special tamales for certain festivals. The fruits are apparently valued for their pigment rather than as food. The red-grained chenopod is known as *chia* in Michoacan and *huauhtli* in the Lake Texcoco area. Both names are highly ambiguous generic terms for various plant species. *Chia* is the usual name for wild and cultivated species of *Salvia* with edible seeds. At the time of the Conquest, *huauhtli*, originally meaning insect eggs or fish-roe, was used with adjectival modifiers for chenopods and amaranths in general, whether grown for potherbs or grain. *Chia* and *huauhtli* were enumerated separately in the lists of grain tribute of the Aztec Empire (Barlow 1949). Several lines of evidence indicate that the *huauhtli* grain was primarily from amaranths (Amaranthaceae).

Although there is no archaeologic evidence of a domesticated grain chenopod in Mexico, there is such evidence from eastern North America. Like the Mexican *C. nuttalliae*, the North American domesticate was evidently derived from the wide-ranging wild *C. berlandieri*. Archaeologists believe domestication took place independently in the two regions. Archaeologic remains of pale chenopod grain from Kentucky have been directly dated at about 1500 B.C. By 1000 B.C., the cultigen was evidently widespread and important in the Ozarks and other midwestern regions. This was during the Woodland period when the economy was still based primarily on hunting and gathering. After about 1000 A.D., when maize became a staple crop, the grain chenopod crop apparently became extinct.

Convolvulaceae — Morning-Glory Family

Containing about 50 genera and 1000 species, the Convolvulaceae are cosmopolitan in distribution but most diverse in the tropics. Most members are herbaceous vines, annual or perennial, colonizing naturally open and artificially disturbed habitats, often spreading vegetatively by rooting at nodes. Several genera include garden ornamentals. *Operculina* species, known as wooden roses, are grown for dried bouquets. Only *Ipomoea* has been domesticated for food crops.

***IPOMOEA* — SWEET POTATO** (Austin 1977, 1978, 1988; Brand 1971; Burkill 1935; Ferdon 1988; O'Brien 1972; Patiño 1964; Purseglove 1968; Sauer, C. 1966; Yen 1971, 1974, 1976)

Ipomoea is a cosmopolitan genus of about 400 species. Some individual species are pantropical, e.g., *I. pes-caprae* and *I. littoralis*, both known as beach morning glories; their seeds are capable of almost unlimited drift dispersal by ocean currents. Other upland pioneer species are also highly disjunct, probably by bird dispersal.

I. purpurea is the common garden morning glory; several other species, including the night-blooming moonflowers, are widely grown ornamentals.

The seeds of several species are used around the tropics in folk medicine and as a narcotic.

I. aquatica and two other species are cultivated in Southeast Asia as potherbs and for fodder for livestock, especially pigs. *I. aquatica*, like rice, can be grown as either an upland or a flooded crop. It is often planted in farm fish ponds in China. *I. batatas* (sweet potato) is of far wider importance.

Ipomoea batatas — Sweet Potato

The story of the sweet potato is still extraordinarily murky, resembling that of many other crops 50 years ago. Its wild progenitor or progenitors are not known. Its presence in aboriginal agriculture in both the New World and Polynesia remains an enigma.

I. batatas does not form a gene pool, but rather an assortment of clones propagated vegetatively. The common cultivars are hexaploid (6x=90). Some wild relatives are diploid (2n=30); a few are tetraploid (4x=60), but chromosomes of most wild relatives remain uncounted. Various hypotheses have been suggested. By analogy with bread wheat, the hexaploid cultigen might have originated as a hybrid between a primitive tetraploid cultigen and a weedy diploid. It is also possible that there are truly wild hexaploids conspecific with the crop. Conceivably, cultivars were independently domesticated in different regions.

There is little hope of understanding origins of the cultivars until the basic systematics of the wild relatives are better known. There are supposed to be about a dozen closely related wild species, but the few taxonomists who have studied them disagree on their definition. The species are highly variable and subtly different. Together, their ranges cover a huge area, not only in the New World but in Pacific and Indian Ocean islands and Australia. Available collections are simply too few to provide a proper sample of variation patterns in these populations.

A special problem with standard herbarium collections is lack of information about tubers. The sweet potato tuber may have been modified by human selection, but it is unlikely that it was created *de novo*. It probably evolved as an adaptation for regrowth after seasonal drought, not fire or winter cold. A wide-ranging species might have both tuberous and nontuberous subspecies. Almost nothing is yet known about the geography of tuber-bearing wild relatives of the cultigen.

One of the most serious deficiencies in the present data base is the lack of good information on ecology of the wild populations. Careful field observations should discriminate between possible wild progenitors in natural habitats and weedy escapes or relics of cultivation. So far, however, taxonomic research on *Ipomoea* has been based mainly on experimental garden plants and herbarium collections lacking good habitat data; thus, the geographic origin of the domesticate remains obscure.

At present, the cultivated sweet potato simply comes out of the blue. Its historical geography begins after domestication. The earliest archaeological remains are from outside the range of wild *Ipomoea* species on the west side of the Andes in Peru and northern Chile. Extremely ancient remains have been reported from caves near Chilca, south of Lima, but the reports are vague and contradictory. Bonavia (1983) notes that neither the identification nor the dating has been confirmed. A fairly coherent archaeological record of apparently domesticated sweet potatoes begins at 2500 B.C. with tubers from Ancon-Chillon on the central coast. The crop was evidently widely grown in irrigated oases of northern Chile and southern Peru during later prehistoric times, as shown by actual tuber remains and accurate ceramic models. Archaeological remains are not available from elsewhere, but there are indirect clues suggesting very ancient cultivation. For example, variation in names for sweet potato in different Maya dialects suggests derivation from proto-Maya in the 3rd millenium B.C.

The crop entered written history in Hispaniola. Columbus was familiar with true yams (*Dioscorea* spp.) from West Africa and at first called the sweet potato a yam, a confusion that persists today in North America where true yams are unknown. However, Spanish explorers soon adopted the Island Arawak names: *batata* for sweeter, moister sweet potatoes with reddish orange flesh; and *aje* for pale, starchy, mealy ones. Sweet potatoes were second only to manioc among Arawak crops in the Greater Antilles. Spanish explorers of the Caribbean mainland coasts reported *batatas* and *ajes* widely cultivated by various Indian tribes. The Indians grew varieties under many names; the twofold naming was soon dropped. *Batatas* and variants became the lingua franca name in the West Indies and Brazil. After the conquest of Mexico, *camote* (Nahuatl: *camotl*) became a common name. By 1550, the Mexican name was replacing all other names as far away as Peru among both Spaniards and Indians.

The sweet potato in aboriginal America was exclusively a crop of low and middle elevations within the tropics. It was grown from Michoacan and Veracruz southward to Peru, Bolivia, Brazil, and adjacent Paraguay and Argentina. In rainy regions, the crop was planted in mounds for better drainage. In arid regions, it was irrigated. In either situation, the crop could be planted and harvested all year round. Propagation was by vine cuttings without storing tubers. Thus natural selection for tuber dormancy during drought was relaxed, and some cultivars were developed that store very poorly. Others still retain the trait of long dormancy and can be stored for many months if properly cured. The quicker growing cultivars could be dug 3 or 4 months after planting. Since seed was not planted, natural selection for sexual reproduction was relaxed, and eventually some nonflowering cultivars developed. However, others continued to flower, and occasional seedlings may have been adopted into cultivation as new clones.

Spanish colonists of the New World found the sweet potato a good source of provisions for slaves and servants. They also adopted it quickly for their own cuisine. In fact, Columbus returned with sweet potatoes for tasting by Isabella and Ferdinand. Whether any of this lot escaped being eaten and was planted is unrecorded, but the crop was established in Spain very soon thereafter. Oviedo wrote in 1526 that he had himself repeatedly brought *batatas* from the Caribbean to Castile. Monardes wrote about 1570 that sweet potatoes were being shipped from Velez to Malaga by the caravel load. By that time, sweet potatoes were being grown in other European countries. By 1600, they were grown in England. *Batata* became *patata*, potato, and other variants, eventually shared with the later-arriving Andean *Solanum tuberosum*.

English colonists in Virginia were growing sweet potatoes before 1650, perhaps obtained from the West Indies rather than via Europe. Before the American Revolution, cultivation was general as far north as New England and had been taken up by Indian tribes in the Carolinas and Florida.

The spread of the sweet potato out of the tropics, in Europe, North America, and elsewhere depended on development of techniques for warm, dry storage of the tubers during winter.

Introduction of the sweet potato to West and East Africa and to India is credited to 16th century Portuguese shipping. The trail of spread among African peoples is marked by the name *batata* and variants, shared among such unrelated language groups as Berber and Zulu. The sweet potato also took over various local names from the aboriginal tuber crops, especially yams. A later introduction to East Africa, probably a British colonial connection, is marked by names attributing the crop to Bombay (O'Brien 1972). In Asia, the Portuguese *batata* trail extends through Ceylon to Java where the crop was cultivated before 1650.

Spaniards are credited with early introduction from Mexico across the Pacific. The trail is marked by the Mexican name *camote*, standard in the Philippines and Guam. Here also, the crop acquired many other names borrowed from the old native crops, such as taro and yams. By the 1590s the crop had been introduced from the Philippines to China and soon after was taken from China to the Ryukyus where it became a staple. By 1674, it had been adopted in Japan. China and Japan became the world's leading growers of sweet potatoes. China alone was growing about 100 million tons a year during the mid-1980s, about ³/₄ of the world harvest (Sokolov 1989). Sweet potatoes have been compatible with various Asian farming patterns from intensive Chinese market gardening to shifting cultivation among hill tribes.

There is one more chapter, the most mysterious one, in the historical geography of this crop. This involves its spread into Polynesia, Micronesia, and Melanesia. The earliest Spanish and Portuguese explorers of these islands repeatedly mentioned *camotes* or *batatas* among the native crops. This happened in such widely scattered places as the Marquesas, Solomons, and Guam. These reports would have been simply dismissed as confusion with the native yams and taro except for what followed. There is solid evidence from the last half of the 18th century that *I. batatas* was an important Polynesian crop, embedded in tradition and ritual. It was clearly reported by French and British expeditions, particularly those of Cook, not only in the Society Islands, Tonga, and the Marquesas, but also in the outermost Polynesian islands: New Zealand, Hawaii, and, less clearly, Easter Island.

In Melanesia, sweet potatoes are a traditional staple crop in the highlands of New Guinea. Palynological evidence of vegetation changes associated with agriculture suggests that sweet potatoes were introduced among the Papuans in the 17th century (Worsley and Oldfield 1988). The French naturalist La Billardiere recorded Melanesian cultivation of sweet potatoes in New Caledonia in 1792.

There are five major hypotheses to explain the introduction of *I. batatas* to the central and western Pacific, none of which can be either summarily dismissed or neatly confirmed: (1) prehistoric introduction by South American Indian rafts drifting downstream and downwind, (2) prehistoric introduction on Polynesian canoes returning from a round trip to South America, (3) historical introduction by the Portuguese from the Atlantic, (4) historical introduction by Spaniards across the Pacific, and (5) natural dispersal by drifting capsules. The massive literature arguing these possibilities will not be reviewed here. The first hypothesis received a dramatic boost by the Kon Tiki expedition. Subsequent archaeological work, however, has shown clearly that all the domesticated animals and crops of Melanesia and Polynesia, except for the sweet potato, were brought from the west, as sketched below in discussions of sugar cane, taro, and bananas. If a South American Indian raft introduced *I. batatas*, it brought nothing else that left a mark. On the other hand, an upwind Polynesian canoe voyage across the Pacific is hard to imagine, but only a little less credible than the known voyages to Hawaii and Easter Island. The Indians of the Peruvian coast were highly civilized and might very well have provisioned visitors from the west for the relatively quick voyage home.

One bit of evidence on this question that has been much discussed is not really very helpful, namely the shared name *kumara* or *kumala*. This is the common Polynesian name for the sweet potato and is also a localized name in the Andes. The common Quechua name *apichu* is unrelated. Linguistic evidence indicates the word *kumara* originated in ancient proto-Polynesian and gradually diverged in different dialects. By contrast, there is no record of the name in Peru before the late 16th century after the Mexican name *camote* had been introduced. O'Brien (1972) suggested Spanish introduction of the name from the Pacific islands.

Another much cited bit of evidence that has not proven decisive is presence of storage pits on the North Island of New Zealand dated to the 14th century. Since sweet potatoes were historically the Maori staple crop and the tubers were stored over winter in similar pits, it was assumed that the archaeologic structures document pre-Columbian cultivation of *Ipomoea*. Ferdon (1988) criticized this assumption, noting that taro and sweet potatoes were both cultivated in the region during Cook's explorations and that the technology of storing could have been developed initially for taro and taken over by sweet potatoes at an unknown date. He suggested sweet potatoes may have been planted in the New Hebrides by the Spanish colonizing expedition from Peru under Quiros in 1605 or 1606, which would have left time for spread to New

Zealand before Cook. Possible early historical introductions by both Spanish and Portuguese are discussed at length by Brand (1971). Transport from the South Pacific to Hawaii in the time available thereafter, however, would have required some very fast, tight scheduling of unrecorded canoe voyages.

Purseglove (1968) suggests introduction of *I. batatas* to Polynesia by none of the above but by natural dispersal: the seeds are viable for more than 20 years; they are hard and dormant unless scarified; they are impervious to salt water; and they are not buoyant, but the capsule is. I doubt that the seedlings could survive in the drift zone on an ocean beach, but conceivably capsules could have been picked up by some Polynesian beachcomber, or seeds might have germinated along the banks of a tidal estuary. A piece of evidence that favors the possibility of long-range natural dispersal is the presence in Australia of endemic *Ipomoea* species closely related to *I. batatas*. One is *I. gracilis*, which was first described in 1810 from Carpentaria Island in tropical Queensland. Its tubers were dug by the Aborigines. Another species closely related to *I. batatas*, *I. littoralis*, has a huge natural range in islands of the tropical Pacific and Indian Ocean region.

The five hypotheses are not necessarily mutually exclusive. Different varieties of the sweet potato conceivably arrived in Polynesia by both drift dispersal and by human transport. Historical introductions undoubtedly also occurred, probably repeatedly. If varieties introduced by missionaries and modern agencies have overlain and displaced the earlier introductions, it may no longer be possible to recognize the original cultivars no matter how much future research is done.

As for modern breeding, extensive collections of sweet potato cultivars for research and breeding were begun very recently. The International Potato Center in Lima, Peru, which was originally devoted to work on *S. tuberosum*, has now added the sweet potato to its program. Research stations elsewhere are working with *Ipomoea*, but among important crops it remains one of the least understood and least affected by scientific breeding.

Cucurbitaceae — Gourd Family

Cucurbitaceae include over 100 genera and over 800 species, nearly all vines, annual or perennial. The great majority of the cucurbits are tropical or subtropical, and most require warm, sunny habitats. Many members are natives of deserts or semi-arid scrub associations. Nearly all have unisexual flowers requiring insect pollination. Some have hard-shelled fruits adapted for dispersal by streams or ocean currents. Others produce edible fruits attractive to birds and other animals, with large quantities of seed. The seeds, like those of legumes, lack endosperm, but store high concentrations of oils and proteins in the cotyledons. The wild species undoubtedly attracted human attention as soon as people encountered them, and they have been widely exploited historically by hunting-gathering groups. In addition to the four genera discussed below, several genera have domesticated members that are widely planted in gardens as minor crops or ornamentals, e.g., *Luffa* (vegetable sponge), *Benincasa* (Chinese preserving melon), *Sechium* (chayote), *Trichosanthes* (snake gourd), *Momordica* (balsam apple).

CITRULLUS — WATERMELON AND RELATIVES (Blake 1981; Dalziel 1948; Hedrick 1919; Neal 1965; Patiño 1969; Watson 1983; Zohary and Hopf 1988)

Citrullus is an Old World genus of three species of vines native to arid regions. *C. colocynthus*, the colocynth, is native to northern Africa, some Mediterranean islands, southwestern Asia, and Australia. Its extremely bitter fruits have been much used as a purgative since antiquity and are grown commercially for medicine.

Citrullus lanatus (=*C. vulgaris*) — Watermelon

Clearly wild populations are reported only from deserts of southern Africa. In the 1850s, David Livingstone described the species as the most surprising plant of the South African deserts; in years with unusually heavy rains, the vines covered vast tracts. The wild populations produce two kinds of melons, bitter ones, which accomplish reproduction *in situ*, and sweet ones, which attract antelopes that disperse the seeds. These *tsama* melons were very important to the nomadic Kalahari Bushmen.

Where and when domestication began is unknown, but by 2000 B.C., watermelons were being cultivated in the Nile Valley. Watermelons were among the fruits of Egypt that the Jews regretted losing in the wilderness. Watermelons were widely grown prehistorically by agricultural peoples of sub-Saharan Africa. They had developed many landraces, varying in fruit size, shape, flesh color, and rind color. Also a spectrum of seed color mutants had been selected. In traditional African cuisines, the seeds and flesh of some watermelon fruits are used in cooking. The seeds are very rich in edible oil and protein.

From Africa, watermelons were introduced to India by 800 A.D. and to China by 1100 A.D.. The seeds are eaten and crushed for their edible oil in India and China, as they are in Africa. Watermelons are grown in other parts of Asia, but are of minor importance in wetter climates, largely because of fungus diseases.

Introduction of watermelons to Europe is credited to the Moorish conquerors of Spain. There are records from Moorish Spain at Córdoba in 961 A.D. and Seville in 1158. The Arabic name became *sandía* in Spanish. Watermelons spread slowly into other parts of Europe, perhaps largely because the summers are not generally hot enough for good yields. However, watermelons began appearing in European herbals before 1600, and by 1625, the species was widely planted in Europe as a minor garden crop.

In the New World, Spanish colonists were growing watermelons in Florida by 1576. By 1600, they were being planted in Panama, Cartagena, and on the coast of Peru. By 1650, watermelons were common in many parts of Spanish America and in Brazil. As in the Old World, the crop was most successful and most appreciated where there is a long, hot sunny growing season. In northeastern North America, watermelons did well in the hot, fairly dry summers. They were abundant there in the British and Dutch colonies before 1650.

A remarkable chapter in the historical geography of *C. lanatus* was its unhesitating acceptance by the North American Indians. Evidently the watermelon was something the Indians had just been waiting for, and seeds were passed from tribe to tribe like smoke signals, outrunning European explorers. The pathways of diffusion cannot be traced and are probably multiple. Before 1600, Spanish explorers penetrating the interior of North America found watermelons already in cultivation by tribes of the Ocmulgee region of what is now Georgia, by the Conchos nation of the Rio Grande Valley, and among the Zuñi and other Pueblo people of the Southwest. Before 1700, watermelons were being grown by the Yumans on the lower Colorado visited by Kino, by tribes of the Great Lakes region visited by Marquette and Cadillac, and by the Hurons of eastern Canada visited by LaHontan.

C. lanatus was also enthusiastically received by the Hawaiians when Captain Cook left seeds on Niihau in 1778. When Vancouver's expedition reached Kauai and Oahu in 1792, the Polynesians there had plenty of watermelons to give away.

Modern breeding has produced some new watermelon cultivars that are grown commercially, mainly because of better disease resistance. However, the size and quality of the fruit are a heritage of folk selection.

***CUCUMIS*— MUSKMELONS, CUCUMBER** (Dalziel 1948; Harlan 1975; Hedrick 1919; Miller and Morris 1988; Patiño 1969; Robinson and Whitaker 1974; Whitaker and Bemis 1976; Zohary and Hopf 1988)

Cucumis includes about 30 species, mostly endemic to desert and savanna regions of Africa, one native to India, and a few wide-ranging. *C. anguria*, the so-called West Indian gherkin, is widely naturalized in warmer regions of the New World. The gourd or melon-like fruits of various wild species are gathered for their edible seeds, for pickling, and other purposes. Two species are widely cultivated crops.

***Cucumis melo* — Muskmelons**

Modern cultivars of this species are very diverse, including some with netted rinds, e.g., cantaloups and Persians, others with smooth rinds, e.g., casabas and honeydews. Some of the more extreme variants, such as the Chinese serpent melon, are rarely grown in western countries.

In Africa and southwestern Asia, the probable homeland of the cultigens, the line between domesticated and wild populations is hazy. Along with large, sweet cultivars, native farmers sometimes grow primitive varieties with fruits no larger than a hen's egg, some of which are grown primarily for the edible seeds, others cooked like a squash. It is not always clear whether these are genetically distinct from wild populations.

Populations of truly wild melons, growing in natural habitats, have been reported from desert and savanna regions of Africa, Arabia, southwestern Asia, and Australia. These have been given various names, e.g., *C. melo* var. *agrestis*, *C. trigonus*, *C. callosus*. Hybridization and isozyme studies (Bates et al. 1990) suggest that all these behave as a single species, at least in an experimental garden. Thus it is unclear where *C. melo* was domesticated. It is conceivable that it was independently domesticated from different wild populations in Africa and southwestern Asia. Zohary and Hopf (1988) suggested two centers of domestication within Asia. Some of the diversity of the cultigen species may derive from differences between geographically separate wild subspecies. The archaeological record consists mostly of seeds, which do not help in tracing the development of the different fruit types.

Archaeological seeds identified as *C. melo*, but not as to variety, are reported from Persia in the 3rd millenium and from Greece in the 2nd millenium B.C. Pictures of melons from Egypt in the 2nd millenium are identified as the variety *Chate*, which is used like a cucumber.

There are apparent references in classical Greek sources to *C. melo* under the name *pepon*. It does not seem to have been grown much or highly regarded and could hardly have been anything like the muskmelons we know today. The first evidence of improvement comes from Rome in the 1st century A.D. when Pliny wrote that a new kind of *melopepo* had recently appeared in Campania; his description of the fruit, including a peculiar trait of detaching itself from the vine when ripe, clearly refers to a cantaloup type. (In the casaba group, the fruit remains attached to the vine until harvested.) Reports of melons in Medieval Spain indicate introduction of new varieties by the Arabs.

The variety and quality of melon cultivars were evidently greatly increased by selection in Medieval gardens, the locales unknown. With the rebirth of scientific botany in the 16th century, European herbals clearly recorded many excellent melon varieties of both the summer or cantaloup type and the winter or casaba type. They already had various shapes, sizes, rind patterns, and flesh colors.

High European esteem for muskmelons is evident in the immediate introduction to colonies in the New World. Columbus planted muskmelons on Hispaniola on his first voyage and ate the harvest on his second. Other successful introductions followed rapidly. The story closely parallels that sketched above for watermelons and will not be recounted in detail. Again, acceptance by American Indian groups was astoundingly rapid. For example, before 1540 Indians in New Mexico and eastern Canada were growing muskmelons when visited by explorers Lopez de Gomara and Jacques Cartier, respectively. As with the watermelon, muskmelons were adopted by Pacific Islanders as soon as seeds were introduced by Captain Cook and other Europeans.

In modern times, *C. melo* has become ubiquitous in regions with sufficiently hot, sunny growing seasons. Whether by mutations or by crossing among varieties, the species is quite plastic genetically, and new cultivars have been selected in many countries. Most cultivars are suitable for home gardens or local markets, but are too perishable to ship if properly ripe. In California, some unusual and delectable varieties are restricted to ethnic markets, including Korean and Japanese. Some cultivars have moved from ethnic to general markets, e.g., the Persian melon, brought to California by Armenian immigrants.

At present, the largest *C. melo* crops grown for shipment to distant markets are produced under irrigation in the hot, dry interior valleys of California and Arizona. The leading commercial cultivars are partly old varieties, e.g., honeydew, a new name for the old White Antibes, essentially unchanged genetically since its introduction from France in 1911, and the Crenshaw, a new name for one of the old Middle Eastern casaba type melons, introduced in 1929. Some commercial cultivars have, however, originated in California, e.g., the Golden Pershaw, a hybrid between the Persian and the Crenshaw. In the cantaloup group, a series of important new cultivars began in California with selection by a Japanese grower of an intervarietal hybrid, the foundation of Hale's Best, released in 1924. Problems with powdery mildew led breeders at a California experiment station to try to combine all the desirable qualities of Hale's Best with resistance to the fungus, which was found in an otherwise unimpressive variety imported from India in 1928. A single dominant gene for resistance was successfully transferred to a Hale's Best background to produce PMR 45, released to growers in 1936. By good luck, PMR 45 differed from all older cantaloup varieties in having firm but nonrubbery flesh. For the first time, growers had a high quality cantaloup that could be harvested when fully ripe and shipped to distant markets; consumer demand led to great expansion of melon acreages. Other improved derivatives of PMR 45, e.g., Campo, have been bred subsequently, and the harvest season greatly extended.

Cucumis sativa — Cucumber

C. sativa has a different chromosome number (2n=14) than C. melo, and the two species cannot interbreed. *C. sativa* is a cultigen species; its only close wild relative and putative progenitor is *C. hardwickii*, native to the foothills of the Himalaya and other parts of India and to Arabia. Fruits of the wild species contain bitter terpene compounds called cucurbitacins; the mode of seed dispersal is unclear.

Cucumbers were presumably domesticated in India at an early but unknown date. India has a great many cultivars selected for various fruit shapes, sizes, and reduced bitterness. Some are eaten cooked, others raw and pickled. Archaeological remains of cucumbers are reported from Iran before 2000 B.C. Written records are reported in China by the 5th century A.D., and various cultivars have evidently originated there.

Cucumbers were grown in classical times in Greece and Rome. In the 1st century A.D., Pliny reported cucumber production in greenhouses. The Emperor Tiberius was reputed to have them on the table every day of the year. Charlemagne demanded a supply for his table in the 8th century. Several varieties were being grown in Europe by the time of the Renaissance herbals.

Spaniards were growing cucumbers in Hispaniola by 1494, and within 50 years, they had been taken into cultivation by Indian tribes visited by De Soto in Florida and by Cartier in Quebec. The species went with early European colonists elsewhere much like the watermelon and muskmelon.

Traditional dill pickles are made by fermentation in brine by the same lactic acid bacteria used for sauerkraut. A mediocre imitation with a different taste is commonly made by packing cooked cucumbers in vinegar (acetic acid).

European greenhouse production of cucumbers utilizes a modern mutant that produces parthnocarpic fruits, valuable when pollinating insects are not present. Another mutant that results in purely female plants is used for crossing two inbred lines to produce commercial F_1 hybrid seed, like hybrid maize.

***CUCURBITA* — SQUASHES, PUMPKINS, SOME GOURDS** (Andres 1990; Cutler and Whitaker 1967; Decker 1988; Decker-Walters 1990; Hedrick 1919; Heiser 1989; Merrick 1990; Nee 1990; Paris 1989; Puchalski and Robinson 1990; Smith, B. 1989; Weeden and Robinson 1990; Whitaker 1983; Whitaker and Bemis 1976; Whitaker and Cutler 1967; Whitaker and Robinson 1986)

About 20 wild species are known, all natives of the New World. All are diploids, but interspecific hybrids are usually nonexistent or sterile. The species are all vines and ecological pioneers in open sites. Those native to desert regions are xerophytic perennials with enlarged storage roots. The species that have been domesticated are mesophytes, mostly annuals, commonly native to riparian habitats.

The vines are monoecious with separate male and female flowers. In nature, pollination is by so-called squash bees, solitary bees closely dependent on the plants. The fruits are small, indehiscent gourds, generally with bitter flesh and hard, dormant seeds, presumably adapted for dispersal by flotation within the dried gourd.

The wild gourds were attractive to prehistoric peoples as containers and rattles. The seeds are rich in fat and protein and quite tasty, especially if toasted.

There are five cultigen species with fruit and seed characteristics dramatically different from their wild progenitors. The technical differences by which experts distinguish the species are subtle and not impressive to laymen, but the five are completely discrete gene pools incapable of producing viable hybrids. Several species have often been grown together in gardens, e.g., by the Pueblo peoples of southwestern North America, with no interbreeding. Nevertheless, the species are not generally recognized in folk taxonomy.

Common names have often been applied to different cultigen species indiscriminately, including the European names, gourd, pumpkin and calabash, which were transferred to *Cucurbita* from other genera. Also, the names derived as corruptions of American Indian names became generic rather than specific, e.g., squash in eastern North America, *ayote* from Nahuatl, *zapallo* from Quechua. The confusion in folk names is partly due to convergent similarity in unrelated cultivars. For example, round, ribbed, orange-colored winter squashes, all called pumpkins, are produced by several cultigen species. On the other hand, summer and winter squashes belonging to the same species may differ strikingly.

Cucurbita pepo

C. pepo had the most northerly distribution as an aboriginal crop. At the time of European arrival, it was grown from the St. Lawrence region through eastern, midwestern, and southwestern North America and in the highlands of Mexico and perhaps Guatemala.

It has two close wild relatives: *C. texana* of the south central U.S. and *C. fraterna* of northeastern Mexico. The two are geographically and morphologically discrete, but are capable of interbreeding with each other and with *C. pepo*. Long before they began deliberate planting of cucurbits, Indian peoples evidently gathered the small, hard gourds of these wild species and traded them far outside their natural ranges. In kitchen middens and other artificial sites, volunteer seedlings may have spread as camp-followers, providing useful gourds and edible seeds.

The earliest archaeological remains attributable to *C. pepo* rather than a wild progenitor are from Mexico: a high, dry valley at Guila Naquitz in Oaxaca, dated about 8000 B.C. and the Ocampo caves in Tamaulipas, dated about 5500 B.C. Selection evidently was initially for larger seeds and larger fruits rather than edible flesh. Eventually, as larger fruits evolved offering more fleshy tissue, mutants were selected for nonbitter and less fibrous flesh and for thinner rinds. Primitive Mexican winter squashes or pumpkins were cut into strips and dried. The beginnings of selection leading to summer squashes, which are picked while very immature for cooking as a green vegetable, are obscure. Such crops leave practically no trace in the archaeological record because hard rinds and seeds were not left as remains.

A possible independent domestication of *C. pepo* in eastern North America has been much discussed. Archaeological remains of rinds and seeds, presumably from wild *C. texana* dating from about 5000 B.C. onward, have been reported from Illinois and other eastern North American regions in nonagricultural context. The first finds of apparently cultigen *C. pepo* in eastern North America begin about 500 B.C. Later finds of *C. pepo* in agricultural context are widespread in eastern and midwestern North America and abundant in the Pueblo region of the Southwest. It seems likely that Mexican cultivars of *C. pepo* spread into eastern North America from the Southwest, along with maize, beans, and grain amaranth. Hybridization with any indigenous *C. pepo* cultivars derived from *C. texana* would account for the diversity of the crop at the time of European arrival.

Cartier in the St. Lawrence region in 1535, Hariot in Virginia in 1587, and many other early observers consistently noted that the Indians grew gourds, summer squashes for boiling, and winter squashes or pumpkins. They were struck by the variety of colors, an inheritance carried on today in the bright, multicolored ornamental gourds, the *C. pepo* cultivars most closely related to *C. texana*. The varieties grown as squashes rather than gourds were said to have a delighful flavor. Evidently they were all rather small. Champlain in New England in 1605 said they were the size of a fist; others compared them to an apple.

Introduction of *C. pepo* cultivars to Europe began very early, considering that they had to come from Mexico and North America rather than the earlier explored regions of the Caribbean. By 1550, European herbalists were familiar with both summer squashes, such as scallops, and winter squashes, such as pumpkins. By the late 16th century, various additional cultivars were recorded in European herbals and in Flemish and Dutch paintings. Some of these may have originated in Europe by hybridization between imported Indian cultivars. For example, the acorn squash, which was recorded by 1590, may have been a hybrid between a scallop and a pumpkin. Mexican pumpkins and North American ornamental gourds may have crossed in Europe to initiate the vegetable marrow and cocozelle squashes. From Europe, vegetable marrows spread to North Africa and the Near East, becoming the most popular summer squashes grown there, with a center of diversity in Anatolia. The zucchini group was selected out of the cocozelle type in southern Europe and spread through the temperate regions of the world as the most common summer squash.

A distinctive group of *C. pepo* summer squashes, the crooknecks and straightnecks, did not enter the historical record until after 1800, perhaps because they were developed by Indians in the interior of North America. Nuttall in 1818 reported that the Arikara Indians in the headwaters of the Missouri cultivated crooknecks. This group of squashes was widely disseminated in North America during the late 19th century.

All the *C. pepo* cultivars are interfertile when brought together. New hybrids can be quickly stabilized by isolation and inbreeding. Thus hundreds of named cultivars have arisen in modern gardens. Deliberate breeding has concentrated on the summer squashes, particularly the zucchinis. A new zucchini cultivar developed in the Far East is known as vegetable spaghetti. In the 1980s, it has become popular in North America.

Commercial seedsmen now supply growers with F_1 hybrid *C. pepo* seed, produced by interplanting selected inbred lines. As in commercial hybrid maize, the seed is harvested from plants from which the male flowers have been removed before shedding pollen or in which male sterile genes have been incorporated. Such seed is relatively expensive, but gives more vigorous and productive plants.

Cucurbita argyrosperma **(incl.** ***C. mixta*****)**

C. argyrosperma (incl. *C. mixta*) is another ancient, complex cultigen species, including some forms of gourds and squashes and cushaws. The wild progenitor is evidently *C. sororia* (incl. *C. kellyana*), a bitter-fruited wild gourd native to riparian habitats in semi-arid lowland regions of much of Mexico and Central America. A probable derivative is the weedy, camp-follower *C. palmeri* also with bitter fruits. Fruits of the wild and weedy species have been gathered since ancient times for use as gourds and for their edible seeds. Before 3000 B.C., archaeological remains in the Tehuacan caves in Puebla suggest selection had already developed cultivars with larger seeds. Definite cultivars are known from later archaeological sites in Mexico. By about 300 A.D., *C. argyrosperma* was being grown by Basketmaker people in the Southwest, and from 1000 A.D. onward, it was common in Pueblo sites, although less abundant than *C. pepo*.

The species is widely cultivated in Mexico and Guatemala today for its large, tasty seeds, which are roasted and sold as a snack. The species includes cultivars grown as squashes, e.g., Tennessee Sweet Potato and Japanese Pie, but the flesh is stringy and watery and less popular than other *Cucurbita* squashes.

Cucurbita moschata

C. moschata was the main pumpkin type cultigen of aboriginal agriculture in the humid lowlands of tropical America from southern Mexico to northern and western South America. The wild progenitor has not been identified, but a nameless bitter-fruited wild gourd in the Amazon headwaters is reported to hybridize with the cultigen.

The earliest known archaeological remains of *C. moschata* are from Huaca Prieta in the Peruvian coastal desert, where it would have required irrigation. The record of the crop there begins in preceramic, premaize context at about 2000 B.C. Later archaeological records in Peru include superb life-size models of the fruits by Moche potters. By 1400 B.C., the species was being grown far to the north in Tamaulipas, Mexico. A few archaeological specimens are known from the Pueblo region of the southwestern U.S.

Spaniards introduced *C. moschata* to the West Indies, where it is now the main squash grown. They probably also introduced it to Florida, where it was adopted by the Seminoles. By the late 17th century, *C. moschata* pumpkins were being cultivated in New England.

In modern times, various *C. moschata* cultivars, including Butternut and Golden Cushaw, have become favorite winter squashes throughout much of the U.S. mainly in regions with a hot, moist summer growing season. Nearly all the pumpkin pies are made from this species, with C. pepo pumpkins left uneaten as jack-o-lanterns.

C. moschata is now widely grown in the Far East with diverse cultivars selected in China and Japan.

Cucurbita maxima

C. maxima was the aboriginal summer squash of temperate South America. The wild progenitor is evidently *C. andreana*, a weedy pioneer native to the Plata River region of Argentina and Uruguay. The cultigen is still interfertile with the wild species, but has diverged radically in fruit structure.

What may be the oldest archaeological remains of the cultigen are undated but in preceramic context in dry caves in Mendoza, Argentina. Intact fruits dated about 500 A.D. have been found in Salta, Argentina. The earliest dated archaeological remains are from across the Andes in the coastal oases of the Viru Valley, Peru, dated about 1800 B.C. Later archaeological remains are known from other places on the Peruvian coast. Prehistoric cultivars of *C. maxima* were diverse, including both summer squashes, eaten when immature, and winter squashes. Among the latter were the largest of all pumpkins. A 16th century Spanish account from the Amazon headwaters in Bolivia described one of these Indian pumpkins as so large that a man could scarcely move it. Pumpkins of this species are the world's largest fruits, sometimes weighing 300 kg. They are commonly grown in all temperate regions. Other common *C. maxima* cultivars include Hubbard, Turban, Buttercup, and Banana squashes.

A Japanese seed company recently began offering interspecific *C. maxima* and *C. moschata* hybrid seed. The cross is not easy to make, and the seed is expensive. Plants grown from it combine desirable qualities of both parents and also have hybrid vigor. Fruits produced by this F_1 hybrid seed contain no viable seed.

Cucurbita ficifolia

C. ficifolia is a cultigen species; the wild progenitor is unknown. It evidently originated at fairly high elevations in the Andes and has remained a crop of cool temperate tropical mountains. Unlike the other squash species, *C. ficifolia* vines are perennial. Also the fruits are smooth and are remarkably similar to a watermelon in size, shape, and rind color. However, the flesh inside is stringy and eaten as a cooked vegetable. The fruits can be kept for over a year after harvest.

The archaeological record of *C. ficifolia* begins about 3000 B.C. at Huaca Prieta on the north coast of Peru in a preceramic, premaize context. There are remains from another Peruvian coastal site in the Casma Valley beginning about 2000 B.C. The fruits may have been brought down from the mountains.

At the time of the Spanish Conquest of Mexico, *C. ficifolia* was grown in the highlands of Central America and Mexico. The Spanish name *chilacayote* was derived from the Nahuatl *tzilacayotli*, probably meaning smooth squash. The Aztecs admired the jade-like appearance of the rind and made offerings to the rain gods of *pulque*, agave wine, in bowls made from the *tzilacayotli* fruits.

Spaniards introduced *chilacayote* to the Philippines, where it is cultivated in some mountain regions. It is also grown in the Himalaya, sometimes for feeding yaks. The species is occasionally grown elsewhere, but is of little importance.

LAGENARIA — BOTTLE GOURD (Burkill 1935; Cutler and Whitaker 1961; Dalziel 1948; Heiser 1973, 1989; Patiño 1964; Smith, B. 1989; Whitaker 1964, 1983; Whitaker and Bemis 1976; Whitaker and Carter 1954)

Lagenaria includes six wild species in Africa, one of which has been domesticated.

Lagenaria siceraria (=*L. vulgaris, L. leucantha, L. bicornuta*) — Bottle Gourd

In contrast to *Cucurbita*, which has yellow, dawn-blooming, bee-pollinated flowers, *Lagenaria* has white, night-blooming flowers. *Lagenaria* produces gourds that are much thicker-shelled (up to $^1/_2$ cm), tougher, harder, and more durable than those of *Cucurbita*.

Little is known about the native range of *L. siceraria*, but apparently truly wild populations occur in both South Africa and East Africa. The species evidently evolved in riparian habitats and relied on flotation for seed dispersal. The fruits are bitter and unattractive to animals other than man. The available archaeological record in Africa begins with Egyptian tombs about 3500 B.C., perhaps merely a curiosity traded from far away rather than locally grown. Finds are reported from Zambia and South Africa dated about 2000 B.C. and from Kenya about 800 B.C. These are not identified as being cultivars rather than wild types, but it seems likely that the species was an ancient domesticate. By the time it entered history, it was cultivated over most of tropical Africa by both agricultural and pastoral people.

In traditional African cultures, diverse *Lagenaria* cultivars produced gourds of many shapes and sizes, and additional forms were obtained by tying up developing fruits to produce constrictions and bends. After curing, the gourds were cut to produce bowls, stoppered bottles, plates, ladles, milking pails, and butter churns. They were lighter and less fragile than pottery. Also, the rind provided an ideal surface for engraving decorative patterns. *Lagenaria* was also important for native musical instruments, both wind and percussion. The gourds provided resonators for xylophones. Folk selection also produced nonbitter fruits that were cooked much like squashes. African cultivars also were used for edible seeds, some containing about 50% oil, which was extracted for cooking.

L. siceraria was introduced prehistorically to India. Dispersal to India by oceanic drift during the summer monsoon is conceivable, but it seems likely that the gourds were carried on ancient ships, like melons, sorghum, and other African crops.

The species was taken eastward bearing a Sanskrit name into Malaya and the East Indies. It was in New Guinea by 350 B.C. It was in China by 100 A.D. Use and cultivation of *Lagenaria* spread prehistorically through the Pacific islands all the way to Hawaii.

In Asia as in its homeland, *Lagenaria* played a multiple role: cooked as a vegetable and cured for a variety of utensils and musical instruments, including the *sitar* and pails for tapping palm sap and fermenting palm toddy. New cultivars were selected, some producing great gourds over a meter long. As a group, Asian and Pacific Island cultivars (*L. siceraria* subsp. *asiatica*) differ in seed and fruit characters from the original African stock (*L. siceraria* subsp. *siceraria*).

Comparative studies show that New World bottle gourd cultivars are derived from the African subspecies, not *asiatica*. The puzzle of how *L. siceraria* arrived in the New World was solved long ago by Whitaker and Carter (1954). Suspecting that the species had been introduced by ancient seafarers, they experimented with flotation of gourds in seawater to see if the possibility of natural dispersal could be excluded. Unexpectedly, they found the gourds could float and retain viable seed for many months, no limits being established, time enough for westward drift from Africa to tropical America or eastward drift in higher latitudes to the Pacific coast of South America. Such dispersal alone would not explain spread of the species, because it is not adapted to grow on a seashore, nor is it known to grow wild in any other habitats in the New World. The likeliest hypothesis is that drift gourds were found by ancient beachcombers, carried inland out of curiosity, and that volunteer seedlings came up in kitchen middens and other artificial habitats.

L. siceraria was evidently one of the very first species domesticated in the New World. Its archaeological record there begins long before any evidence of its planting in Africa. As usual, the record is biased in favor of arid regions and does not pinpoint the place of first domestication. The oldest dated finds are from caves in the interior of Mexico: Guila Naquitz in Oaxaca about 7200 B.C., Ocampo in Tamaulipas about 7000 B.C., and Tehuacan in Puebla about 5000 B.C. In North America, finds are dated at about 5300 B.C. at the Windover site on the Atlantic Coast of Florida, at 2300 B.C. in the interior at Phillips Spring, Missouri, and later prehistoric finds widely scattered from the Dakotas to Kentucky. In the southwestern U.S., archaeologic remains began about 300 B.C. and become ubiquitous in later cliff-dwelling and other Pueblo sites.

In South America, archaeological remains begin on the Peruvian coast about 4000 B.C. and are abundant by 3000 B.C. at Huaca Prieta, where the gourds were not only important for utensils but for floats for fishnets. Gourds are widespread in later archaeological sites in Peru and in northwestern Argentina.

Thus, in both North and South America, planting of bottle gourds began long before maize agriculture and long before pottery making. The gourds were probably priceless to Indians lacking pottery, especially in desert regions. Their cultivation did not cease with pottery-making, however. As in the Old World, they continued in folk use for light, nonfragile utensils, often beautifully decorated with engraved patterns.

The geography of *Lagenaria* planting in the New World at the time of Columbus is unclear, because Europeans confused *Lagenaria* with the unrelated tree calabash, *Crescentia cujete*. However, in western South America, Lagenaria can be clearly distinguished under its Quechua name, *mate*. *Mate* was reported in early Spanish accounts from Peru and Ecuador as widely traded in Indian markets. Besides many utensils, uses included floats for rafts to ford rivers and manufacture of musical instruments. The fruits were said to be bitter and inedible. During the colonial period, intricately engraved bottle gourds continued to be produced in Peru as art objects, as they are today.

In the U.S. and western Europe, would-be ethnic musicians and folk artisans are sufficiently numerous to support commercial growers of bottle gourds. One grower in the Central Valley of California ships thousands of gourds nationwide and internationally.

Ericaceae — Heath Family

About 70 genera belong to the Ericaceae, some with hundreds of species, ranging from the equator to very high latitudes in both the northern and southern hemispheres. The family includes shrubs, vines, epiphytes, and small trees. The family is remarkable for its ability to grow in infertile soils, commonly highly acidic. Several genera, including *Rhododendron* and *Arbutus*, are commonly grown as ornamentals. Only *Vaccinium* has provided domesticated crop plants.

VACCINIUM (INCL. *OXYCOCCUS* AND *CYANOCOCCUS*) — CRANBERRIES, BLUEBERRIES, AND RELATIVES (Galletta 1975; Sokolov 1981; Whealy 1989)

Vaccinium is a large genus with many species native to North America, Asia, and Europe with a few in Mexico, the Andes, and Madagascar. It includes evergreen or deciduous shrubs or subshrubs and some trailing vines, generally growing on acid sands and bogs, often in nearly pure stands. *Vaccinium* species are ecological pioneers of naturally open sites and also invade logged and burned sites and abandoned fields. Natural interspecies hybrids are common, the species generally being interfertile. The species form a polyploid series with diploid ($2n=24$), tetraploids ($4x=48$), and hexaploids ($6x=72$). *Vaccinium* berries, usually blue or red, are attractive to birds, and the seeds are thus widely dispersed, facilitating colonization of scattered habitat patches. Wild *Vaccinium* berries have long been gathered by people of many regions, even Greenland, under a variety of names: whortleberry, blue huckleberry, bilberry, whinberry, lingonberry, sparkleberry, cowberry, deerberry, to mention only a few. In addition to being flavorful, they are an excellent source of vitamin C.

Cultivation of *Vaccinium* leading to the domesticated cranberries and blueberries began remarkably recently in eastern North America.

Vaccinium macrocarpon — Cultivated Cranberry

There are several closely related *Vaccinium* species known as cranberries, both diploid and polyploid, with a combined circumboreal range. All have similar red berries that are gathered wild, but the only species cultivated as a crop is *V. macrocarpon*, a diploid native to northeastern North America. Its range is from Minnesota to Newfoundland and south to Tennessee and North Carolina. It grows only in acid bogs that are periodically flooded. The plants grow as creeping subshrubs, the stems rooting at the nodes. The plants are capable of photosynthesizing for long periods while submerged or covered with ice, unless shaded by deep snow cover. Artificial bogs are produced by dikes for controlled flooding. Cuttings from wild stands are covered with a thin layer of sand. Flooding is used to provide frost protection and also to float berries during harvest. Sand is added every few years to encourage rooting of trailing branches.

The first commercial planting was established about 1815 on Cape Cod, Massachusetts, by Henry Hall. The modern commercial crop, occupying about 10,000 ha, is still mostly grown within the native range of the species, in Massachusetts, New Jersey, Wisconsin, Quebec, and Nova Scotia. The crop was introduced to Oregon in the 1880s and has spread to Washington and British Columbia. About 900 North American farms grow cranberries commercially today; the great majority belong to a farmers' cooperative that markets the fresh and processed berries under the Ocean Spray brand.

The clones grown commercially are essentially wild genetically. Nearly all the crop is produced by four varieties, three of which were selected from the wild in Massachusetts, one in Wisconsin. During the mid-20th century, several experiment stations attempted to breed superior cultivars by deliberate hybridization and seedling selection. A few cultivars were released in the 1960s, mostly first generation crosses between the four main commercial varieties, but they have not had wide acceptance.

V. macrocarpon is being grown experimentally in several countries of central and eastern Europe.

Vaccinium angustifolium (incl. *V. lamarckii*) — Lowbush Blueberry

There are about 10 species of lowbush blueberry, all natives of eastern North America with much overlap in ranges and much hybridization. Of these, *V. angustifolium* is the most important commercially. It is a tetraploid (4x=48), supposedly an autopolyploid derivative of *V. boreale* (2n=24). The two often grow together. They range from Minnesota east to Newfoundland and Labrador and south to the mountains of West Virginia and Virginia. They are natives of bogs and open sandy and rocky upland sites. Bird dispersal allows colonization of new openings after fires or agricultural clearing. After arrival of Europeans, habitat opportunities were undoubtedly expanded with logging and mowing of meadows.

Berries of *V. angustifolium* are considerably larger than those of *V. boreale*, but still very small. The bushes are only ankle-high, and the berries are not easy to harvest in quantity. Commercial harvesting began in the late 19th century with the invention in Maine of a special hand rake with long, closely spaced tines that comb the berries off the vine-like bushes. Berries are often bruised by raking and winnowing and are seldom shipped as fresh fruit. Most are used for muffins and pancake mix or in other processed forms. However, something like 10,000 tons are harvested annually in New England and eastern Canada.

The entire crop comes from spontaneous stands, initiated by natural seed and clonal propagation. However, the stands are now generally managed to favor *Vaccinium* and discourage grasses and other competitors. The blueberry stands are burned over after harvest, which eliminates other shrubs and tree seedlings that would assume dominance. The blueberry rhizomes survive underground and resprout. Two years after burning, another berry crop is harvested. Some manipulation with herbicides and fertilizer is also involved. Genetically, the plants remain essentially wild. Experimental breeding has begun at various government stations, but successful cultivars have not yet been developed.

Vaccinium corymbosum and *V. australe* — Highbush Blueberries

There are about a dozen highbush blueberry species in eastern North America, diploid, tetraploid, and hexaploid, but the commercial crop comes from two tetraploids and their hybrids. *V. corymbosum* ranges from Michigan to Nova Scotia and south to New York. *V. australe's* northern border overlaps the southern margin of *V. corymbosum*; it ranges southward to Florida. Both grow in marshes, bogs, and wet sandy areas. *V. corymbosum* is a compilospecies incorporating genes from at least six other tetraploids. *V. australe* is supposed to have originated as an autopolyploid of the diploid *V. caesariense*.

Domestication of highbush blueberries was begun shortly after 1900 by Frederick Coville, a U.S. Department of Agriculture botanist. He was assisted for decades by Elizabeth White, an amateur blueberry enthusiast, on whose New Jersey farm the pioneer breeding was carried out. Selected wild clones were propagated vegetatively, and hybrids, produced by controlled pollination, were tested by the thousands. The first crop of hybrid blueberries was marketed from the White farm in 1916. In 1920, the first highbush cultivars were released to the public — Pioneer, Cabot, and Katherine, all three first generation *V. corymbosum* and *V. australe* hybrids. By 1959, the U.S. Department of Agriculture had released a total of 30 cultivars developed by Coville and his successors. Nearly all of these were *V. corymbosum* and *V. australe* hybrids, which traced back to a single wild *V. corymbosum* parent. By 1965, experiment stations and private growers in a dozen states were cooperating in highbush blueberry breeding. Highbush cultivars are planted on thousands of hectares in many parts of the eastern U.S. and are being tried in various European countries.

Vaccinium angustifolium and *V. corymbosum* — Lowbush and Highbush Hybrids

Coville crossed a wild New Hampshire lowbush selection with his wild New Jersey highbush selection, and this led to U.S.D.A. cultivar releases. In the 1970s and 1980s, other highbush x lowbush cultivars were released by the University of Minnesota and Michigan State. These are widely grown in the northern parts of the U.S., especially in home gardens and in "pick it yourself" farms.

Vaccinium ashei — Rabbit-Eye Blueberries

V. ashei is a hexaploid (6x=72) compilospecies, believed to incorporate germ plasm from at least five highbush tetraploids, including *V. australe.* It is even taller than the tetraploids, but the berries are relatively small. Rabbit-eye blueberries presumably originated as riparian pioneers, but they entered into a mutualism with human occupants of their native region as colonizers of old fields. The species range is restricted to Florida and adjacent regions in the far southeast of North America. In the late 19th century, wild seedlings were transplanted into Florida gardens and offered by commercial nurseries, sometimes with extravagant claims and disappointing production. Successful domestication of the rabbit-eye blueberry followed scientific breeding, begun in the 1930s by cooperative programs between the U.S. Department of Agriculture and state programs in Florida, Georgia, and the Carolinas. Several cultivars have been successfully introduced in the southeastern region for commercial production, "pick your own" farms, and home gardens.

Euphorbiaceae — Crown-of-Thorns Family

Euphorbiaceae are a family of hundreds of genera and thousands of species native to the tropics and temperate regions in general. Growth forms are extremely varied from tiny prostrate annuals to tall evergreen trees. Many are succulent, spiny, and resemble cacti. Most have milky juice. Some genera, especially *Croton* and *Codiaeum*, are commonly planted for hedges in the tropics. Other garden ornamentals include *Poinsettia*, the coral plant and other *Jatropha* species, the red-hot-cattail and other *Acalypha* species. Some euphorbs are planted commercially as sources of special oils, e.g., *Ricinus communis* (castor-bean) and *Aleurites moluccana* (tung-oil tree). Only the two most important crops will be discussed here.

MANIHOT — MANIOC (Burkill 1935; Cock 1985; Jennings 1976; Jones 1959; Miracle 1967; Nye 1991; Patiño 1964; Rogers 1963, 1965; Roosevelt 1980; Sauer, C. 1969; Ugent et al. 1986)

Manihot includes about 100 species of trees and shrubs natives of the New World from the southwestern U.S. to Argentina, with the greatest diversity in western and southern Mexico and in northeast Brazil. Many of the species have been experimentally hybridized, but in nature they are fairly discrete and segregated ecologically and geographically. Most species grow in arid or seasonally dry regions and drop their leaves during drought. They are adapted to riparian or other open habitats. *M. glaziovii*, the Ceará rubber tree, has been widely planted for its latex, but has been abandoned in favor of *Hevea* (see below). Only one species of *Manihot* is a major crop.

Manihot esculenta (=*M. utilissima, M. palmata, M. dulcis*) — Manioc, Cassava, Yuca

M. esculenta is the leading starchy root crop of the tropics. It supplies far fewer calories than rice and maize but far more than yams, sweet potatoes, taro, and other noncereals. Manioc roots contain about 65% water and 30% starch. Apart from some vitamin C, other nutrients are negligible. Nevertheless, Africa now harvests about 45 million tons, wet weight, a year; Tropical Asia and the Pacific harvest about 45 million, and the Americas about 30 million tons a year. Outside of the Americas, manioc is a remarkably recent crop, and much of its ascendancy is a 20th century phenomenon.

M. esculenta is a cultigen unknown in natural habitats, although often surviving after a field is abandoned. It is widely assumed, perhaps correctly, that domestication took place in northeastern South America, the center of manioc's aboriginal distribution as a staple crop. No particular wild species has been identified as its probable progenitor. Rogers (1963, 1965) has suggested that it is a compilospecies with genes pooled from perhaps a dozen wild relatives. The crop is always propagated by cuttings of the woody stems, not by seed. If the plants are allowed to mature, however, they flower and produce seed. They retain the wild type characteristics of explosively dehiscent pods and hard, dormant seed. Thus hybridization between *M. esculenta* and its wild relatives has taken place, and hybrid seedlings may have occasionally been adopted as new cultivars and propagated as clones. It is also possible that the crop was initially independently domesticated in different regions and later merged as cultivation spread. In any case, the crop includes an uncounted array of very heterogeneous clones.

Manioc's archaeological record reveals little about its ancient history. In Mexico, cave deposits dated at 1st millenium B.C. in Tamaulipas and at Tehuacan contain bits of plant tissue that have been tentatively identified as from manioc roots, but they might have come from wild plants rather than a cultigen. In South America, there is indirect archaeological evidence of possible manioc cultivation in the Orinoco Basin from about 2000 B.C. onward. This is based on finds of ceramic griddles and stone chips that match artifacts used by the aboriginal peoples in historical time to grate manioc roots and bake the meal into flat bread. Actual remains of the roots have not been preserved in the humid Orinoco-Amazon region.

In the Peruvian-Chilean desert where manioc was grown under irrigation, the earliest archaeological remains are from the Ancon-Chillon site dated about 2500 B.C. (Hawkes 1990) and from the Azapa Valley before 1500 B.C. (Rivera 1991). Starting about 1000 B.C. and continuing into the late period of Inca and Spanish conquests, abundant remains of manioc roots and other identifiable organs have been found in various Peruvian oases. Also there are superbly realistic ceramic models of manioc tubers attached to the base of the stem from the Moche and Chimu potters. There is no possible wild progenitor of the cultigen in the flora of the Peruvian desert. How far away and how long ago it was domesticated remains a mystery.

At the time of arrival of the first Europeans, manioc was cultivated almost throughout the New World tropics and to nearly 30° S.L. in eastern South America. There were many different clones of both so-called sweet and bitter manioc. Only sweet manioc was grown on the Peruvian coast, in the Andean valleys and eastern slopes, and in Central America and Mexico. It could be eaten raw after peeling or prepared by simply baking or boiling. It was usually interplanted with other crops in dooryard gardens and harvested after a single growing season. It was a moderately significant food, but not a staple.

From the Magdalena River in Colombia eastward through the Guianas and from the Amazon Basin to the Atlantic coast of Brazil as well as in the West Indies, the manioc crop was more complex. In addition to sweet clones, again as minor garden vegetables, bitter manioc was generally the staple crop. The roots contain a glucoside in quantities that vary with plant age and environment as well as between clones. The glucoside is converted to hydrocyanic acid when the roots are injured or dug, perhaps a defense against herbivory shared with various unrelated plants. The Indians retained this defense mechanism in the clones grown in shifting cultivation as staple field crops. There was no compelling motive for selecting nonpoisonous cultivars because detoxification was accomplished automatically during the complex processing required to convert fresh roots to cassava flour or tapioca starch. This usually involved grating the roots, squeezing out the juice, cooking down both the meal and the juice, and drying the meal, which could be stored indefinitely. Although dug up roots are highly perishable, they could be left in the ground for up to 3 years and dug as needed, meanwhile growing larger and more fibrous. The tall woody shrub competed well with weeds and needed little attention after the initial planting.

Wild *Manihot* species have somewhat thickened roots and substantial starch storage, but prehistoric Indian selection produced much larger, tuberous roots. A single plant commonly produces 5 to 10 tubers, some over 1 m long. Under optimal conditions, yields are phenomenal, over 30 tons of fresh roots per hectare. Manioc is usually grown under far from optimal conditions, and yields are usually 5 to 15 tons/ha. It could provide such yields on leached, acid soils with nutrient levels too low and aluminum levels too high for other staple crops. Also, although manioc has ordinary C_3 type photosynthesis, it yields well even where it has to survive long and irregular drought seasons. It does this by dropping most or all of its leaves under moisture stress and then rapidly growing new shoots when the rains come.

Without irrigation, the crop needs a good rainy season. It is grown in some areas, e.g., northeast Brazil, with less than 750 mm annual rainfall, but is mostly grown where rainfall is over 1000 mm/year. It can be grown in rain forest regions, but requires good drainage. Most cultivars are adapted to constantly hot lowlands; a few Andean sweet manioc varieties are grown in temperate elevations up to about 2300 m.

Manioc entered written history when Columbus landed on Hispaniola, where it was the major Arawak staple. The Arawak names *yuca* for the roots and *cassava* for the bread made from them became *lingua franca* words in Spanish America. The Portuguese first encountered the crop among the Tupi-Guarani of Brazil and derived from them the names *mandioca* for the roots and *tapioca* for the starch extracted from them.

The Spaniards were much impressed with the productivity of manioc in Arawak agriculture in the Greater Antilles. Bartolomé de las Casas calculated that 20 persons working 6 hours a day for a month could plant enough *yuca* to provide *cassava* bread for a village of 300 persons for 2 years. With abundant fish and sea turtle for protein and with their other crops for variety, the Arawak on Hispaniola had a well-nourished population of over a million people. They had abundant leisure for ball games, dances, music, and crafts. Columbus and his henchmen figured that the Indians' time could be better used as forced labor for gold mining and other enterprises. It took about 30 years after Spanish colonization of Hispaniola for the Arawaks to become extinct, partly by destruction of the native society, mostly by introduction of Old World diseases to which they had no immunity. Extinction of other West Indian native peoples took a little longer. By the time the Indians were gone, Spaniards had learned to grow *yuca* and make *cassava* bread, and the crop has been inherited by the African people brought to the West Indies to replace the natives as forced labor.

In eastern South America, bitter manioc also survived as a staple among whatever Indian cultures survived and was adopted by the immigrant ethnic groups. The Spaniards introduced bitter manioc and cassava making to Central America and the Mexican Gulf coast. It is now grown by the Maya in Yucatan along with sweet manioc, which they had grown before the Conquest.

Because of its cheapness and keeping qualities, bread made from manioc meal became a staple in provisions for the African slave trade early in the colonial period. In 1593, a Portuguese ship laden with manioc meal was captured off South America by Sir John Hawkins, himself one of the leading English slave traders. Hawkins reported that the ship had been bound for Angola to supply slave ships returning to South America. Evidently, manioc planting had not yet become established in West Africa.

Introduction of manioc planting to the Old World tropics in general was surprisingly slow. It seems likely that occasional undocumented early introductions failed because of episodes of cyanide poisoning. Old World tropical agriculture had a good roster of traditional staples: bananas, yams, taros, rice in the wetter regions, sorghum, and millets in the drier regions. There was little initial incentive to add a blander, pure starch source that was tedious to process and sometimes lethal.

The first records of establishment of manioc planting in the Old World were in West Africa in the mid-17th century. Some of these were on islands in the

Gulf of Guinea — Sao Tomé, Principe, Fernando Po — where the Portuguese had long established trading posts and plantations. The Portuguese and other European traders seldom ventured onto the mainland Guinea coast except at a few fortified ports or actual castles. Manioc was first reported being grown on the mainland in 1635 at the Portuguese post at Bissau. By 1682, it was an important crop at Warri in the Niger Delta, where Portuguese missionaries had gained a foothold and converted the local king to Christianity. By 1750, manioc cultivation had begun spreading inland from the Gulf of Guinea, but slowly and evidently only as a garden vegetable using nonpoisonous clones. Finally, late in the 18th century, bitter manioc planting and processing were established on the Guinea coast by freed slaves returning from Brazil. The black Brazilians became prominent middlemen in the slave trade between African sellers and European buyers. Eventually bitter manioc cultivation and manufacture of cassava bread and tapioca starch spread northward into the interior but did not penetrate through the forest region into the Sahel-Sudan savanna until the 20th century.

A separate introduction of manioc from Brazil to Africa was made south of the mouth of the Congo River, a region of seasonal forest and savanna less hostile to Portuguese colonization than the wet forests of the Guinea coast. Portuguese cultural and trade contacts with the native kingdoms there were far more extensive. By the 1680s, manioc was an important crop in Angola near Portuguese colonies. In the Congo Basin east of Angola, oral history of the Bushongo Kingdom recounts how manioc and peanuts were added to the traditional crops of yams, bananas, and millets. Manioc was initially grown as a vegetable to be sliced and boiled, but by about 1650, the Bushongo had learned how to make meal for baking bread. Adoption by other tribes of the Congo Basin came very irregularly. Adoption was often grudging and came only after failure of traditional crops through drought, plagues of locusts, or predation by baboons.

When the written history of the interior of tropical Africa began in the mid-19th century, manioc cultivation was widespread but patchy and still rapidly spreading. For example, Livingstone saw manioc among only certain tribes along the chain of great lakes from Lake Victoria to Lake Nyassa. East of there manioc was still almost totally absent in 1850 except right along the coast.

Introduction of manioc to the East Africa coast and other parts of the Indian Ocean remains quite obscure. It seems likely that viable stem cuttings would have been brought repeatedly by Portuguese and other early voyagers. The first definite record from the region was not until 1739 when the French colony on Reunion, then Ile Bourbon, received cuttings from Brazil. From there and the neighboring island of Mauritius, then Ile de France, manioc introduction to Madagascar, Zanzibar, and the East African mainland supposedly took place.

Acceptance of manioc as an indispensable staple crop was thus long delayed in much of tropical Africa, and its ascendancy has been mainly a phenomenon of the late 19th and the 20th century. The revolutionary economic development and huge population growth during this period demanded a food supply beyond the capacity of the well-balanced traditional agricultural systems. These systems were enormously diverse, including a wide spectrum from intensive, polycultural gardening and irrigated rice to shifting cultivation with brief cropping followed by a long fallow period. During the fallow, savanna or forest vegetation was allowed to regrow until enough humus and ash could be provided to support another crop. Manioc was eventually fitted into various phases of these systems, sweet manioc in gardens, bitter manioc in shifting cultivation mainly. As population growth outran the food supply, fallow periods were shortened, and crops were planted on less fertile soil. In crops such as yams and bananas that require highly fertile soils, yields declined rapidly. Even in hardy crops such as millets, shortening fallows brought declining yields. Manioc, being less demanding of soil nutrients, was increasingly the last resort. Sometimes manioc was replanted immediately after harvest, increasing the cropping period and even becoming a continuous monoculture. This was, of course, an expedient that was not a solution. In many regions, manioc yields have begun declining, partly because of soil exhaustion but largely because of bacterial and virus diseases and insect pests, problems that were absent when the crop was planted less continuously in time and space.

In Africa, manioc processing techniques have not been simply transferred from the New World. Africans have developed new procedures that are highly unstandardized and seem to be still in a state of flux and experimentation. They include soaking of the roots for brief or long periods, chopping, pounding, drying, boiling, and fermentation, in various permutations. The end product may be a dry meal or a moist paste. Sometimes there are problems with cyanide toxicity. Manioc foliage, considered toxic in the Americas, is often boiled as a potherb in Africa.

I cannot offer a history of the spread of manioc in Asia and the tropical Pacific. Early introduction by Spaniards to the Philippines seems likely, and spread from there by Chinese traders to Southeast Asia and the East Indies has been suggested. As in Africa, there may have been repeated early introductions that failed or led to only minor garden planting. Various recorded introductions during the 19th and 20th centuries, e.g., from Surinam to Java in 1851 and from Brazil to India by the British in 1886, added new clones. As in Africa, the ascendancy of manioc as a staple came during the late 19th and the 20th century as population growth overtaxed the traditional agriculture. Again, expansion was concentrated on marginal and deteriorating lands. Manioc is now a major food, particularly among poorer people in Southeast Asia and the South Pacific in general. It is a staple in such different places as southern India, Indonesia, Fiji, and Tonga.

Most manioc is still grown as a subsistence crop or for local markets by small land holders. Much of the crop is processed on the family or village level for local sale or for regional urban markets. The labor required for processing is greater than that used in growing the crop. Simple mechanization of processing is increasing rapidly in small factories dotted all around the tropics. A few large factories, which require large plantations to keep busy, are operating in Brazil, Venezuela, and Thailand. Thailand is unique in being an important manioc growing country whose crop is mainly processed for export. In addition to food products, such as tapioca, Thailand sells high quality manioc starch for use in sizing paper and in the textile industry.

Because of its low cost, manioc starch is often discussed as a basis for alcohol fermentation, and Brazil actually produced some ethanol used for motor fuel from manioc during the 1970s. This was part of the same program that produced fuel alcohol from sugar cane on a much larger scale. Neither process is economic with current low petroleum prices.

***HEVEA* — PARÁ RUBBER TREE** (Brockway 1979; Burkill 1935; Purseglove 1968; Webster and Baulkwill 1989; Wycherley 1976)

Hevea includes about 12 species of trees native to lowland rain forest and seasonal forest of the whole Amazon Basin and adjacent regions. All are diploid and interfertile; some natural hybridization occurs, but the species are fairly discrete and segregated ecologically and geographically. Most grow in habitats subject to brief seasonal flooding, but the species vary in tolerance of flooding; some grow on dry uplands. The flowers are entirely insect-pollinated, and individual trees can set seed by both self- and cross-pollination. The mature pods dehisce explosively and scatter the seeds to about 15 m, but dispersal by floating seeds can be very wide. All *Hevea* species have latex secreting ducts throughout the plant body. Outflow and coagulation of the latex presumably serve as a defense against boring insects and other injury. Latex is a watery suspension whose main organic component is natural rubber. A natural rubber molecule is an enormous branching polymer of thousands of isoprene (C_5H_8) building blocks. Elastic rubber can be obtained from the latex of several *Hevea* species, other genera of euphorbs, *Ficus* and *Castilloa* in the mulberry family, and other unrelated angiosperm families.

Elastic rubber can be made by simply drying successive layers of latex over a smoky fire. When the first Europeans arrived in the West Indies and Mesoamerica, the Indians had bouncy balls and other rubber artifacts. It seems likely that the South American Indians used *Hevea* similarly, but the first reports of such use within the range of *Hevea* were in the mid-18th century. The roster included balls, figurines, bracelets, waterproof capes, boots, and syringes (*Hevea=seringa* in Brazil), suggesting a combination of native and European inventions.

During the late 18th century, European surgeons began using rubber tubing and gloves made at Pará in Brazil from *Hevea* latex. By then, however, European chemists had discovered that rubber was soluble in turpentine, and soon they found better solvents. This opened the way for manufacture of rubber goods in Europe based on imports of raw rubber from Pará. During the early

19th century, Mackintosh and other inventors put on the market not only tubing and rubber gloves, but air mattresses, raincoats, rollers, and other useful but minor products. Their utility was limited by the fact that unvulcanized rubber is brittle and hard when cold, soft and tacky when hot. This problem was solved in 1839 by vulcanization, i.e., heating rubber with molten sulfur to produce sulfur bonds between the natural rubber molecules. By varying the amount of sulfur and other ingredients, all sorts of modern rubber products could be made: tough hoses, including air hoses for railroad brakes, drive belts, gaskets, electrical insulation, tires. We now have substitutes for natural rubber in some of these uses, but there were none at first. Wild rubber, mainly *Hevea* from the Amazon Basin, played a leading role in the industrial revolution. The demand exceeded the supply; the price skyrocketed.

A trickle of commercial rubber came from a lot of different plant genera growing wild in various tropical regions, but Amazonian *Hevea* provided far more than all the rest combined. In the 1850s, the Amazon rubber boom began like a gold rush, first in the lower basin, then up river at Manaus and eventually in the remote headwaters in country claimed by several countries. The rubber tappers led a miserable life of daily rounds to collect latex from trees scattered through the forest. Many of them came from impoverished drought-ridden farms of northeast Brazil. Formerly isolated Indian peoples were overrun and forced into virtual slavery. Large areas were ruled by the armed henchmen of lawless entrepreneurs. Demand for rubber continued to exceed supply, and the average price in 1910 was $2/lb.

Meanwhile, the British had been trying to start cultivation of *Hevea* in some of their colonies, a project directed by the Royal Botanic Gardens at Kew. In 1876, Henry Wickham, later Sir Henry, took a chartered British ship to the Rio Tapajoz, which flows into the Amazon near its mouth, and with the help of Tapuyo Indians loaded about 70,000 *Hevea* seeds gathered in the forest. Wickham's seeds provided the stock for the rubber plantations of Asia. Like most rain forest species, *Hevea* produces nondormant seeds, and most of Wickham's were dead before they could be planted. However, some 2800 seedlings were raised at Kew. Batches of seedlings were shipped to the British colonial gardens of Peradeniya in Ceylon and Singapore and to the Dutch botanic garden at Buitenzorg, Java. Soon a few other Hevea introductions were made to Kew and also directly from Brazil to Asia, but they are not known to have survived.

The *Hevea* species that was introduced to Southeast Asia was *H. brasiliensis*, the best source of rubber of several *Hevea* species that were being tapped in the Amazon basin. *H. brasiliensis* has a huge range in Brazil and neighboring countries and is quite variable in morphology and ecology. Wickham and his Indian seed collectors evidently chose a uniform, highly productive grove on an upland site.

During the 1880s, the species began flowering and producing abundant seed in Asian botanic gardens and on a few private estates. Commercial planting had not yet begun. A long period of trial and error intervened. Some plantings failed because of a misconception that the species grew in swamps. Tapping at first was very timid and rubber yield correspondingly unimpressive. Eventually, largely through experiments in botanic gardens, methods of planting and tapping were developed that promised profitable rubber yields. The first record of a serious commercial plantation was in 1898 in Malacca; the planter was not a European colonist but Tan Chay Yan. Meanwhile, the price of rubber kept rising, and coffee plantations began dying from a fungus disease, particularly in lowland regions. By 1905, *Hevea* planting had begun in earnest in Ceylon, Malaya, Sumatra, and Borneo. By 1914, Asian plantations were yielding more rubber than American forests, and new plantations were being laid out at a rate of 100,000 ha/year, mainly in Malaya and Sumatra. In some areas, rubber replaced coffee or tea; most of the planting was in former lowland forest that had previously been occupied by a sparse native population. Indentured Tamil labor was recruited by planters in Ceylon and Malaya; Chinese were also recruited for Malaya. The Sumatran plantations relied heavily on Javanese labor. The pattern of ownership of plantations was a complex mixture of large corporate monocultures, private European colonial estates, local Asian entrepreneurs, and family-operated small holdings. The large plantations could sometimes get better yields of higher quality rubber, but the small operations had important advantages. They could adjust tapping to fluctuations in the price of rubber and to demands of seasonal labor for their other crops. They could save their best land for rice and put a few rubber trees on poorer, drier sites. By 1914, the Malayan *Hevea* hectarage was about evenly divided between estates and smallholders.

Between World Wars I and II, *Hevea* production in Southeast Asia increased more than tenfold, mainly from Malaysia and the Dutch East Indies, but some also came from the Malabar Coast, Burma, Indochina, and Siam.

Between World Wars I and II, selection and breeding of *Hevea* were begun in earnest in the British and Dutch colonies in Asia, primarily to increase yields and disease resistance. Contrary to textbook theory, the narrow genetic bottleneck of Wickham's seed collection did not lead to lack of variation in the newly founded population. By keeping track of yields of individual trees, superior trees were identified. Wide variation was found within the first generation of commercial plantation trees, and budwood was taken from selected trees to graft onto seedling rootstocks for the next generation. During the 1920s, hand-pollination came into general use in *Hevea* breeding to produce crosses between the best and the best for new hybrid clones. In the 1930s, some breeders established seed orchards of the best clones to provide progeny for commercial planting. By then, some new plantations were yielding about a ton of rubber per hectare per year, nearly quadruple the initial yield. The best third generation clones reportedly could yield 5 tons/ha. There has been considerable lag in adoption of the new clones, not only because of the long time needed to evaluate and distribute them, but because a *Hevea* plantation is still in full production after 30 or 40 years and planters do not want to wait several years before getting any yield from a replacement planting.

Since 1950, fresh introductions of *H. brasiliensis* and other *Hevea* species have been sent from Brazil to Asia and West Africa to augment the original Wickham gene pool for experimental breeding. These have not yet contributed to commercial plantations.

By the 1930s, *Hevea* rubber was in such plentiful supply that the price was less than 20 cents a pound and planters were seeking to restrict production. The oversupply problem was solved during World War II when the Japanese occupied most of the Asian *Hevea* regions. A second boom in Amazon wild rubber ensued along with many expensive attempts to start plantations of *Hevea* and other rubber sources, most nearly or quite fruitless. In the long run, the main result of the World War II rubber crisis was the development of a huge synthetic rubber industry.

After World War II, as European colonial governments of Southeast Asia were replaced by independent nations, *Hevea* planting became increasingly the domain of small landholders, often with diversified crops. The former estates in Sri Lanka are now owned by 150,000 or so farmers, each with a hectare more or less in *Hevea*. Malaysia has about 400,000 rubber producers; Indonesia has perhaps 2 million, and of course, the Michelin plantations in Vietnam are now divided up. Generally the raw rubber is marketed by government supported cooperatives. China recently began planting rubber, mainly on Hainan Island and in Yunnan, using special clones selected for resistance to cold and wind. Some of the plantations have an understory of tea bushes. By 1984, China was fourth in the world in *Hevea* production. Between 1945 and 1985, *Hevea* production in Southeast Asia increased more than fourfold, supplying nearly 95% of the world's natural rubber. About 5% comes from West Africa, and less than 1% comes from South America.

African production is mainly from Liberia, where Firestone plantations of *H. brasiliensis* were established in the 1940s. As in Asia, these have been converted to small holdings.

Brazil has become a net importer of natural rubber and also is a heavy consumer of synthetic rubber. Early in the 20th century, *Hevea* plantations were established by the European colonial powers in their Guiana colonies. In the 1920s and 1930s, several *Hevea* plantations were started on the Tapajoz River, where Wickham's seed was collected, and elsewhere in Brazil and tropical America. Both local *H. brasiliensis* seed and improved clones from Asia were tried. The plantations suffered severely from a parasitic fungus *Microcyclis ulei*, a South American leaf blight, which the scattered and genetically diverse wild rubber trees could live with. It devastated large monocultures. So far the blight has not spread outside the Americas. In 1937, the Ford Motor Company, which had large, failing plantations in the Tapajoz, began a program to breed resistant trees, later continued by the Brazilian government. Blight resistance was found in populations of *H. brasiliensis* in the upper Amazon and also in other species of *Hevea*. Hybrids between high yielding and resistant populations, however, have so far not been successful, partly because new races of the fungus have evolved capable of attacking the new hybrids. In the 1980s, evidently despairing of being able to breed resistant cultivars, the Brazilian government announced a scheme to plant 200 million *Hevea* trees to be protected by fungicidal spraying involving huge ground-based mist blowers or aerial spraying. The fungicide will presumably have to be reapplied following each rainfall. Brazilian plantations are being extended south into drier regions in the hopes of reducing leaf blight.

Fabaceae (Leguminosae; Papilionaceae + Caesalpiniaceae + Mimosaceae) — Legume Family

Fabaceae are an enormous family of 600 or so genera and 15,000 or so species, extremely versatile in morphology and ecology. Legumes are important members of tropical, subtropical, and temperate vegetation throughout the world. Common growth forms include tall hardwood rain forest trees, lianas, low shrubs of savanna parklands and semi-desert scrub, and soft grassland herbs. Some belong to undisturbed communities, but many are ecological pioneers of burns and other open sites. The seeds of many genera have very tough coats and are capable of extremely long dormancy in the soil. Legumes are uniquely adapted to colonize infertile soil by their special relationship with *Rhizobium* bacteria, which inhabit root nodules and convert abundant, inert atmospheric nitrogen to compounds useful in metabolism of the host plants. Lavish use of nitrogenous compounds is especially obvious in legume seeds. These usually lack endosperm, but have rich stores of protein in the cotyledons. This is precious food for herbivores, and the legume seeds commonly are also loaded with alkaloids and other toxic nitrogenous compounds. Wild legumes typically have very hard seeds that, if simply swallowed, can pass through an animal undigested. Some genera produce bright red seeds, evidently attractive to birds. Others attract browsing animals by producing edible pods that hide indigestible seeds, e.g., *Prosopis* (mesquite) and *Ceratonia* (carob bean).

Most legumes produce the familiar pod, a single chambered fruit sutured along both margins and containing a row of seeds. Perhaps the basic, primitive pod type was one that dehisced explosively when ripe, scattering the beans away from the parent plant. In riparian species, the ejected seeds are buoyant and further dispersed by streams or ocean currents. There are many variants of the basic legume pod, such as inflated, buoyant pods, pods that disarticulate in one-seeded segments, pods that adhere to animals. The effectiveness of these adaptations is evident from the wide ranges of many legume genera, which span different continents and remote islands.

Legume flowers are typically insect-pollinated, which permits coexistence of a diverse flora occupying different niches in a complex community. The attractive flowers have led to cultivation of a great many genera as ornamentals.

A few legume genera have been domesticated as special purpose crops, e.g., *Indigofera* spp. prehistorically domesticated in several regions, producing indigo, a major commercial dyestuff until development of synthetic dyes; *Crotalaria*, sunn hemp, a low-grade fiber crop in India; *Glycyrhiza*, cultivated since ancient times in Eurasia for its extremely sweet roots, the source of liquorice; *Pachyrrhizus* spp., jicama or yam-bean, domesticated independently in ancient Mexico and South America for juicy, edible roots. Another odd legume crop, grown as a source of a useful poison rotenone, is briefly discussed at the end of this chapter.

The bulk of this chapter is devoted to legumes grown as high-protein food for humans and livestock. There are far too many of these for complete treatment here. Some are declining relics of mainly ethnographic interest, e.g., *Lupinus* species that were independently domesticated in the Mediterranean and the Andes, or *Canavalia* with four species independently domesticated in Central and South America, Asia, and tropical Africa. Several other genera that will be merely mentioned here are important subsistence crops in certain regions: *Cajanus*, the pigeonpea of India; *Vigna*, with several domesticates, including the cowpea and Bambara ground-nut of tropical Africa and the grams of Asia (mung, adzuki, moth, and rice beans); and *Vicia faba*, the broad bean of Eurasia.

As for livestock feed, an incalculable number of legumes have been managed and planted as pasture and hay crops, including several genera called clovers. These commonly serve a double purpose as soil building and green manure crops. Only *Medicago*, alfalfa, will be treated here.

PISUM — COMMON PEA (Gritton 1986; Hedrick 1919; Patiño 1969; Simoons 1990; Zohary and Hopf 1988)

Pisum is a small genus with three wild species native to the Mediterranean and the Near East. All three species are diploids and partly interfertile. The primary progenitor of the domesticate was evidently *P. humile* (=*P. syriacum*, *P. sativum* var. *pumilio*), a twining winter annual, native to oak parklands from Anatolia through the Fertile Crescent, the same region as the wild ancestors of barley and the early wheats. Later on, the cultivated pea is believed to have hybridized with *P. elatius*, a wild climber in the *maquis* scrub vegetation in moister regions of the Mediterranean.

Pisum sativum (incl. *P. arvense*) — Field and Garden Pea

Like other grain legumes, the domesticated pea differs from its wild ancestors in having indehiscent pods, which retain the seeds for shelling, and also in having seeds with thin, smooth coats that are permeable to water and thus have lost the wild-type dormancy. Unfortunately, the earliest archaeological finds are mostly carbonized seeds that lack coats. After 6500 B.C., however, seeds with smooth coats and thus clearly domesticated became common in most Neolithic sites in the Near East. It is evident that the pea was a regular companion of wheat and barley during the so-called Neolithic Revolution. Although the legume crop yields comparatively few calories, it contains over 20% protein, which the early farmers must have appreciated. Gradually, they selected larger seeded cultivars.

Archaeological finds show that by 5000 B.C. the cultivated pea had spread into Egypt, Cyprus, Crete, and through Greece into the Balkans. By 4000 B.C., the crop was widespread in central Europe and had reached the Rhine. By 3000 B.C., it was across southern France and into Spain. By 2000 B.C., it had spread west to Portugal, north to Denmark, and east to the Indus Valley.

The crop is mentioned in various classical Roman works, but was apparently not highly regarded.

There is the usual dearth of records for the Middle Ages, but the crop evidently continued to be grown over Europe in general. The crop is first mentioned in England after the Norman Conquest as grown by monasteries. The Renaissance herbals describe several varieties, including the field pea with small, round starchy seeds, a type familiar today in split-pea soup, and the garden pea with larger sweeter seeds, usually eaten green, and wrinkled if allowed to mature and dry out. Also peas with edible pods were already grown in northern Europe in the mid-17th century.

According to Chinese tradition, *P. sativum* was introduced from the west, possibly as early as the 2nd century B.C. There is clear evidence of cultivation in China by 600 A.D. It was adopted in medieval Japan.

European colonists introduced peas overseas early. Some of the early explorers' accounts mention peas as grown by the American Indians, obviously a confusion with the indigenous beans. By the 17th century, however, cultivation of the European pea seems to be definitely recorded in Spanish America, particularly in the northern Andean highlands. In North America, sowing of European peas seems to be clearly recorded in the early 17th century in the English colonies.

Both in Eurasia and overseas, *P. sativum* has adapted to a wide range of climates, from the subtropical, dry summer Mediterranean type in which it originated to subarctic cool summer and tropical humid highlands. This is partly due to being a quick growing ephemeral that can evade coping with unfavorable seasons. Also, the 8500 or so generations since *P. sativum* was domesticated have been enough for evolution of a diverse array of cultivars. The domesticate inherited the self-pollinating habit of its progenitor, along with occasional cross-pollination and infrequent hybridization events. As a result, whatever viable mutations occurred could quickly give rise to new pure-line cultivars. For the last 200 years, the number of cultivars has been rapidly increasing through deliberate hybridization. This was initiated in England by Thomas Andrew Knight in 1787, perhaps the first time that artificial cross-pollination was practiced to breed new varieties in any crop species. Scientific interpretation of *P. sativum* genetics began a century later in Moravia in Gregor Mendel's pea patch. Twentieth century breeding of the crop has been largely concerned with traits desirable for large-scale machine harvesting for commercial canning and more recently, for freezing.

LENS **— LENTIL** (Smartt 1990; Zohary 1976, 1989; Zohary and Hopf 1988)

The genus has four wild species native to the Mediterranean and Southwest Asia. All are small ephemerals, inconspicuous but locally common on barren rocky and gravelly slopes, usually at elevations between 1200 and 1600 m in the general oak parkland. They also volunteer on edges of wheat fields and on stone heaps. The tiny pods contain one or two seeds about 3 mm in diameter. All lentil species are mainly self-pollinated, but are interfertile. The primary progenitor of the cultigen is evidently *L. orientalis* native to the Near East; the other species may have contributed some genetic diversity to the crop.

L. culinaris (*L. esculentus*), the domesticate, differs from the wild species in having indehiscent pods, due to a single recessive gene, and nondormant seeds. There are many landraces of the crop, some with seeds no larger than the wild, some with diameters up to 9 mm.

Between 9000 and 7500 B.C., prefarming village sites in the Near East contain tiny lentil seeds presumably gathered wild. During the 7th millenium, the earliest Neolithic village sites contained lentils that may have been incipient domesticates. During the 6th millenium, almost all Neolithic sites in the Near East contained abundant lentils, clearly domesticated. By then, the archaeological lentil finds were no longer confined to the range of wild *L. orientalis.*

The spread of the lentil crop so closely paralleled that of *Pisum* that it need not be separately narrated. Like peas, lentils are a relatively low yielding but extremely high quality source of nutrition for people and their livestock. The seeds contain about 25% protein, filling a void in a diet of wheat and barley. Lentils went along with wheat and barley to Egypt and Ethiopia, to India and China, through Europe and with the European colonists overseas. The only notable difference in global geography from peas is that lentils have been less successful outside the warm, semi-arid, Mediterranean type climates. Their history departs from that of peas also in that lentils have remained simply a grain legume, like split peas, usually stored dry and not eaten as a green vegetable. For about 8000 years, lentils were evidently more important to mankind than peas, but for the last 500 years, they have declined in relative importance.

CICER **— CHICKPEA** (Patiño 1969; Ramanujam 1976; Zohary and Hopf 1988)

Cicer includes about 40 wild species of annuals, perennial herbs, and shrubs in southwestern Asia and southeastern Europe. Some are pioneers of scree and rock outcrops. The only wild species that interbreeds with the crop is *C.*

reticulatum, a self-pollinating annual growing on limestone in southeastern Anatolia.

Cicer arietinum — Domestic Chickpea

The historical geography of the chickpea starts out much like that of the common pea and lentil, but ends up differently. A few archaeological finds in and near the range of *C. reticulatum* probably represent gathering. In the 7th millenium, sites at Jericho and several other places in the Levant have larger seeded *Cicer*, probably domesticated. Bronze Age sites in the Fertile Crescent have smooth-seeded *Cicer*, definitely domesticated. The crop did not spread into northern Europe, but did spread westward to Greece by the 6th millenium, southern France by the 4th millenium, and Iberia by the 3rd millenium. The crop is important and apparently ancient in North Africa and Ethiopia. Several varieties were grown in the Roman Empire.

In the Indian subcontinent, the first available archaeological record of the crop is from about 2000 B.C. in north central India. Two completely different sets of folk names for the crop suggest separate introductions: one to northern India, perhaps overland, and one to southern India, perhaps by sea from Egypt. Northern India and Pakistan became by far the world's most important chickpea region. *Cicer* is the common gram (i.e., grain legume) especially but not exclusively among the poor, eaten as *dhal*, similar to split-pea soup, and in various other forms. The grain not only contains over 20% protein, high in lysine, but also about 55% carbohydrate and 5% fat. The value to the subcontinent is enhanced by being grown on marginal land, in the cool season, without irrigation. Chickpeas become especially vital in years with below average rainfall.

In Spain at the time of Columbus, three kinds of chickpeas or *garbanzos* were distinguished with white, golden, and dark seeds. They were standard provisions for Spanish voyages. The first crop in the New World was harvested in 1494 at Isabela on Hispaniola. The crop was introduced by early Spanish colonists in various places on the mainland, but did not do well in the humid tropical lowlands. During the Spanish colonial period, *garbanzos* became established as a temperate highland crop from Mexico to the northern Andes and on the Peruvian coast. *Garbanzos* are today standard fare in Mexico, although not the staple that *Phaseolus* beans are.

***GLYCINE* (INCL. *SOJA*) — SOYBEANS** (Burkill 1935; Dovring 1974; Hymowitz 1976; Hymowitz and Harlan 1983; Hymowitz and Newell 1981; Simoons 1990)

Glycine includes about half a dozen wild species, if broadly defined, mostly perennial, native to Southeast Asia, Pacific islands, and Australia. One species *G. soja* (=*G. ussuriensis*) is exceptional in being an annual native to temperate eastern Asia. It is a twining vine with dehiscent pods and small, hard, dark seeds. The immature, green seeds are edible when cooked. The mature seeds have a strong, unpleasant flavor and are rather indigestible unless elaborately processed. They did, however, offer a potentially very nutritious food, with protein excellent in quantity and quality and a remarkably high oil content.

G. max, the cultigen soybean, was evidently domesticated in northeastern China in antiquity. Clear evidence of its cultivation is not available until about 1000 B.C. By the 3rd century B.C., it was regularly reported as one of the two most important food crops of northern China, the other being millet. Before 100 A.D., soybean cultivation was general in central and southern China and Korea and may have already spread to Japan, through Indochina to Burma and India and into the East Indies. By 100 A.D., Buddhism was spreading out of India into China. Its vegetarian tenets must have been hard to follow as it left dairy cultures behind. The timely arrival of the soybean may have provided a crucial substitute for dairy products.

The soybean is normally self-pollinated, with about one seed in a hundred the result of cross-pollination. As a result, innumerable landraces, nearly pure genetically, were selected, differing in adaptation to latitude and photoperiod, in seed color, and many other ways. As a group, however, the cultivars differ from the wild *G. soja* in being larger, more erect plants with reduced shattering of the ripe pods, larger seeds with higher oil content, better flavor, quicker cooking, and more digestible. The cultigen species does still offer some problems of digestibility. For example, the ripe seeds contain tryptophan inhibitors which limit availability of essential amino acids in the seed. The inhibitors can be inactivated by proper cooking, but overcooking reduces protein quality in other ways.

G. max became a vital pillar supporting the enormous human populations of the Far East and Southeast Asia, including Java. In central and eastern Java, soybeans are regularly grown as a dry season crop, rotated with wet season rice. The importance of soybeans in Asian cuisines depends on diverse and sophisticated methods of preparation. In addition to using green beans and bean sprouts, the mature seeds are processed by infinitely varied combinations of soaking, cooking, milling, liquefying, coagulating into curd or *tofu*, pressing, drying, and pickling. The most elaborate procedures involve inoculation with fungal or bacterial cultures, e.g., in Indonesian *tempeh*, Japanese *miso*; microbial fermentation is also involved in making soy sauce, an ancient Chinese *ketchup*. In addition to providing nutrition, the results are often tasty and have become globally appreciated in ethnic Asian cuisine, but preparation has generally remained a mystery to non-Asian cooks.

Starting in the 18th century, Europeans who contacted the crop in Asia repeatedly brought seed back home or to the colonies. Seed was brought to eastern North America by British sailors who had served with the East India Company. One of these men Samuel Bower began planting soybeans in Georgia in 1765; he tried to start manufacturing soy sauce, noodles, and other Oriental recipes. Benjamin Franklin sent soy seed from London to William Bartram. Commodore Perry brought back Japanese seed in 1854, but continuous planting began only after introduction of three other Japanese varieties in 1900. Initially, U.S. farmers grew soybeans mainly as a minor hay crop. By far the most important introductions came with 4000 or so seed lots collected in 1929 to 1931 in northeastern China, Korea, and Japan by two U.S. Department of Agriculture plant explorers.

The U.S. soybean crop was of little importance until World War II when a shortage of butter created a strong market for soybean oil for margarine. Soybean oil was soon found excellent for salad oil, soap, and many other purposes and rapidly displaced other vegetable oils. The residual cake was rapidly accepted as a high-protein feed for livestock. Farmers in the midwestern Cornbelt found that rotating soybeans with maize, usually once every 4 years, reduced the need for nitrogen fertilizer. The crop proved to be suited to mechanized planting and combining. The soybean crop skyrocketed. By 1973, U.S. farmers were planting over 20 million ha of soybeans and harvesting 45 million tons, about 3/4 of the total world crop. Soybeans have surpassed wheat, maize, and all other crops in cash value and in the value exported. Soybeans are grown commercially in over half the U.S. and several Canadian provinces, but Illinois and adjacent Cornbelt states produce most of the crop.

As far as the North American diet goes, *G. soja* is nearly invisible. A fair quantity of the beans is eaten disguised in mixtures with wheat and other ingredients for enhanced protein. The dominant role of the species as a source of vegetable oil is not generally appreciated, nor is its role in producing pork, eggs, and other farm products.

Soybeans have been planted experimentally since the 19th century in other parts of the world, but as in the U.S., were slow to become established. In Europe, the crop temporarily became important for human food in Germany and Austria when those countries were starving during and after World War I. The crop has since been used mainly for oil and feedstuff and is grown in moderate quantities in eastern Europe. Soybeans are grown commercially in several Latin American countries with Brazil by far the most important. Brazil's first significant crop was in 1927 in Rio Grande do Sul, and for 20 years, the crop remained minor and confined to that state. In 1968, soybeans began replacing coffee in much of southern Brazil. By 1978, Brazil was harvesting 12 million tons of beans in seven states and exporting most of the crop. The beans are usually rotated with wheat, reducing the need for fertilizer.

Attempts have been made to establish the crop in tropical Africa, without notable success. It seems unlikely that soybeans have much future in humid tropical regions, but their success in parts of Java shows they might do well in seasonally dry tropical climates. Unfortunately, the whole history of soybean introductions, except for the one systematic U.S.D.A. collecting expedition in 1929 to 1931, has been of rather casual seed transfers, without regard to the narrow adaptations of particular races to latitudinal belts and photoperiod. Most attempted introductions in Latin America came from quite inappropriate latitudes.

The U.S. soybean germ plasm banks, mainly at the University of Illinois, contain only a fraction of the genetic diversity developed over centuries by Asian farmers. Over 2/3 of the 4000 seed lots collected in Asia in 1929 to 1931 have been lost. Many may be irreplaceable because old landraces have undoubtedly been lost in China, Korea, and elsewhere as the traditional peasant farms were converted to large-scale, centrally planned collectives and other more modern systems. Nearly all the commercial soybean cultivars in the U.S. are derived either directly or through experiment station breeding from just 11 original Oriental landraces.

PHASEOLUS — NEW WORLD BEANS (Brücher 1988; Carter 1951; Evans 1976; Gentry 1969; Gepts 1988; Hedrick 1919; Kaplan 1956, 1965, 1981; Kaplan and Kaplan 1988; Mackie 1943; Martin and Adams 1987; Patiño 1964; Silbernagel 1986; Singh, S. 1989; Smartt 1990)

Until recently, *Phaseolus* was defined to include about 200 species native to the tropics and subtropics of both the New and Old Worlds. The genus is now narrowly defined to include only about 30 New World species. All the other species have been reassigned to *Vigna* or *Macroptilium*.

Four species of *Phaseolus* were prehistorically domesticated by the American Indians for their high protein seeds, which were an invaluable complement to the staple starch crops and to lysine deficient maize protein. Two of these species — *P. acutifolius*, the tepary bean, and *P. coccineus* (incl. *P. polyanthus*), scarlet-runner bean — will not be discussed; they are important crops only in restricted regions near their original homelands. The other two species, the common bean and the lima bean, have much more complex histories.

Phaseolus vulgaris — Common, Kidney, Navy, String, and Wax Bean, *frijol*

Linnaeus based the binomial on plants cultivated in the Old World, and no likely wild progenitor was identified until very recently. Since 1960, several botanists have independently discovered apparent close wild relatives of the cultigen in many places in the western Cordilleran system of the Americas. The wild populations differ from the cultigen in having explosively dehiscent pods; seeds half as big as in the cultivar with hard, impermeable coats; and delayed, irregular germination. The wild seeds are dark brownish and often mottled.

The wild and cultivated beans can be crossed artificially to produce fertile F_1 hybrids, and apparent spontaneous hybrids are found as weeds. However, both the wild and cultigen plants are normally entirely self-pollinated, the pollen being shed before the flowers open. Presumably crossing is due to bees forcing their way into unopened flower buds.

Wild *P. vulgaris* populations have a phenomenally wide latitudinal range, extending with some gaps from north of the Tropic of Cancer in Chihuahua to far south of the Tropic of Capricorn in Cordoba, Argentina. They grow at medium to fairly high elevations in the mountains, usually as annual vines that grow during summer rains and survive a long dry season as dormant seed. Habitats are rather diverse. In the Andes, the wild beans grow mainly in broadleaf deciduous forests. In Mexico, they grow in more open xerophytic woodlands and naturally open sites, such as arroyo banks and rocky slopes; they also grow in secondary vegetation in clearings and abandoned fields. Where domestic livestock are present, the vines are usually restricted to thickets of thorny, unpalatable shrubs. In some Mexican sites, the wild beans twine on stalks of *Zea mexicana*, teosinte, the wild progenitor of maize. There is no evidence, however, that the two species were domesticated and spread as a partnership.

These wild bean populations were named first in South America as *P. aborigineus* or *P. vulgaris* subsp. *aborigineus*. In Mexico, they have been called *P. vulgaris* var. *mexicanus*. A case could be made for assigning the Mesoamerican and South American wild beans to separate subspecies or even species. They differ in both morphological and biochemical traits and have internal barriers to interbreeding. Experimental F_1 hybrids between these two groups are largely inviable or sterile. However, formal taxonomic separation is better left undone. As we shall see, cultivars of *P. vulgaris* came from separate domestications in Mesoamerica and South America. Therefore, formal splitting of wild *P. vulgaris* would require parallel splitting of the cultigen derivatives; this would open a nomenclatorial Pandora's box. This can be sensibly avoided by using informal names for subdivisions of *P. vulgaris*, both wild and domesticated.

Independent domestications are strongly supported by electrophoretic analysis of seed proteins in wild and cultivated beans (Gepts 1988; Gepts et al. 1986; Singh et al. 1991). Eight genetically controlled types of phaseolin, the main seed storage protein, have been identified in different races. In the wild populations, the types are geographically segregated. Two of the types found in wild populations have not been found in any cultivars. The other phaseolin types show strong regional correspondence between wild and cultivated landraces. The great majority of the cultivars belong to two types: the S-type, evidently originating in Mexico and Central America, and the T-type, evidently originating in the northern and central Andes. In northeastern South America, in Brazil, and in the West Indies, S- and T-type cultivars are roughly equal in frequency, as would be expected if the crop had originally spread eastward from Colombia and then been taken into the islands. The geographic patterns of phaseolin in conjunction with isozyme and morphological patterns suggest that the cultigen *P. vulgaris* has six major races derived from independent domestications between northern Mexico and the southern Andes; subsequent hybridization has only slightly blurred the discreteness of these major races.

There are no known archaeological remains of *P. aborigineus* or of forms transitional to the domesticated *P. vulgaris*. In South America, this would be expected because nearly all the archaeological work has been done outside the range of the wild populations. In Mexico, however, some archaeological sites with plant remains are within regions where the wild species grows. Presumably the wild beans were too rare there to provide a regular food source. The dark, mottled seeds are well camouflaged after being scattered in the litter from an exploding pod. It seems remarkable that the wild beans ever attracted enough human attention or were plentiful enough to initiate planting.

The archaeological record so far has not revealed any stage of incipient domestication. All the seeds recovered are much larger than the wild. The earliest of all, from levels dated about 5500 B.C., is from Guitarrero Cave in the Callejon de Huaylas, an inter-Andean valley in northern Peru. This was long before the introduction of maize or pottery making. In Mesoamerica, the earliest *P. vulgaris* find, like those of maize, is from the Tehuacan caves in Puebla; a single pod valve was found in levels dated 5000 to 3500 B.C. and another in levels dated 3500 to 2300 B.C. Thus, by 5000 B.C., *P. vulgaris* had been independently domesticated in Mexico and the Andes. The only other very ancient archaeologic finds are from even more remote regions: Ocampo, Tamaulipas, Mexico in levels dated earlier than 2000 B.C., and Pichasca on the west side of the Chilean Andes in levels dated about 2700 B.C.

In Peru, *P. vulgaris* did not spread from the highlands into the coastal oases, with their great civilizations and rich archaeologic plant record, until about 500 B.C. In North America, the crop did not spread into the maize-growing cultures of the Southwest until about 300 B.C. nor to those east of the Mississippi until 1000 A.D. Even slower than introduction of the crop was its acceptance as a staple food. Everywhere there was a lag of centuries before the beans became common in the archaeological crop record, and in some places, it took far longer. For example, in the famous Tehuacan and Ocampo sequences in Mexico, *P. vulgaris* became abundant only after 100 A.D., in the Pueblo region of the Southwest after 1000 A.D., and in eastern North America after 1200 A.D. Maize and beans did not advance initially as the crop complex they formed later. Perhaps people felt no need to grow beans until hunting and fishing failed to provide enough animal protein to feed their rapidly increasing populations. Nevertheless, before 1492, beans had caught up with maize over most of the New World and had become a staple.

Moctezuma's hieroglyphic tribute list shows that just before the Spanish Conquest, 18 provinces were contributing about 230,000 bushels of beans as annual tribute, compared to about 280,000 bushels of maize and 200,000 of amaranth grain (Barlow 1949). No doubt some of the beans were from other *Phaseolus* species, but the bulk was certainly *P. vulgaris*.

Almost every European explorer from Columbus on reported beans wherever indigenous agriculture was practiced. Instead of using Indian names, the explorers used names of Old World grain legumes, such as *Pisum sativum*, the common pea, or *Vigna unguiculata*, the cowpea. (The New World beans eventually usurped the Greco-Roman name of the cowpea, *Phaseolus*, as well as Spanish derivatives, *frisol, frejol.*) The beans were clearly recognized as different from and often as better than the European ones. Many accounts noted the abundance of the Indian bean crop and the diversity of seed colors. Among such accounts written during the early 1500s were Cartier in the St. Lawrence, Verrazzano in New England, Narvaes in Florida, De Soto in the Mississippi Valley, Cabeza de Vaca in the Southwest, Oviedo and a multitude of others in tropical America.

The diversity of seed colors noted by these explorers has remained characteristic of the bean crop of peasant agriculture in various parts of Latin America, particularly among highland Indians. In markets, mixtures of half a dozen or so strikingly different *P. vulgaris* seed colors are commonly offered in a single sack. A vendor with nothing better to do may sort out a particular cultivar for sale at a slightly higher price, but normally the medley is used and planted together. The diversity does not represent a gene pool. The cultivars are self-pollinating pure lines that rarely, if ever, interbreed. These mixtures presumably give the crop flexibility, spreading out seed germination and varying response to weather. Grain ripening and harvest are thus extended, but this is acceptable when pods are handpicked. By comparing medleys of beans in modern Mexican markets with those recorded archaeologically and historically, Kaplan (1981) found a complex situation. At Tehuacan, mixtures offered in the modern market are essentially identical to those found in archaeological caches from about 1000 A.D. Such stability is not found in other parts of Mexico, however. For example, some very common modern cultivars have sulfur-yellow seed in northwest Mexico, black seed in the southeast, and a rabbit-eye pattern in the central plateau, none of which are represented in the well-preserved archaeological records of the region. Obviously evolution and diffusion of cultivars did not cease with the colonial period.

The New World beans were rapidly taken around the world as ship stores, and innumerable introductions overseas went unrecorded. The Old World acquired a grand mixture of cultivars from different regions, e.g., the navy bean, acquired from the northeastern Indians but ultimately of Mexican origin; the pinto bean, another Mexican cultivar; the large red kidney bean of Andean origin. Both the Mexican and Andean lineages contributed edible-podded string and stringless cultivars, the more slender ones being Mexican. Illustrations and descriptions in European herbals show that various *P. vulgaris* cultivars were being grown in Europe by the mid-1500s. Their diversity, particularly in seed colors, kept European taxonomists totally confused for 200 years after that.

East Africa has a remarkable pattern of *P. vulgaris* planting as mixtures of cultivars, about 80% Andean, 20% Mexican in origin. The species is now well established in subsistence agriculture from Ethiopia down through the eastern highlands to South Africa. Multiple introductions were probably begun from Brazil by the Portuguese in the early colonial period. The crop spread inland in native cultivation ahead of European exploration. In the 1850s in what is now Malawi, David Livingstone found a whole roster of American Indian domesticates abundant in native gardens: maize, pumpkins, tobacco, sweet potatoes, and beans. Currently farmers in Malawi grow *P. vulgaris* in even more complex mixtures than those used in Latin America. Often over 20 recognizably different seed types are harvested together. This suggests a preference for crop insurance through diversity rather than selection for quality.

Acceptance of the introduced bean in Africa was probably facilitated by the similarity in cultivation and use with the indigenous cowpea *Vigna unguiculata*. In some regions, particularly the cool highlands of Uganda and adjacent countries, *P. vulgaris* is the most important staple, accounting for over half the calories in the total diet. How much of the diversity stems from original introductions is unknown.

The species was introduced to Java during the Dutch colonial period and accepted into native gardens. In Asia in general, it has not been able to displace the old legume crops, such as the cowpea, mung bean, and soybean.

About 75% of the *P. vulgaris* cultivars grown in Europe and eastern North America today are ultimately derived from the central and southern Andes, about 25% from Mexico. They have, of course, undergone mutation and selection in their peregrinations, but modern cultivars remain mainly the product of natural and folk selection.

Scientific breeding during the 20th century has produced commercially important new cultivars. These are preferred by growers largely because they are adapted for mechanical harvesting. For example, canners and freezers of green beans want a bush rather than a pole bean plant form and also a crop that can be harvested once rather than by repeated handpicking. Breeders were able to transfer the high quality pods of the Blue Lake pole bean type to cultivars suitable for machine harvest. Unfortunately, breeders have also been able to produce new green bean cultivars that are tough enough to look attractive after shipment, handling, and a long time in a fresh produce market or on a steam table.

Phaseolus lunatus **(incl.** ***P. limensis*****) — Sieva, Butter, Lima Bean**

The binomials *P. lunatus* and *P. limensis*, as well as others belonging to this group, were all bestowed on cultivars, and the wild relatives have no Latin name of their own, although they are morphologically quite discrete. Seeds of the wild limas are less than half the dimension and therefore less than $^{1}/_{10}$ the weight of even the smallest cultivar seeds. They also are dark brown with black mottling and contain, at least in some cases, 10 to 20 times the glucoside concentration found in cultivars. The glucoside breaks down to hydrocyanic acid when the seeds are bruised or chewed, and the wild beans are generally considered inedible, although they can be detoxified by sufficient boiling in changes of water. The content of glucoside in cultivars, particularly those with white seeds, is well within the tolerable level. The wild plants are strongly photoperiodic and mostly do not flower until around the fall equinox, a trait which has been bred out of some tropical and all temperate cultivars. Most cultivars differ from the wild in being grown as annuals, and in later prehistoric time, the Indians developed erect, quick maturing bush beans.

Wild lima beans are common in Central America, particularly Guatemala, at low to medium elevations on the seasonally wet and dry Pacific slope. The plants usually have perennial roots. The vines grow during the rains, twining on thickets and roadside hedges, and die back during the dry season. The range extends northwestward on the Pacific slope of Mexico through Oaxaca and Guerrero to Jalisco, with an outlier on Socorro Island. The range is poorly delimited in the Caribbean, but is reported to include Belize, islands of the coast of Venezuela, and Puerto Rico. Wild limas are reported rather vaguely from the east slope of the Andes in Peru, Argentina, and Brazil. There is no archaeological record of domesticated *P. lunatus* anywhere within the reported range of the wild progenitor, nor are transitions or incipient domesticates known. Experimental crosses have shown, however, that the wild and cultigen are fully interfertile.

The archaeological record strongly suggests independent domestication in Central America and northwestern South America. From the outset, the beans of the two regions were quite different in size. The Central American type, although far larger than the wild, were small-seeded compared to the South American type, a difference that is still easily recognizable in modern cultivars. Sieva beans, butterbeans, and baby limas belong to the Central American type, the large limas to the South American type. The groups are quite interfertile.

The South American large *pallar* type was the first to be recorded archaeologically as a domesticate. It appeared about 6500 B.C. in Guitarrero Cave in northwestern Peru about the same time as *P. vulgaris* and long before maize. The site is in a thorn steppe vegetation in a deep valley where the species could not have been naturally present nor would irrigated agriculture be expected at such an early date. Perhaps the beans were packed in from east of the Cordilleran crest. About 3000 B.C., *pallares* entered the archaeological record in incipient agricultural levels at Huaca Prieta on the north coast of Peru. After 1500 B.C., *pallares* were preserved in abundance in the irrigated oases that supported the great prehistoric civilizations along the Peruvian coast. Under the extreme aridity, the color patterning of the ancient beans can still be seen. Also, *pallares* were a motif of the classic pre-Inca pottery and textiles, with the diverse color patterns beautifully depicted. Patiño (1964) reproduces the pattern on a Mochica pot with both realistic glyphs of the beans and anthropomorphic figures of lima bean warriors. Archaeologists have suggested that the *pallar*, with its varied colors and markings, served as a model for ideographic glyphs found on ceramics and textiles of several pre-Inca cultures. It seems that the species was of more than prosaic dietary interest to the anicient Peruvians.

The other type, the smaller sieva bean, entered the archaeological record much later, but again in arid sites outside the range of the wild progenitor.

It showed up roughly simultaneously at levels dated before 800 A.D. in widely scattered parts of Mexico: Tehuacan caves in the central east, Ocampo in the far northeast, and Dzibilchaltun, Yucatán. This has been interpreted as diffusion from Central America northward following domestication in Guatemala, which is the present center of diversity of sieva type limas in folk cultivation.

By about 1300 A.D., the sieva had been introduced to what is now Arizona, presumably by the Hohokam. Perfectly preserved sieva beans of that age were found in a cliff dwelling in Tonto National Monument. Very similar sieva cultivars survive in the so-called Hopi Limas. It is possible that the crop was introduced in pre-Columbian time to eastern North America, either from the Southwest or from the Gulf Coast of Mexico. Hariot's famous report "on the new found land of Virginia", published in 1588, described Indian crops in what is now North Carolina. Carter (1951) suggests that one of them seems to have been *P. lunatus*. Hariot (1590) distinguished two Indian crops, one resembling English peas, i.e., *Pisum*, and the other resembling English beans, probably *Vicia faba*. This does indeed suggest the presence of both *P. vulgaris* and *P. lunatus*. About 1700, a so-called bushel bean being grown in the Carolinas was described as very flat, white, and with a purple figure, almost certainly referring to *P. lunatus*. Eastern North America is a possible source for a *P. lunatus* bean grown in Holland clearly shown in drawings published by Lobel in 1591. Very similar beans were described later as growing spontaneously in abandoned Indian gardens in Florida.

In tropical America, *P. lunatus* may have entered written history when Columbus reported two kinds of beans grown by the Indians of Cuba, one under the name of *fava*, which implies resemblance to *Vicia faba*. Unfortunately, as noted above, most early Spanish explorers seem to have lumped all the American *Phaseolus* crops under *frijoles*. Only in Ecuador did some early accounts compare Indian bean crops to *habas*, i.e., *V. faba*. These accounts suggest that *P. lunatus*, undoubtedly of the *pallar* type, was grown not only in the Peruvian oases but also in moister climates of Ecuador, both on the coast and east of the Cordilleran crest. By the 18th century, Spaniards had adopted the Peruvian name *pallar* for the crop. In Spanish America, the name *haba* has now been reclaimed by *V. faba*, which is now commonly grown in highlands above 2000 m elevation, above the usual zone of lima beans.

Like other grain legumes, limas made superb stores for sailing ships and were presumably disseminated worldwide on innumerable unrecorded voyages. The earliest clear proof of introduction to Europe was in the drawings published by Lobel in 1591, which depicted not only the sieva, as already noted, but the large *pallar* type already under the name lima. Limas were probably first introduced to Asia on the Manila galleons. They were in Cochin-China by 1790 and India soon after. By the early 19th century, the *pallar* type had been introduced to the Cape of Good Hope, Madagascar, and the Mascarene Islands.

Both sieva and *pallar* limas are now worldwide garden crops, mainly in low rainfall tropical and warm temperate regions. The Hopi and some other sieva types are especially heat and drought resistant. The *pallar* types retain preference for mild, foggy coastal regions like their Peruvian homeland. The largest scale, mechanized field cultivation of both types is in California. Commercial lima bean farming was founded on a *pallar* variety introduced in the early 19th century. According to tradition, it did not come directly from Peru but in a Yankee clipper ship from Madagascar, where the Malagasy farmers of the arid southwest coast had selected a quick-maturing variant. Growing of large, *pallar* type limas, such as the Ventura and Fordhook, remains concentrated along the southern California coast. In California, growing of sieva type baby limas, such as Henderson and Wilbur, developed later in the 19th century in the interior Great Valley. The Mackie, a hybrid between a sieva parent obtained from the Hopi Reservation and a *pallar* type similar to the Fordhook, was released in 1955. It was selected to provide a commercial strain combining the large seed size of one type with the heat resistance of the other. Modern scientific breeding, however, has done far less to change the genetics of the lima bean than was done by prehistoric Indian farmers. Outside the U.S., the crop remains essentially a subsistence crop.

ARACHIS — PEANUT (Burkill 1935; Gregory and Gregory 1976; Krapovickas 1969, 1973; Patiño 1964; Smartt 1990)

Arachis is a South American genus of about 35 species of annual and perennial herbs native to seasonally dry savannas and scrub forests between the Amazon and La Plata rivers. The plants are mostly prostrate or trailing over the ground, rooting at the nodes. All the species have tough, indehiscent pods that ripen underground. The flowers usually do not open and are self-pollinated. After pollination, a meristem below the young pod, called the peg elongates and pushes the pod down into the earth and away from the parent plant. In the wild species, another meristem in the middle of the pod pushes the halves of the pod, each with a single seed, apart, so the seeds do not germinate right together.

The pods are not known to be dug up by any native animals, although swine root them out enthusiastically, and the seeds lack hard coats that would resist digestion. Burying the pods seems to be primarily an adaptation for survival *in situ* by a soil seed bank that evades the seasonal drought. Nevertheless, *Arachis* seed must have occasionally served the more ordinary functions of genetic recombination and dispersal. Variation patterns of the populations indicate that some cross-pollination and even interspecies hybridization have occurred. The geographic ranges of the species are evidence of effective dispersal, most likely by washing out of pods by running water and later burial under alluvium. The plants require loose soil and often grow in pure sand.

With the seeds hidden, *Arachis* was able to dispense with inedibility, and the South American Indians found this out long ago. Two species were domesticated prehistorically, *A. villosulicarpa* and *A. hypogaea*. The former is a perennial diploid (2n=20) cultivated by a few Indians in Mato Grosso. The latter is a major world crop.

Arachis hypogaea **(incl *A. africana, A. asiatica*) — Peanut, Groundnut, *maní***

The peanut is unusual in its genus in being a polyploid (4x=40). It can interbreed only with one other species, *A. monticola*, the probable wild progenitor of the crop (also 4x=40), which is native to the eastern foothills of the Andes in northwest Argentina and southern Bolivia. The domesticate differs from *A. monticola* in lacking the meristem in the middle of the pod that pushes the two seeds apart in the soil. Some cultivars differ from *A. monticola* in having more than two seeds per pod, larger nondormant seed, and in being erect rather than prostrate. Analyses of the wild seeds are not available; the cultivars generally have about 45 to 50% oil and 25 to 30% protein, and the protein has a desirable amino acid makeup; peanuts also supply some vitamins. On the downside, under certain conditions, peanuts can cause liver cancer in humans due to a fungal aflatoxin. This is dangerous only if fungus infected peanuts are consumed in large quantities, a problem mainly in parts of Africa.

The crop entered the archaeological record only after it had been taken across the Andes to the Peruvian desert oases. It was being grown there before 2000 B.C. and became common during the 1st millenium B.C. In addition to abundant remains of intact pods, the fruit was depicted in textile patterns and on the elegant pottery produced in the coastal city states before the crushing Inca Conquest.

A. hypogaea entered written history with the Spanish entry into Hispaniola, where the Arawak cultivated it under the name *maní*. A cognate name, *mandubi*, was reported from the coast of Brazil about 1550. These names and variants are still used for the peanut among Arawak peoples from the Caribbean across the heart of South America to the region where the crop originated. Early Spanish and Portuguese accounts record the presence of the crop through most of the West Indies and South America. The range extended northwestward into the temperate valleys of the northern Andes but not into the Pacific or Caribbean lowlands of Colombia, nor into Panama. There are no accounts from European contact times of Indian cultivation of peanuts anywhere in Central or North America. The first Spanish mention of the crop in Mexico states that Spaniards had brought it from the West Indies. It was given a Nahuatl name meaning ground-cacao, now corrupted to *cacahuate*. The only report of archaeological peanut material in Mexico is from the Tehuacan caves where some peanut shells were found in the top levels, which may date from after the Spanish Conquest. One seed was found in a lower level, dated before 750 A.D. Standing alone, this must be suspected of being intrusive through disturbance.

A. hypogaea is usually divided into four major varieties, all traceable to prehistoric American Indian selection. Unfortunately, the names of three varieties are quite misleading. The so-called Virginia variety may have been developed in Amazonia and later in Indian cultivation in the West Indies. The Peruvian variety is aptly named; it is the common type in archaeological sites in the oases of Peru. The Spanish variety was grown by the peoples of northeastern Brazil. The Valencia variety may have been developed by the Guaraní peoples of the Paraguay-Paraná basin. The Virginia and Peruvian varieties are prostrate, have seed dormancy, and require a 5- to 10-month growing season. The other two varieties are erect, have nondormant seeds, and mature in 3 to 5 months. The Spanish variety is especially rich in oil.

Krapovickas (1969) postulated a history of introductions since 1500 that can be summarized as follows:

- The "Virginia" variety (*A. africana*) was taken from the Antilles to Mexico soon after 1500 and was probably quickly introduced to West Africa by slaving ships. Peanuts were being cultivated in Senegambia in the 1560s and by 1600 were a widespread West African crop and used to provision ships. The variety almost certainly was introduced to eastern North America from both the West Indies and West Africa in the 17th century.
- The "Peruvian" (*A. asiatica*) was taken to the Philippines by Spanish galleons and from there to southeastern China before 1600. It gradually spread throughout China and to Japan. Chinese traders and settlers were probably largely responsible for introduction to the rest of Southeast Asia and Indonesia.
- The "Spanish", an erect variety, was almost certainly taken from Brazil to Africa very early to mix with the prostrate "Virginia" and produce the great diversity of African landraces. Introduction of the "Spanish" race to Spain evidently occurred in the late 18th century, from Brazil via Lisbon. From Spain, it was introduced to southern France and finally, in 1871 to the U.S.
- The "Valencia" was probably introduced to Spain from Cordoba, Argentina, about 1900 and was introduced from Valencia to the U.S. about 1910.

This sketch has only attempted to trace initial, successful introductions. There have been innumerable modern exchanges of varieties and improved cultivars worldwide.

The peanut is now a significant source of human nutrition almost everywhere with suitable climate and soils. It needs hot summers, alternating wet and dry seasons, and sandy soils. The nuts are valuable for a great many uses in addition to simply being roasted and eaten or ground for butter. Different ethnic groups discovered independently that the fine quality cooking oil could be easily extracted in various ways and the residual high-protein cake could then be used as an ingredient of tasty cooked dishes. There are a great variety of these in Africa, Southeast Asia, and especially China. Some recipes involve microbial fermentation, as in soybeans. Commercial mills crushing peanuts for oil were developed independently in Asia and Europe early in the 19th century. In Europe, crushing was pioneered in Marseille when the demand for olive oil could not be met, the supply of nuts coming mainly from West Africa. After the opening of the Suez canal, the French mills imported large quantities of peanuts from the Madras region of India, via the port of Pondicherry. In Europe, the residual cake is mostly fed to dairy cattle. Africa and Asia have continued to supply most of the export crop.

MEDICAGO **— ALFALFA** (Bolton 1962; Hanson 1972; Lesins 1976; Patiño 1969; Wilsie 1962; Wing 1912)

Medicago is a genus of about 50 species of annual or perennial herbs and low shrubs, mostly native to Eurasia, some in Africa. The species are generally pioneers of naturally disturbed or open sites, including grasslands. Some are common weeds of pathways, roadsides, and lawns, e.g., black medic and bur clover. Bur-like pods of several species are apparently well adapted for adhesion to animal fur.

The genus includes diploids (2n=16), tetraploids (4x=32), and hexaploids (6x=48) with different ploidy levels repeatedly found within morphological species. This is interpreted as due primarily to autopolyploidy rather than allopolyploidy involving interspecies hybridization. *Medicago* chromosomes are extremely short, perhaps too short to have crossing over with more than one partner during meiosis. This would preclude weaving together of multivalents by multiple crossovers, the cause of sexual sterility in many autopolyploids. In any case, the polyploids in *Medicago* are sexually fertile and genetically stable.

Medicago is one of several legume genera whose seeds were gathered from the wild for human food in the early Neolithic of the Near East. *Medicago* seeds are abundant in some Fertile Crescent archaeological sites in levels dated to the 8th and 7th millenia B.C. Unlike some other legume genera already discussed, however, *Medicago* was never domesticated as a grain crop, and the archaeological finds diminished as other crops developed. After 6000 B.C., although no longer important for human food, *Medicago* continued to be gathered as hay for livestock. After a long delay, this led to domestication of one species.

Medicago sativa — Alfalfa, Lucerne, Medic

This species, the most valuable of the world's forage crops, is a tetraploid (4x=32). It is not known truly wild, although it commonly escapes cultivation in artificially disturbed sites.

It is derived primarily from *M. coerulea*, a diploid (2n=16) that grows wild in grasslands of southwestern Iran, the Caucasus region, and eastern Anatolia. The tetraploid is a more vigorous grower than the diploid, and some ancient farmers evidently appreciated the difference and began planting selected seed for pasture and hay. It is a perennial and has sometimes been grown as a permanent pasture or hay field, but it usually remains highly productive for only 3 to 6 years before being replanted or rotated to other crops.

Alfalfa was not part of the ancient Neolithic wheat-barley-sheep-cattle agricultural complex. It was probably not domesticated until the Bronze Age and perhaps primarily for feeding horses. Domestication of the horse had begun by about 2500 B.C. somewhere in the steppes of the Ukraine or Inner Asia. By 2000 B.C., nomadic Indo-Aryan warrior horsemen had crossed the Caucasus. Early in the 2nd millenium B.C., they were establishing their rule over the Near East and southwestern Asia, with the horse-drawn chariot playing a crucial role. The care, feeding, and breeding of horses became matters of supreme importance, particularly among kings and aristocracy. Alfalfa has been claimed to be mentioned in a Babylonian text from about the 7th century B.C. This is far outside the native region and would suggest cultivation under irrigation.

Alfalfa growing probably came to Europe in the 5th century B.C. when it was introduced to Greece during the Persian wars. *Medicago* is derived from the classical Greek name attributing it to Medea. The Romans were growing it by the 2nd century B.C. Cultivation in Europe is not known to have survived the fall of the Roman Empire. The crop was reestablished in Spain in the 8th century A.D. during the Moorish rule. Lucerna is a famous horse-breeding center near Cordoba in southern Spain.

Little is known about the ancient and medieval geography of alfalfa as a crop in the hinterlands of Eurasia. By the 18th century, the crop was grown from northwestern Europe to China by people among whom horses or dairy cattle were important and valuable. By Chinese tradition, alfalfa was brought from Turkestan in 126 B.C. by an expedition sent to acquire the much admired Iranian horses.

M. sativa includes a diverse assortment of cultivars adapted to a wide range of temperate and cool temperate climates. Much of this diversity is attributable to hybridization with wild *Medicago* species during spread of the crop across Eurasia. The most important genetic contributions were from the so-called Siberian alfalfa *M. falcata*, the source of extremely winter-hardy cultivars (once given species status as *M. media*, *M. varia*).

Other *M. sativa* varieties are thought to have genes obtained by hybridization with *M. glutinosa*, native to the Caucasus, and *M. glomerata* of the Mediterranean region.

The broad climatic tolerance of alfalfa is partly owing to its remarkable root system. In deep, well-drained soils, the modest shoot system may be supported by roots reaching downward 7 m or more to permanently moist soil. Alfalfa is in no way a xerophyte, however, and needs irrigation on really dry soils. Alfalfa needs unleached, nonacid soils and thrives on soils rich in calcium carbonate. Its historical geography has been closely linked with limestone substrates.

The Spaniards introduced alfalfa widely to the New World in the 16th century, but it was generally a failure in humid lowlands. By 1650, the crop was notably successful in irrigated desert oases around Lima in Peru. Later in the Spanish colonial period, alfalfa and sugar cane largely displaced wheat on the central coast of Peru. By 1800, alfalfa was being grown in the highlands of Ecuador and Colombia. Today in Latin America, alfalfa is a widespread but minor mountain crop in the Andes and Mexico. It is a major crop in Argentina and Chile.

The U.S., now a leading grower of alfalfa, was amazingly slow in developing the crop. It did not spread north from Mexico until the late 19th century. By then, it had been introduced from afar by several pathways. Around 1850, ships coming around the Horn to the California Gold Rush brought alfalfa seed under the name of Chilean clover. Other varieties were introduced from Peru and Argentina. Experimental plantings often thrived, but expansion of acreage was very gradual. California livestock were expected to forage for themselves on unimproved rangeland, not depend on irrigated pasture or expensive hay. It took a while to learn that more intensive livestock raising was profitable. A man who found it most profitable was Henry Miller, who started out in Gold Rush, San Francisco as a butcher and began acquiring ranch land in the San Joaquin Valley to be brought under irrigation. Alfalfa hay, many thousands of tons of it a year, became an integral part of Miller's (later Miller & Lux) ranch, one of California's greatest agribusinesses. With development of large irrigation systems, alfalfa spread over huge areas of California's Central Valley and southern deserts.

During the late 19th century, irrigated alfalfa growing spread through the Great Basin and into the Colorado Rockies. Pioneer cattle ranchers in those regions had depended on open range that soon became overstocked and overgrazed. Catastrophic dieoffs of livestock during severe winters forced development of haying, initially in natural wet meadows along the streams coming down from the mountains. Artificial irrigation works and alfalfa planting soon followed. Alfalfa proved important to the Mormon settlers; it was ideally suited to their pattern of intensive, diversified agriculture and animal husbandry. Also, Utah became a major source of commercial alfalfa seed for planting in much of the West. Before the end of the 19th century, alfalfa growing had spread east of the Rockies into the High Plains, from Nebraska and Kansas to Texas.

In the Middle West, alfalfa varieties introduced from South America via California were joined by varieties brought from Europe via the eastern states. These had been introduced repeatedly over a long time span, often unsuccessfully. The earliest records are from the late 18th century. George Washington and Thomas Jefferson, for example, experimented with alfalfa in Virginia. The crop was called lucerne, suggesting it came from France or England. For a variety of reasons, the crop failed to become important in eastern North America throughout the 19th century. The main reason was that farmers had not yet learned that in this region alfalfa needs applications of agricultural limestone and also subsoil drainage. Also, *M. sativa* had come from regions with milder winters and drier summers than eastern North America, and some evolutionary adaptation was prerequisite for success in the new environment. An important advance in this regard was introduction and subsequent selection of a variety brought from Baden in Germany to Minnesota in the mid-19th century by an immigrant farmer Wendelin Grimm. The Grimm cultivar proved extremely winter-hardy, probably due both to some genes from *M. falcata* in its ancestry and to generations of natural selection in Minnesota during the late 19th century. Around 1900, the U.S. Department of Agriculture imported alfalfa cultivars from interior Eurasia, some of which were also winter-hardy. Some of the new introductions, such as Cossack, Ladak, and Turkestan, were hybridized and selected to produce a major synthetic cultivar Ranger.

Starting about 1900, alfalfa planting spread rapidly in much of the U.S. and parts of Canada. The most spectacular expansion was in the Lake States and High Plains east of the Rockies. In the Cornbelt, alfalfa expansion was rolled back when farmers inserted the soybean into their crop rotation.

Alfalfa planting was also a 19th century development in Australia, New Zealand, and South Africa.

Alfalfa is now planted in many regions where its seed yields are poor and seed is imported from distant sources. Commercial seed from southwestern North America, for example, is planted in the northeast. This obviously entails risk of adaptation loss to severe winters, a problem met by constant testing for trueness to type. Another chronic problem in obtaining alfalfa seed is lack of pollinating insects. The ubiquitous European honeybee eagerly extracts nectar from alfalfa flowers but soon learns to avoid triggering the anthers and releasing pollen. Bumblebees and a variety of other solitary bees are quite effective pollinators, but they are not usually abundant enough to service wide expanses of alfalfa. Progress has recently been made toward domesticating and transporting solitary bees in artificial nesting cavities placed in commercial alfalfa seed fields.

DERRIS* AND *LONCHOCARPUS* — *TUBA* AND *BARBASCO (Burkill 1935; Killip and Smith 1930; Patiño 1967; Purseglove 1968)

Both these closely related genera are pantropical and include numerous species of diverse growth forms and habitats. Those that have interested humans are native to lowland forests. Seedlings start out as erect understory trees, but upon growing up and contacting larger trees, they become climbers, clambering up into the forest canopy where they finally flower and bear pods. The organs of interest to people are the roots, which have an extremely high content of various toxic organic compounds, including rotenone ($C_{23}H_{22}O_6$), supposedly as a defense against soil invertebrates. Native peoples occasionally use the roots to kill lice and ticks, but the roots are far more important as fish poisons. Traditionally, *Derris* and *Lonchocarpus* roots were simply ground or pounded, for example in a dugout canoe, and dumped into streams or ponds. The results are quick and spectacular, with dead and stupefied fish of many kinds floating on the surface to be taken by hand. Rotenone is lethal to some fish in dilutions of 1 part in 10 million. The fish are still edible, the toxins having no appreciable effect on mammals unless ingested in large quantities. This practice was important and apparently ancient among native peoples of the Old World tropics from Southeast Asia through Indonesia to the Philippines, Melanesia, and Australia. It had developed independently in the New World tropics.

In addition to *Derris* and *Lonchocarpus*, several other legume genera were used as fish poisons, as well as some plants in other families. The folk taxonomy commonly lumped all the fish poison plants under one name, e.g., *tuba* in Malay, *barbasco* in Spanish, and various other regional names. Thus, early accounts are usually quite ambiguous as to species involved. The situation has improved only moderately with scientific botanical exploration. Few specimens have been collected with the flowers and pods needed for accurate identification, so the geography of the various species is still not well known. It seems, however, that two species *Derris elliptica* and *Lonchocarpus nicou* are especially important both in folk use for fishing and in commercial insecticide production.

D. elliptica is believed to be native to forests from Burma eastward to Siam, Indochina, and the Philippines and southward into some of the East Indies, where it was noted as a fish poison by Rumpf about 1650. Although rarely seen in flower, it is easily propagated vegetatively by stem cuttings, and the roots can be harvested 2 years later. Folk use as a shampoo for head lice may have suggested use against garden insects, but this usage was not recorded until about 1850. Cultivation of *D. elliptica* for insecticidal use seems to have been pioneered in Southeast Asia in the late 19th century by Chinese gardeners. In the early 20th century, European planters in Southeast Asia, including India, rapidly adopted *Derris* for garden use. Samples were taken to Europe and Japan for chemical analysis, with extraction of rotenone and other compounds not long after.

Commercial planting of *D. elliptica* boomed in Malaya, the Dutch East Indies, and the Philippines until the Japanese occupation in World War II. By then, the species had been introduced experimentally in many tropical countries. The U.S. Department of Agriculture had select *D. elliptica* varieties in cultivation in Puerto Rico, which supplied the stock for wartime emergency plantings in various West Indian and Central American regions, especially on United Fruit banana plantations. New World rotenone production has depended mainly on the following species, however.

L. nicou entered scientific history in 1773 when Aublet described it from French Guiana using the local Indian name for fish poison plants in general as its specific name. Various South American Indian tribes traditionally used not only wild plants but propagated the species vegetatively, planting stem cutings that could be harvested after a few years, sometimes in the same field as their staple starch crop *Manihot*. This was a sensible combination because the fish were crucial in supplying protein for which *Manihot* is a notoriously poor source. A single family would require perhaps 25 to 100 *Lonchocarpus* plants, but villages planted much larger numbers. *L. nicou* was evidently in Indian cultivation over much of the tropical forest region of the Orinoco and Amazon basins from moderate elevations on the east slope of the Andes to the river mouths.

Chemical analysis of South American *Lonchocarpus* roots showed they contained rotenone just like *D. elliptica* from Asia. Smithsonian botanists identified *L. nicou* from Peru as one of the most potent sources. During the early 1930s gathering and planting of the species for export boomed, especially around the ports of Iquitos in the upper Amazon basin and Belem at its mouth. By 1940, even before the supplies of Southeast Asian *Derris* were cut off, the U.S. was getting most of its rotenone from *Lonchocarpus* root imported from South America.

L. nicou was introduced in 1932 to an experiment station in the Virgin Islands. In 1932, living plants of two unidentified kinds of *Lonchocarpus* were introduced from British Guiana to Malaya by way of the Royal Botanical Gardens at Kew, England. Whether the American species can compete with *Derris* is unclear.

More recently, the abandonment of DDT and mistrust of other synthetic insecticides have greatly enhanced the importance of both rotenone sources.

Fagaceae — Beech Family

The Fagaceae have over 500 species in six genera, including *Quercus*, oaks; *Castanea*, chestnuts; and *Fagus*, beeches. The species are all hardwood trees and shrubs with simple evergreen or deciduous leaves. Most have inconspicuous wind-pollinated flowers. Wind pollination is effective due to the enormous success and abundance of the species. Many are widespread in forest, woodland, and scrub vegetation. They are most important in low and middle latitudes of the northern hemisphere. *Nothofagus* emigrated to the southern hemisphere during Cretaceous time.

The fruits of Fagaceae are one-seeded nuts, without endosperm but with much carbohydrate stored in the cotyledons. The fat and protein contents are remarkably low. Many contain bitter tannins. Because of their abundance, however, they are a staple food, especially in fall and winter, for many kinds of birds, squirrels and other rodents, deer, and various other mammals. In spite of much destructive consumption of the nuts, animal dispersal evidently works very well for the Fagaceae.

Ever since humans arrived in Europe, Asia, and North America, Fagaceae have been a familiar part of the environment and an invaluable source of fuel, timber, and food. They were so abundant that even after millenia of heavy exploitation deliberate planting has seldom been considered necessary. Chestnuts, *Castanea* species, have been widely planted as individual trees and groves in Europe and eastern Asia since early historical times.

QUERCUS — OAKS (Parsons 1962a, b)

Among the many oak species that have been exploited as natural stands, two Iberian species are outstanding as semi-domesticates.

Quercus ilex and *Q. suber* — Holm Oak and Cork Oak

These two evergreen oaks are widespread in the Mediterranean region but important as crops in Andalucía, Extremadura, and adjacent regions in southern Spain and Portugal. They form extensive open woodlands on hills and lower mountain slopes. *Q. ilex*, *encina* in Spanish, is valued mainly for its acorns; *Q. suber*, *alcornoque* in Spanish, is the source of commercial cork, with the acorns a secondary product. *Q. ilex* grows on soils generally good enough to be planted intermittently to wheat, barley, or a legume crop. *Q. suber* grows mainly on sandy and rocky slopes that could support a scrub understory, but are kept free of shrubs for pasture grasses. In both cases, the trees are typically pruned by itinerant experts, with the trimmings used for charcoal. The resulting landscape is closer to an orchard than a forest.

The importance of acorns for both human food and mast for swine has been recorded in Iberia since classical Greco-Roman times, when acorn bread was a staple. The medieval Moorish conquerors recorded that Spaniards tended oaks for their high quality acorns, even planting, irrigating, and manuring the *encina*. Herding of pigs to fatten on oak mast was, of course, suspended during the Moorish occupation, but rapidly re-established with the Christian reconquest. After 1492, Iberian pigs regularly accompanied colonists to the New World. De Soto and his Extremaduran companions set out to explore North America driving a herd of hogs in the van.

In the Iberian peninsula, acorn bread and gruel have continued to be eaten by poor peasants. A large part of the modern pork supply is fattened on mast. Pigs are generally herded through the oak groves from October to January. Sometimes the landowner collects a fee from itinerant swineherds, based on weight gain of the pigs during this period. A 25-kg pig in October may emerge a 70-kg hog in January.

Cork tissue is not unique to *Q. suber*. Among woody plants in general, corky bark forms as replacements for the primary water-retaining epidermis, which cracks as the stems expand by cambial activity. What is special about the cork oak is the continuity and homogeneity of the corky tissue. If the outer, dead corky bark is skillfully stripped without cutting into the inner, live bark, the tree rapidly regenerates a thick, solid cork tissue. Cork can be stripped, sometimes in 2-m long slabs, every 10 years or so, while the tree may live for 150 years.

Q. suber bark was first used in antiquity for fishnet floats and stoppers for amphorae and casks of olive oil and wine. The industry was initially diffuse through the Mediterranean range of the tree. It became more geographically concentrated during the 18th century after cheap glass bottles replaced casks for wine. By 1800, Spain and Portugal were emerging as suppliers of cork for Champagne and other wine regions. During the 19th century, about 100 towns in Andalucía and Extremadura had cork cutting factories. In Spain, the industry continues today as a relatively minor and picturesque part of the economy. Cork is far more important in Portugal. During the 1970s, the country had about 600,000 ha of cork orchards, mainly south of the Tagus, and the industry produced over 100,000 tons of cork products annually, the country's most valuable single export.

In spite of cheap plastic bottle stoppers, real corks remain important for better wines. Huge quantities of cork, much of it reconstituted from scrap, are used in tile, insulation, gaskets, and electrical insulation. The bark that protects the oak from hot, dry winds, and fire has proved wonderfully impervious to heat, sound, solvents, shocks, and electrical current.

The notion of establishing *Q. suber* plantations outside the Mediterranean has, of course, been much discussed. The species thrives in California and other parts of the southern U.S., but the skilled, low-cost labor supply for stripping and manufacture has not been developed.

Juglandaceae — Walnut Family

The Juglandaceae are a family of six genera of hardwood trees with pinnately compound deciduous leaves. Flowers are characteristically inconspicuous and wind-pollinated. Fruits generally have a hard shell and are surrounded by a tough husk, evidently defenses against seed eating animals. The seed lacks endosperm, but contains much stored food, especially fats, in the large, wrinkled cotyledons. Some nuts are attacked by rodents and large birds, e.g., jays and woodpeckers. Although the animals usually destroy the seeds, enough are cached or lost to disperse the species. The family evidently evolved in the temperate mixed hardwood forests that were shared between Eurasia and North America in the Tertiary. A few members range into South America and the East Indies. Many wild species have been exploited for their edible nuts and fine, strong timber. *Juglans* and *Carya* each have one domesticated species.

JUGLANS **— WALNUTS** (Forde 1975; Hutchison 1946; Roach 1985; Simoons 1990)

Juglans includes about 12 species native to Eurasia, North America, and the Cordilleran system of Central and South America. The eastern North

American black walnut *J. nigra* is one of the world's most valuable cabinet woods. With depletion of old-growth wild trees, there has been much interest in planting as a salable timber tree, especially by midwestern farmers with woodlots of mixed hardwood trees. However, the species is ecologically demanding, takes a long time to develop a salable log, and is subject to loss by lightning, wind, rot, and theft, so extensive forestry is impractical. By contrast, the Eurasian *J. regia* is an important orchard tree.

Juglans regia — Persian or English Walnut

This *Juglans* species is believed to be native to a huge region of Eurasia, including the Carpathians, Balkans, Crimea, Anatolia, the Near East, Iran, the Himalaya, and western China. The species tolerates, but does not demand extremes of heat and cold. Planting and artificial selection began prehistorically, perhaps independently in different regions. Seedling cultivars are highly variable and not clearly distinct as a group from wild populations. Presumably, thinner shelled forms were preferentially planted, but the trait could not be stabilized until open-pollinated seedlings were replaced by grafted clones.

The first historical records of walnut orchards are from classical Greece and Rome. The walnut was introduced by the Romans to Iberia, Gaul, and southern Germany, where cultivation has continued down to the present. A common name notwithstanding, there is no clear evidence of walnut planting in Britain until the late Middle Ages and then only on a rather trivial scale. Significant planting in northern Europe began during the 16th century after selection of improved varieties in France, but most of the walnut supply continued to be imported from Italy. Planting of walnuts expanded greatly during the 17th century in northern Europe in general, often as much for the timber as for the nuts. The species was in great demand as a cabinet wood and for gunstocks. Grafting of clones selected for superior nuts on seedling rootstocks began in western Europe, especially France, in the 18th century.

According to Chinese tradition, walnut cultivation was introduced to central China from the west, perhaps about 2000 years ago. Cultivation in northern China was recorded about 700 A.D. Walnuts are a traditional ingredient in many Chinese recipes.

In post-Columbian times, *J. regia* has been repeatedly introduced overseas, mainly from Europe. For example, Polish and German walnut cultivars were introduced to eastern North America and Spanish cultivars to Chile and California. The most successful overseas introduction of the species was to California. In the last half of the 19th century, southern California farmers were experimenting with a variety of orchard trees. By the end of the century, they had developed superior, soft-shelled walnuts by selection of seedlings of the old Spanish cultivars. By 1900, they were switching to grafting, using the wild California black walnut species as rootstocks. Not until after 1900 was there much planting in California outside the southern coastal region, evidently due to a mistaken belief that *J. regia* required a mild climate. Finally, between 1910 and 1930, walnut orchards were greatly expanded in the north and the interior, and California became the world's leading producer of the crop, as it has remained. The orchards are now mainly French cultivars, grafted on native rootstocks. Generally they are medium-sized, family-run properties with co-operative marketing.

CARYA — HICKORIES, PECAN (Madden and Malstrom 1975; Peattie 1950)

Carya includes about 15 species, native to North America and Asia. Generally, they are shade-intolerant pioneers growing with other deciduous hardwoods. Timber of some *Carya* species is used where extreme toughness is wanted. The seeds are generally quite edible. The nuts are hard to crack, and the nut meats are pried out in fragments. Only the pecan has become an orchard crop.

Carya illinoensis — Pecan

The wild species produces thin-shelled nuts which, unlike ordinary hickories, attract animals other than squirrels. The nuts are commonly carried by jays, crows, and other birds. The species has a huge range in the middle and lower Mississippi Valley, including the lower Ohio River, westward to about the 100th meridian and southward through eastern Mexico. Its habitat is open woodlands on river-bottom alluvium, briefly flooded but well drained. The trees can live for centuries, reach 50 m in height, and 2 m in trunk diameter. Pecan groves are most continuous along the rivers of Texas and Oklahoma; elsewhere, they are generally scattered disjunctly.

Many Indian tribes gathered and stored quantities of pecan nuts, and archaeological sites are commonly associated with pecan groves. This has led to speculation that the Indians expanded the range of the species, but this is doubtful.

De Soto encountered wild pecans in the Mississippi Valley in 1541, and they were reported by innumerable later explorers and frontiersmen, especially in Louisiana and Texas. Occasionally they carried off viable nuts, and some trees planted in the 18th century as far eastward as Virginia still survive. Pecans at Monticello were planted by Thomas Jefferson and at Mount Vernon by George Washington.

Much more important were plantings in the Pecan Belt of the South, both within the native range west of the Mississippi and east of that range in Georgia and neighboring states. By 1850, pecan trees were common in the dooryards of houses throughout this belt. Handsome, giving good summer shade, and a shower of nuts for festive occasions, the species remains a part of the rural and suburban human environment. Part of the commercial supply still comes from dooryard trees and is funneled through local pecan buyers. Until the mid-19th century, only seedlings were planted, but by 1879, a New Orleans pecan nursery was selling grafted stock. Since then, many orchardists have sought improved varieties to be propagated by grafting on seedling rootstocks. Between 1905 and 1920, hundreds of cultivars, some produced by controlled hybridization, were named. Only about two dozen cultivars are planted in commercial orchards, having stood the test of time. Recently commercial growers and government breeders have developed cultivars suited for mechanized harvesting and shelling. Outside the old Pecan Belt, pecan orchards are proving successful in California, particularly in the semi-arid San Joaquin Valley, which has hot and dry weather for the harvest period, highly desirable where irrigation is possible. By the late 1980s, over 100,000 ha of pecan orchards had been planted in California.

The species has been introduced to all temperate regions of the world. The first overseas introduction recorded was to Spain from Mexico in the early 17th century. However, most of the world supply is still from North America.

Lauraceae — Laurel Family

The Lauraceae include about 50 genera and over 1000 species native to many of the warmer regions of the earth. Most are evergreen trees and shrubs. The family characteristically has aromatic wood, bark, and foliage, which mankind has used in various ways. *Cinnamomum* provides cinnamon and camphor wood; *Laurus* and related genera provide bay leaves. The leaves of other genera, including *Sassafras*, are used locally for spice. Only *Persea* has been domesticated as a food crop.

PERSEA — AVOCADO AND CONGENERS (Anderson 1950; Bergh 1975, 1976; Hutchison 1946; Patiño 1963; Popenoe 1950; Purseglove 1968; Schroeder 1967; Smith, C. 1966; Towle 1961)

The genus was widespread in the northern hemisphere in the Tertiary, but has retreated to disjunct relicts in the Canaries and other Atlantic islands, the Americas, and eastern Asia. Most of the species are not closely related to the avocado and have small fruits of interest mainly to birds.

Persea americana (incl. *P. gratissima, P. drymifolia, P. guatemalensis*) — Avocado

The domesticated avocado is evidently a compilospecies descended from various wild progenitors that were taken into cultivation independently but have since coalesced into a single breeding population. There is no clear line of demarcation separating cultivars from the wild avocados; the main effect of human selection was a gradual increase in fruit size and quality. The wild species had already evolved a fairly large fruit, 4 to 5 cm in diameter, with a thin edible pulp surrounding the single seed, which averages about 4 cm in length and 2 cm in diameter. Such a large seed, although exceptional in the

Lauraceae, is common in unrelated trees of dense, mature tropical forests. A large endowment of stored food allows seedlings to survive longer in dim light, thus increasing their chances of inheriting a gap in the forest canopy. The main habitat of wild *P. americana* is cloud forests on the great volcanic peaks from eastern Mexico through Central America to the northern Andes; it also grows at lower elevation on windward slopes in patches of well-drained rain forests. Such habitats have probably always been highly disjunct. The success of *P. americana* in colonizing these places indicates a nice balance between natural selection of larger seeds for survival vs. smaller seeds for dispersibility. Fruit-eating birds were presumably the usual agents in prehuman times.

The first archaeological record of human use of avocados, as in the case of several other species, is from caves near Tehuacan in Puebla, Mexico. Fortunately, prehistoric human use of avocados is recorded in dry sites by the seeds, which are loaded with toxic defenses. The earliest finds at Tehuacan date from about 7000 B.C. They are of a size suggesting gathering from wild trees, which do not grow in the local thorn-scrub vegetation but in mesic canyon forests higher up the nearby mountains. Avocado seeds were few and far between in the cave deposits until about 500 B.C., when a sudden increase in abundance and in average seed size suggests planting under irrigation and selection for larger fruits. In contrast, no increase in size above the wild is shown by much later avocado seeds, dated about 700 A.D., found in caves in the dry Valley of Oaxaca near Mitla, although they may also have been grown under irrigation.

In irrigated oases of the Peruvian coastal desert, archaeological avocado seeds have been reported from widely scattered sites, the earliest dated about 750 B.C. in the north at Huaca Prieta.

The Spanish explorers found no avocados in the West Indies but they reported them among innumerable Indian groups in Mexico, Central America, and in northwestern South America. From the Aztecs, the Spaniards picked up the Nahuatl name *ahuacacuauhitl*, meaning testicle tree, shortened it to *aguacate*, later corrupted to avocado in English. This has long been the lingua franca name in Latin America, although there are many other local names.

At the time of the Conquest, the avocado was clearly an important food in tropical America, as it remains today. The fruit is available over a long time span, especially from heterogeneous seedling orchards, stores well on the tree, and ripens only slowly after picking, if kept cool. It is very rich in fats, mostly unsaturated, and in vitamins and minerals, and is a fair source of protein. The highly digestible oil is similar in composition to olive oil.

It is not always easy to distinguish between wild and domesticated groves in the early colonial accounts. Even today, it is often unclear whether trees in an Indian garden are volunteers or planted and whether trees in a forest are wild or relics of former gardens. There were, however, definite references in 16th century sources to avocado planting by the Indians and also to different varieties with fruit larger than the wild. The Indians did not practice grafting nor could they develop pure lines because avocado trees are outbreeders, cross-pollinated by insects. Neverthless, by long selection among heterogeneous seedlings, the Indians had bred cultivars with fruit of size and quality that has not been significantly bettered by modern breeding.

Indian cultivars have been loosely classified into three geographical races: Mexican, Guatemalan, and so-called West Indian, actually native to southern Central America and Columbia. The Mexican race is the most cold tolerant and has the smallest fruit with the highest oil content and the thinnest skin. (Such generalizations are for the group as a whole; there is great diversity within each group.) The West Indian race is the least cold tolerant and tends to have large fruits with low fat content. The Guatemalan race is intermediate in some characters, but has the thickest, roughest skin and rounder, less pear-shaped fruits.

Early Spanish colonists learned to like avocados surprisingly fast, considering the slow subsequent acceptance by most ethnic groups. Regard for the fruit was no doubt helped, except among the clergy, by a notion that it was an aphrodisiac. There are hints that this belief was shared by Indians and Spaniards from Mexico to Peru. Patiño (1963) quotes a Jesuit account of the extirpation of idolatry in Peru, which decried orgiastic Indian rites preceded by feasting on avocados. The avocado was conspicuously absent in the gardens of some Spanish missions, including those of the California Indians.

During the colonial period in tropical America, the three geographic races were intermixed regionally, but tended to be segregated altitudinally: the South American race succeeding at lower elevations, the Mexican at high elevations, and the Guatemalan in between. Where the races were planted together, they interbred freely, producing very diverse hybrids. The most famous case is the hacienda of the Rodiles family near Atlixco, Puebla, in eastern Mexico. For generations, the family planted seeds of the best varieties from local markets and from the progeny in their own orchards. Atlixco became a Mecca for horticulturists hunting avocado cultivars. One of the seedlings originating there, believed to be a hybrid between the Mexican and Guatemalan races, is the Fuerte, the world's leading commercial cultivar today.

Introduction of the avocado outside its aboriginal region was begun early in the Spanish colonial period and was easily accomplished because seeds remain viable for weeks even after removal from the fruit. Spaniards took what is now known as the West Indian race from the mainland to the Greater Antilles, whence it was taken to other islands. Spaniards introduced the Mexican race to central Chile where it flourished. Early Spanish introductions, presumably mainly of the Mexican race, also succeeded in the Canaries and other Atlantic islands and in Spain, but failed in the Philippines where more tropical cultivars were recently successfully introduced.

Outside the Spanish sphere, avocado planting spread very slowly. During the 18th century, the heyday of much tropical plant introduction, the species was finally recorded in West Africa, Mauritius, and India, but not as a significant crop. During the 19th century, avocados were introduced to Hawaii, other Pacific islands, and the Old World tropics in general, but were only slowly adopted as a crop. One reason was perhaps the poor fruit quality of the majority of trees grown from seed of even the best parent trees. Until clonal propagation by budding and other modes of grafting was developed, the only way to develop an orchard of high quality was to plant lots of seeds and eliminate the inferior progeny when they came into bearing.

Clonal propagation, with buds or twigs of selected scions being grafted onto seedling rootstocks, was initiated in Florida shortly before 1900 and soon became the basis of commercial orchards in California and elsewhere. In Florida, the scions were initially selected from the so-called West Indian race, which had been grown there casually since the 1830s. An important one of these is Pollock, a large fruited, low oil content cultivar. Later hybrids between West Indian and Guatemalan races, such as Lula, became predominant. These and other Florida selections have been taken back to the West Indies and lowland tropical America for commercial planting.

In California, casual planting of Latin American avocado seeds brought back by travelers began in the 1850s shortly after statehood. By the late 19th century, occasional individual trees, widely scattered through coastal southern California, were bearing acceptable fruit. By 1910, the possibility of successful commercial avocado orchards had dawned on a few California growers. Soon private nurserymen and the U.S. Department of Agriculture obtained collections of budwood in the avocado districts of Mexico and Guatemala. Budwood of the great Fuerte clone, mentioned above, was introduced to California in 1911. Other named cultivars originated in California orchards as seedlings, including the Hass. In the 1930s, Rudolph Hass planted a batch of Guatemalan avocado seed to be used as rootstocks for grafting in his orchard near Los Angeles. On one grafted tree, the scion died, but sucker shoots from the rootstock grew up and bore fruit of remarkable quality. This provided budwood for the Hass, which became a major cultivar. The fruit is egg-shaped with a rough black skin, less handsome than the Fuerte, which is pear shaped with a smooth green skin. However, the Hass has an excellent taste, keeping quality and yield, and comes into bearing during a different season than the Fuerte.

During the first half of the 20th century, 500 or so named avocado cultivars had been selected in California orchards. The orchards were generally relatively small operations with resident owners personally interested in selecting new, superior varieties. Seldom, if ever, has a new crop region had such a fund of genetic diversity to experiment with. Those days are now past, and the crop has undergone a dramatic standardization. Of course, most of the named cultivars were not actually of high quality. Also, the diversity created marketing problems. As late as the 1970s, something like 200 different cultivars were being delivered annually to the packing sheds of the major cooperative association, which marketed under the Calavo brand. The odd kinds tended to be passed over by consumers. Growers stopped propagating them and began top-working the existing orchards with the most popular cultivars, especially

Fuerte and Hass. Planting of new orchards boomed. Bearing area increased from 10,000 to 30,000 ha between 1961 and 1981, when the market became glutted. Much of the new acreage was speculative, owned by consortia of absentee investors and corporately managed. Such expansion of avocado orchards in a region with limited farmland and limited water depended on techniques of microirrigation, i.e., drip and sprinkler systems. The technology had originally been developed for avocado orchards in Israel, where land and water are even more precious. It allows planting on steep, previously unusable slopes.

Since mid-20th century, commercial orchards in other temperate regions — Israel, South America, South Africa, Australia, and New Zealand — have followed California in standardizing on Fuerte, Hass, and a few closely related cultivars. The same is true of new commercial orchards in tropical highlands, e.g., Kenya. This remarkable narrowing of the genetic base of an expanding crop seems to stem from a general realization that the best avocado cultivars in Latin American folk cultivation cannot be easily improved by breeding. In fact, current breeding of avocados, particularly in California, Florida, and Israel, has rather modest objectives, such as improving yield and disease resistance, and is not attempting radical genetic change.

It is remarkable that a cloud forest and rain forest species, domesticated in the tropics by simple selection of open-pollinated seedlings, had broad enough ecological tolerances to become a productive orchard crop in environments so different from its native habitat. Given water and freedom from frost, the species is amazingly tolerant of climatic and edaphic variation. It thrives in tropical deserts and subtropical Mediterranean regions with unleached, nonacid soils, strong light, and dry air.

Persea has an unusual mechanism promoting cross-pollination at the cost of lower fertility; the flowers have a diurnal rhythm with stigma receptivity preceding pollen release. In avocado orchards, few flowers are fruitfully pollinated. In California, for example, an average of about 1 Fuerte flower out of 5000 sets fruit. Fortunately, this is enough because the tree bears inflorescences with astronomical numbers of tiny flowers.

Malvaceae — Hibiscus Family

The Malvaceae include about 100 genera and 1500 species native to tropical and temperate regions worldwide. Members are generally adapted to special, local habitats rather than dominating regional vegetation formations. Having scattered, thin populations, they rely on insect and bird pollination and have showy flowers. Genera domesticated as ornamentals include *Hibiscus*, *Alcea* (hollyhock), and *Abutilon* (flowering maple). A species domesticated for the edible seed pods is *Abelmoschus esculentus* (okra, gumbo). Only one genus includes major commercial crop species.

GOSSYPIUM **— COTTON** (Dalziel 1948; Fryxell 1979; Hobhouse 1986; Hutchinson 1977; Patiño 1967; Phillips 1976; Purseglove 1968; Santhanam and Hutchinson 1974; Stephens 1967a 1967b, 1973, 1975, 1976; Towle 1961; Watson 1983; Wendel et al. 1989)

There are about 50 living wild species native to the tropics and subtropics of America, Africa, Asia, and Australia. Most of the species are diploids (2n=26). The diploids belong to six genome groups, conventionally labeled A through F, with chromosomes so different that diploid hybrids between any two groups are sexually sterile. Five wild allopolyploid species (4x=52) have AADD genomes. They are capable of interbreeding among themselves but not with diploids.

Of the 50 broadly defined *Gossypium* species, only four have been domesticated, two diploids (AA) and two polyploids (AADD). Various wild cotton species produce short seed hairs, but only the A-genome bears genes for true lint, i.e., long seed hairs that are convoluted and flattened enough to adhere firmly when spun into thread. What the initial adaptive value of lint was for the ancestral African AA cotton is unknown; attraction of birds as seed dispersers by taking the lint for nesting material is one possibility; adhering lint would lead to inadvertant longer range dispersals. At any rate, this seemingly trivial characteristic had profound consequences for the natural history of AA and AADD species and later for human history.

The origin of the AADD species presented a fascinating puzzle. Clearly they resulted from ancient hybridization between AA and DD diploids, but wild populations of the AA cottons are known only in tropical Africa and possibly southwest Asia; wild populations of DD species grew only in tropical America. Hypotheses to explain contact ranged from sympatry in Gondwanaland a hundred million years or so ago to transport of AA cottons across the ocean by prehistoric raftsmen. A less extreme hypothesis is now generally accepted: natural trans-Atlantic drift of AA cotton seed a few million years ago followed by hybridization with a tropical American species related to *G. raimondii* (DD). Crucial to development of this hypothesis was the discovery that seeds of wild AA and AADD species are impermeable to sea water and remain viable after prolonged immersion.

From perhaps a single original AADD stock, five wild AADD species eventually diverged during extensive migrations in tropical America and through widely scattered islands in the tropical Pacific. Three of the five (*G. mustelinum* of northeast Brazil, *G. tomentosum* of the Hawaiian Islands, and *G. darwinii* of the Galapagos Islands) have not been domesticated and will not be discussed. It is interesting, however, that although isozyme and morphological analyses suggest that *G. darwinii* deserves to rank as a species distinct from its close South American relative *G. barbadense*, after the introduction of cultivated *G. barbadense* to the Galapagos, the two interbred (Wendel and Percy 1990).

Fryxell (1979) has suggested that the wild AA species evolved in the interior of Africa in arid, open habitats but that it was preadapted to colonize sea beaches and for sea dispersal. (This sort of ecological switch between seashore and dry, open inland habitats has probably occurred in many unrelated genera.)

Lint of the AA and AADD species would have inevitably attracted attention of various prehistoric peoples. From crude cordage, some of them eventually progressed to spinning fine thread and weaving. Domestication of each species probably occurred repeatedly in different places.

In addition to lint, cotton seed has economic value as a source of vegetable oil and protein. However, the raw seed is defended by a polyphenol that is toxic to most animals. Ruminants are an exception, and the cotton seed or the cake after oil extraction can be fed to camels, cattle, sheep, and alpacas. Archaeological evidence suggests wild cotton seed was collected for stock feed in Egyptian Nubia ±2500 B.C. by people who did not practice weaving. The oil can be made edible by chemical treatment, but the process is costly and in commercial cotton growing, the seed is a by-product.

In *Gossypium*, unlike many crops, domestication did not involve abrupt qualitative genetic change resulting in clearcut dichotomy between wild and cultigen. Rather there was gradual divergence of the crop. Primitive cultivars remained perennial shrubs, bearing small bolls a few at a time, with hard, slow-germinating seed and sparse, brown lint. Casually planted dooryard cottons retain the ability to survive and reproduce as escapees from cultivation, so distinguishing truly wild from feral cottons is often speculative. More advanced cultivars are quite distinct and totally dependent on human cultivation, but the whole array of forms is a continuum. Consequently, cotton taxonomists generally use broadly defined species with informal races as subspecific groups.

Gossypium herbaceum (incl. *africanum, acerifolium, persicum*) — African-West Asian Cottons (Genome AA)

The species name is unfortunate because the wild and primitive cultivar members of this species are shrubs. Wild forms (*africanum*) are native to large regions of sub-Saharan Africa in semi-desert and savanna vegetation on both sides of the equatorial forest region. The most primitive cultivar probably originated in Ethiopia or southern Arabia. It may have been the wool-bearing tree recorded by Theophrastus on the Persian Gulf about 350 B.C. As the crop spread northward into the interior of Persia, Afghanistan, and Turkey, annual forms were selected. The crop spread westward in medieval Islam, being taken across North Africa to Moorish Spain. Cotton growing and weaving declined with the Christian Reconquest of Spain, with only a few vestiges in the far south at the time of Columbus. Other annual derivatives of *G. herbaceum* spread into the Ukraine and Turkestan and were cultivated in western China by 600 A.D. These are of minor commercial importance on a world scale, but allow cotton cultivation in regions with short, hot summers and long, cold winters at very high latitudes.

Gossypium arboreum **(incl.** ***indicum, soudanese, bengalense, burmanicum, sinense)*** **— Pakistani-Indian Cottons (Genome AA)**

This is another unfortunately chosen species name; some cultivars are tall perennial shrubs, others short annuals. The species is closely related to *G. herbaceum*, but the two differ by a major chromosomal translocation that hinders interbreeding. There has been much speculation about the possibility that this mutation occurred after domestication of *G. herbaceum*, but *Gossypium* geneticists now generally prefer to postulate ancient natural divergence in the wild and independent domestication of the two species and perhaps even multiple domestications of different wild forms of *G. arboreum*.

The earliest archaeological evidence of cotton textiles in the Old World is from the Harappan civilization of the Indus Valley about 2000 B.C. The Harappan textile craft was sophisticated and does not document incipient stages in use or cultivation of cotton. Wild *G. arboreum* has been reported from Gujarat and other parts of northwestern India, but it is not necessary to assume that domestication occurred close to wild stands. Bolls and unginned seeds may have been carried long distances by traders as curiosities or for minor uses before the seed was planted. Spinning and weaving technology probably was adapted from wool and other fibers.

In any case, cultivation of *G. arboreum* probably began around the north end of the Arabian Sea in the Indian subcontinent. The most primitive modern cultivars of the species are still found in western India. Another relatively primitive group, perennials with sparse lint, may have been taken from India to East Africa in antiquity and were grown 2000 years ago by the people of Meroe in Nubia, the first cotton weavers known in Africa. This race spread westward in sub-Saharan Africa in pre-Islamic times. Kano in Nigeria has been a cotton manufacturing center since the 9th century.

A very diverse group of more advanced cultivars, both perennial and annual, were developed in northeastern India; some of them spread widely through southeast Asia and the East Indies. Annual forms of *G. arboreum* were spread through southeastern China between 900 and 1300 A.D. and eventually reached the Ryukyus.

In pre-Columbian times, *G. arboreum* was the Old World's main cotton crop, for both home use and commerce. From ancient through medieval times, the Indian subcontinent remained pre-eminent in manufacturing fine calicos and muslins. Traded widely by caravan and ship, these were luxury goods, competing with silks in China and with fine linens in classical Greece and Rome. Europeans knew cotton only as trade goods and the plants only from travelers' accounts. They visualized the fiber as borne on a fantastic plant-animal chimera, the vegetable lamb. Importation of raw *G. arboreum* cotton from India for manufacture in Europe began in the Renaissance. The cotton came through the Near East to be carried on by ships of Genoa, Venice, and the Hanseatic League. Weaving was initially concentrated at Brussells and Ghent, but spread to Manchester by immigrant Flemish weavers. The fall of

Constantinople to the Turks in 1453 broke the chain of cotton supply, along with that of spices and dyestuffs, and stimulated Portuguese efforts to reach India by rounding Africa. The Portuguese, Dutch, and later British East India companies initially concentrated on importing Indian muslins rather than raw cotton. Eventually cheap cotton textiles manufactured in Britain from American cotton proved disastrous for the old Indian weaving craft, and India became an exporter of raw cotton for manufacture in the Far East. Not until about 1900 did British and Indian capital begin large textile mills within India. Until about 1950, almost all the cotton for these mills came from diploid *G. arboreum*. Since then, the New World tetraploids (see below) have gradually taken over most of India's 8 million or so hectares in cotton. Elsewhere in Asia and Africa, the Old World diploids are mere relics in folk cultivation, and commercial cotton is produced by the two following American tetraploids.

Gossypium barbadense (incl. *G. peruvianum, G. brasiliense*) — South American Cotton (Genomes AADD)

Clearly wild populations of this species are known from sea cliffs on both sides of the Guayas River estuary in southern Ecuador. It is quite likely that there were disjunct wild populations on both Pacific and Atlantic coasts of South America. Most tropical seashores have been heavily modified by human disturbance, often including planting of cotton in dooryard gardens, so that former wild populations would have been eliminated or mongrelized.

The oldest cotton textiles reported in South America, presumably made from *G. barbadense*, are from 3600 B.C. at Quiani in the northern Chilean desert; this is a premaize, preceramic fishing site in the Chinchorro Tradition, believed to have its roots east of the Andes (Rivera 1991).

On the Peruvian coast, *G. barbadense* also enters the archaeological record in a preagricultural fishing village. Cotton bolls from about 2500 B.C. in northern Peru show transitional forms between the wild form, which has small bolls with hard seeds bearing sparse brown lint, and the improved cultivar forms. The earliest Peruvian cotton artifacts were fish nets and other cordage. About 1000 B.C., cotton planting began expanding on the Peruvian coast as spinning and weaving developed and irrigation works were constructed. By then, remains of Peruvian cotton bolls were indistinguishable from modern *G. barbadense* cultivars. The pre-Inca city states of the Peruvian oases were among the highest civilizations of the ancient world. Elaborate cotton textiles are normal components of mummy bundles and other grave goods. Weavers used cottons of various natural shades, from brown to white, as well as dyed cotton. Fabrics often combined cotton with alpaca wool and other fibers. After the wreckage of the classical civilizations by the Inca and Spanish conquests, commercial cotton plantations became extensive in Peru. Other forms of *G. barbadense*, especially *G. brasiliense*, were cultivated prehistorically east of the Andes over a huge expanse of the Amazon and Orinoco basins. *Gossypium* cannot survive in undisturbed forest and requires dry weather for flowering, fruit set, and picking of the crop. However, a little was planted even in rain forst regions by tribes who went naked. Enough could be harvested for their modest needs for fishing lines and other cordage.

By 1492, *G. barbadense* was a common dooryard garden plant throughout the West Indies. On his first voyage, Columbus encountered cotton, presumably *G. barbadense*, in abundance among the Lucayo of the Bahamas and the Arawak of the Greater Antilles. Later he found cotton in Carib gardens of the Lesser Antilles. The Caribs used cotton sails for their sea-going canoes. The presence of cotton textiles and cordage was a factor in confirming Columbus' belief that he had reached Asia.

Prostitution of *G. barbadense* to become a commercial slave plantation crop came rather slowly. The Spaniards admired the fineness of the fiber and the West Indian textiles, but exploitation was usually by extracting tribute of cotton goods from the Indian villages and the take was short-lived because of extinction of the Island Arawak. Spaniards had begun planting cotton with African slave labor on Jamaica by the time the British took the island in the 1650s. By then, Barbados had become the first British West Indian colony to export cotton. The Barbados colonists inherited *G. barbadense* from the Arawak and grew it on small estates with a few African slaves. During the later colonial period, many British, French, and Dutch West Indian colonies exported modest quantities of cotton.

In British eastern North American colonies, planting of *G. barbadense* began about 1670 with the arrival in the Carolinas of Barbadian planters.

Sea Island cotton is an annual *G. barbadense* type that arose in the coastal plain and offshore islands of the Carolinas and Georgia in the late 18th century. It differs from primitive *G. barbadense* in being able to flower outside the tropics in long summer days. It is a premium cotton with very long, fine lint. Sea Island is very late maturing, compared to *G. hirsutum* (see below), and vulnerable to the boll weevil. Its cultivation was abandoned in the U.S. in the early 20th century because of the advent of the boll weevil, but continued until about 1960 in some of the West Indies, where it had been introduced from North America.

In the Old World, perennial *G. barbadense* was probably introduced repeatedly by both Spaniards and Portuguese to the Mediterranean and West Africa and then spread through the Sudan to Egypt. Late in the 18th century, annual Sea Island lines were introduced from the southeastern U.S. to Egypt. So-called Egyptian cotton was developed in the early 19th century from crosses between the old perennial forms of *G. barbadense* and annual Sea Island forms. The productive Ashmouni variety was developed in Egypt about the time that the American Civil War cut off cotton supplies to the mills at Manchester. The English fostered Egyptian cotton planting to keep their mills running; Egyptian cotton has been important ever since. Some of the improved Egyptian lines were introduced to Arizona and adjacent states, where they contributed to the Pima varieties of *G. barbadense*.

Gossypium hirsutum (incl. G. punctatum, G. marie-galante, G. taitense) — Mexican and Related Cotton (Genomes AADD)

The best known wild population of the species is the race *yucatanense*, a perennial shrub prominent in natural beach-ridge scrub vegetation along the north coast of Yucatan. It has small bolls and hard seeds with sparse, brown lint. Fryxell has collected similar wild *G. hirsutum* on the barrier islands of Tamaulipas and on the Florida keys. Similar cottons are reported at Todos Santos Bay in Baja California, on Socorro Island off the Mexican Pacific coast, and on the dry Caribbean coast of South America. Wild forms of *G. hirsutum* (*taitense*) with sparse, brown lint were encountered on Tahiti on James Cook's first voyage and also have been found in the Marquesas, Fiji, Wake Island, and Saipan. No aboriginal use of the fiber for textiles was reported except in the Americas. Wild *G. hirsutum* is reported from some West Indian islands, but these islands have heavily disturbed seashores, and the cotton bushes may be escapees from gardens. Even setting aside the West Indian colonies, it is clear that *G. hirsutum* has a curious native range, with extremely disjunct and highly localized colonies. This makes sense if we recall that the ancestral polyploid hybrid is believed to have originated on a tropical American seashore and from the outset was probably adapted to dispersal by both ocean currents and birds, perhaps including wide-ranging frigate birds, which nest in coastal scrub. Before domestication, it was probably a rather rare species, restricted to a narrow littoral zone and to semi-desert climates. It is not a true xerophyte nor can it compete in dense coastal thickets in humid climates, but does have an elevated tolerance of salinity.

The only wild colonies that are likely sources of the domesticate are those on the Mexican Gulf coast and the Caribbean coast of South America. *G. hirsutum* has two major cultigen varieties, which were apparently independently domesticated. The shrubby Mexican variety was taken inland westward and northward from the Mexican Gulf coast; the tree-like *marie-galante* variety was perhaps domesticated on the Caribbean coast of South America and spread northwestward along both sides of Central America. *Marie-galante* may have been taken into the Lesser Antilles by the Caribs. During the Colonial Period, the West Indies became a melting pot for varieties of *G. hirsutum* as well as *G. barbadense*, with planters acquiring cultivars from many sources. Also, although gene flow between the two species is restricted, some introgression apparently occurred in the West Indies.

The archaeological record of *G. hirsutum* begins in the famous Tehuacan caves in Puebla, Mexico. An early report on the excavations estimated the date of two cotton boll segments as 5500 B.C. Stephens noted (1967b) that the bolls were large and not from a primitive cultivar and concluded that they were probably of younger age and intrusive in a disturbed cave deposit. More securely dated cotton remains begin about 3500 B.C. in the Tehuacan caves, and from about 200 B.C. on, there were numerous bits of string and sizable cotton fabric remains.

At the time of the Spanish arrival, cotton was in general cultivation throughout the lowlands of Mexico and Central America, presumably all forms of *G. hirsutum*. Cotton was grown in dooryard gardens for domestic use and also as a commercial crop. Trade was usually in the form of *mantas*, i.e., woven textiles as they came off the loom, and in cotton garments. The Spaniards were impressed by the colorful patterns and fineness of these products, not only among the high civilizations of the Maya and Aztec regions, but among various Central American tribes as well. Cotton artifacts were clearly not merely utilitarian, but were items of beauty, value, and ceremonial importance. Moctezuma, the last Aztec emperor, collected cotton *mantas* as major tribute from 34 of the 38 provinces of his empire. Often the color and patterning of the *mantas* were specified. Some provinces were required to submit several thousand bundles of *mantas* each year. Also tribute was levied in cotton garments and warriors' costumes, including quilted cotton armor. The geographic extent of the levy indicates that there was an extensive trade in cotton fiber and that weaving was more widespread than cotton cultivation (Barlow 1949).

G. hirsutum cultivation spread among maize growing peoples northwest through Mexico into what is now the southwestern U.S. in prehistoric times. Annual forms were selected that are able to flower outside the tropics in long summer days. The trade in cotton cordage and textiles and perhaps also in cotton fiber extended beyond the range of the crop; thus, archaeological finds of cotton offer interpretive problems. Archaeological cotton seeds first appear in the record from the southwestern U.S. at a Hohokam site in southern Arizona. They occur in all occupation levels from about 100 A.D. until abandonment in 1200 A.D., suggesting but not proving local cultivation (Bohrer 1970). In Anasazi sites in northern Arizona, at higher elevations and with a very short growing season for cotton, cotton seeds, bolls, and stem pieces appear in many cliff dwellings and other ruins starting about 1100 A.D. By the time of Spanish contact, cotton was being cultivated by various pueblos in Arizona and New Mexico. It was especially important among the Hopi. In the 16th century, Hopi cotton was grown not just in gardens but in extensive fields, and the Hopi pueblos wove quantities of cotton goods for trade to other pueblos and also for *Agave* products obtained from the Apache. Balls of cotton fiber, representing rain clouds, and cotton garments are still of great ceremonial importance among the Hopi.

Since Columbus, the annual Mexican lines of *G. hirsutum* and their Upland derivatives have been spread throughout the world's cotton growing regions. The first overseas introduction was the early Spanish to the Philippines. The so-called Cambodian cotton is derived from a much later introduction to Asia of a Mexican *G. hirsutum* cultivar.

Worldwide the most important commercial cottons during the last 200 years have been the Upland lines derived from old Mexican cultivars. These were developed further in British colonies in southeastern North America during the mid-18th century. During the late 18th and early 19th century, cotton varieties were exchanged repeatedly between planters in the Carolinas, West Indies, and Old World tropics, so the pathways of germ plasm introduction became too tangled to trace in detail.

Whitney's invention of the cotton gin in 1793 initiated great expansion of planting with a feedback between lower cost and greater demand. During the next 100 years, cotton rose from <5% to >75% of the world's textile fiber. In the early 19th century, cotton became the main export of various tropical colonies that were unable to produce sugar, even such backwaters as the Cayman Islands and the Seychelles. Slave-grown cotton collapsed in the British Empire in 1835 with Emancipation.

Meanwhile, in the southeastern U.S., the pattern of cotton planting was changing dramatically. Before 1820, cotton had been grown mainly on the Atlantic coastal plain and the Piedmont. It had typically been a crop of poor whites on mixed farms, not of slave monocultures. After 1820, extensive planting was pushed west of the Appalachians as new land was cleared. By 1850, the Old and New South were supplying 80% of the booming demand for English cotton mills. Slavery, which had been moribund around 1800, was revitalized. Although further import of slaves into the U.S. was banned after 1808, the slave population grew from less than half a million to over 4 million by the outbreak of the Civil War.

When the Civil War shut down cotton exports from Dixie, there was a boom in Upland cotton planting in many tropical and subtropical countries. In some places, such as Fiji, the boom in cotton was brief before U.S. production resumed, mainly under sharecropping.

Meanwhile, King Cotton had led the Industrial Revolution, entraining coal mining and steel manufacture. During the 19th century as mechanization of spinning and weaving developed, cotton mills blighted first England and then New England.

During the 20th century, this tyrannical crop moved westward in the U.S. through Texas, New Mexico, and Arizona to California, leaving the old Cotton Belt in the east a legacy of soil erosion, farm abandonment, and poverty. Meanwhile, many old New England mills were closed as the industry moved southward to exploit cheap, nonunion labor. On the national level, cotton became a problem during the depression of the 1930s when federal price supports nearly destroyed the export market, resulting in huge surpluses until drastic acreage reductions were enacted. Another economic and social upheaval began in the 1950s with the mechanization of both planting and picking, eliminating the vast majority of the field labor jobs.

Since 1950, commercial cotton has been largely grown in arid regions under irrigation, as was some of the aboriginal crop. Whether in Peru, Mexico, or the southwestern U.S., however, the modern pattern is a mechanized monoculture. One of the most extreme examples developed in the San Joaquin Valley of California. In order to standardize plant growth, harvesting, and quality of the fiber, only a single cultivar could be planted by law, namely Acala, selected from Upland landraces introduced from Chiapas, Mexico, and

Guatemala. Great machines plant multiple rows of fungicide-treated seed at each pass across the field. The fields are heavily fertilized with phosphate and ammonia, with additional fertilizer injected into the irrigation water. Herbicides and pesticides are sprayed repeatedly, usually by airplane. Other sprays thin the cotton and defoliate it when the bolls are ripe and ready for the mechanical pickers. The landscape is devoid of people and dwellings. As in other crops, this monoculture generated serious pest problems, which in the long run were aggravated by constant use of pesticides. Recently, emphasis has shifted to so-called integrated pest management, with much more emphasis on natural predators of cotton pests.

Oleaceae — Lilac Family

The Oleaceae include about 20 genera native to many tropical and temperate regions. Genera cultivated as ornamentals include *Syringa* (lilac), *Ligustrum* (privet), *Jasminum* (jasmine), *Forsythia*, and *Fraxinus* (ash). Only one has been domesticated as a crop.

***OLEA* — OLIVE** (Boardman 1977; Hutchison 1946; Patiño 1969; Zohary and Hopf 1988)

There are about 45 wild olive species, whose combined ranges extend from New Caledonia, New Zealand, and Australia through southern Asia and through most of Africa to the Near East and southern Europe. The species are isolated primarily ecologically and geographically and are believed to be capable of hybridization. *O. europaea* conceivably originated as an ancient natural hybrid between two species no longer in contact: *O. ferruginea*, now in the Iran-Afghanistan region, and *O. laperinii*, now in mountain ranges of the Sahara.

O. europaea includes the domesticate and the wild olive or oleaster. (The name oleaster is shared with the superficially similar but unrelated *Eleagnus angustifolia*, also known as Russian olive, which is planted for hedges.) The wild and domesticated *O. europaea* intercross, but hybrids tend to be eliminated in both natural habitats and orchards. The oleaster's natural habitat is the sclerophyll scrub the *maquis* or *garrigue*, from Morocco and Spain eastward along the mainland nondesert coasts and islands to the Near East. It differs from the cultivar olives in being shrubby, having spiny trunks, and bearing smaller, less oil-rich fruits. The fruits, like the domesticate's, contain a bitter glucoside, which may repel mammals but not birds, including gulls and doves, which are probably responsible for the broad, disjunct natural range. Humans learned prehistorically to make olives palatable, usually by soaking them in alkali solutions and pickling them in brine, often flavored with herbs.

There is little difference between stones of wild oleaster and domesticated olives. Finds in archaeological sites of the Near East and on Cyprus dating from Natufian and Neolithic times, i.e., from about 9000 to 4000 B.C., may all come from wild gathering. During Chalcolithic time, i.e., between 3500 and 3000 B.C., however, abundant olive stones began to be deposited in Near Eastern sites too dry for oleaster, e.g. in villages practicing irrigated agriculture in the Jordan Valley and nearby arid regions. Between 3,000 and 2,000 B.C., Bronze Age farmers in the Near East and on the Greek islands and mainland evidently regularly practiced oleiculture. By the 14th century B.C., Greek palaces contained huge magazines for storing the specialized jarfuls of olive oil, as well as tablets with written records of the trade. Boardman (1977) illustrates Athenian vases from about 500 B.C with exquisitely detailed paintings showing harvesting and processing of olives. Oleiculture did not spread to Egypt, which regularly imported olive oil from the Levant. Olive oil and pickled olives became, like wine, important articles of daily diet and commerce throughout the ancient Mediterranean.

Many different local cultivars were developed, all presumably originating as chance seedlings found to bear superior fruit. Olives were propagated vegetatively as clones, both by rooting of cuttings and by grafting onto seedling rootstocks. Sometimes cultivars were grafted on oleasters in the wild. Individual olive trees can live for centuries, and a clone may be propagated for millenia, so there have been few sexual generations in most cultivars since initial domestication. Very little evolution of cultivars has been noted in historic time or under scientific breeding.

The geography of oleiculture has remained amazingly stable since classical times. The vast bulk of the crop is still grown in the Mediterranean basin. Successful introductions elsewhere have been essentially limited to closely similar climatically regions. Spaniards attempted to grow olives in the New World from the beginning of their colonization. During the 16th century, olive cuttings brought from Spain had been planted in the Greater Antilles, Mexico, Central America, and as far as Peru. Generally the trees grew well, but bore little or no fruit. Peru was an exception, with fruitful olive orchards widely established in the coastal desert by the early 17th century. Eventually, Chile became more successful in oleiculture in its Mediterranean climatic region where irrigation was unnecessary.

The most successful introduction in the long run, however, proved to be through Mexico. Plantings were desultory until the 18th century when Spanish missions on the Pacific coast pushed into another Mediterranean climatic region in Baja and Alta California. All the missions and many Spanish and Mexican ranches developed successful olive groves, extending north to central California. After statehood, olive planting boomed in California. In the last quarter of the 19th century, olive trees were planted by the millions in commercial orchards. Over 150 cultivars were brought to California from the Mediterranean for trial, and a few were planted commercially. However, the original Mission variety remained strongly dominant both for oil and for canned ripe olives.

Proteaceae — *Grevillea* Family

Proteaceae are primarily a southern hemisphere family of over 50 genera and over a thousand species, the vast majority native to Australia and South Africa. Proteas are very distinctive in growth habit and inflorescences, and many members are cultivated as ornamentals in subtropical regions of the world. Only one genus is planted commercially.

MACADAMIA — MACADAMIA NUT (Fukunaga 1967; Neal 1965; Smith, P. 1976; Storey 1965)

Macadamia includes 10 species of evergreen, medium-sized trees native to subtropical eastern Australia, Indonesia, and New Caledonia. Two species bear large crops of hard-shelled edible nuts. The round embryos, consisting mainly of hemispherical cotyledons, have a fat content of about 70% and are rich in vitamins and minerals. They are delicious raw or roasted, and the wild nuts were gathered from the ground by the Australian aborigines.

M. integrifolia (=*M. ternifolia* var. *integrifolia*), the smooth Macadamia nut, is native to rain forests of subtropical Queensland. *M. tetraphylla* (=*M. ternifolia* var. *tetraphylla*), the rough macadamia nut, has a range overlapping the southern margin of the former species and extending farther south into northern New South Wales. The two species differ in many characteristics, including ecological adaptations, but are fully interfertile when grown together. Orchard plantings of both species were begun in eastern Australia about 1860, and the species were introduced to the Hawaiian Islands in the 1880s and 1890s. Named cultivars were soon selected in both regions and propagated by grafting. Often interspecies hybrid cultivars were used as scions, grafted onto *M. tetraphylla* seedling rootstocks. The hybrids may combine the preferred nut quality of *M. integrifolia* with seasonal flowering and harvest characteristics of *M. tetraphylla*. Selection has, of course, also aimed for higher yielding, larger-fruited, thinner-shelled cultivars, but unfortunately, large nut size seems to be linked with thick shells.

In Hawaii, many *Macadamia* trees were planted as ornamentals between 1910 and 1915. The tree is beautifully proportioned, with handsome foliage, and makes a fine shade tree. On the island of Hawaii, the *Macadamia* was chosen in 1916 for reforestation of areas destroyed by cattle. In 1919 and 1920, when coffee prices declined, Macadamia was planted in the Kona district to replace coffee. By the early 1920s, commercial value of the delicious nuts had become clear, and the territorial legislature gave a 5-year tax moratorium on land planted to the trees; planting began in earnest on all the major islands. Currently plantations totalling several thousand hectares provide a significant, high-value export. Harvesting is expensive, especially if ever-bearing cultivars are grown, which means repeated picking of fallen nuts from the ground. Attempts to mechanize harvesting with huge vacuum machines have not been very effective.

Both species and hybrid cultivars are currently being grown on a smaller scale in southern California, Central America, and South Africa. Their climatic and soil requirements are similar to those of the avocado orchards. In places where the avocados have been killed by soil fungus disease, *Macadamia* trees have shown resistance.

The crop currently looks like a potentially successful incipient domesticate that has not yet filled its potential geographic range.

Rosaceae — Rose Family

The Rosaceae are a family of over 100 genera and thousands of species native to most of the vegetated world but with the greatest diversity in temperate Eurasia and North America. The rose family contains many genera of herbs, vines, shrubs, and trees that are familiar garden ornamentals grown simply for their flowers or colorful fruits. Many genera have fruits attractive to birds and mammals, more by their flavor than by their food value, although they do offer some vitamins and calories. Some genera, including *Rosa*, *Pyracantha*, and *Cotoneaster*, get by with marginally attractive fruits by storing them on the plant until winter when birds have little choice. Such nonperishability contributes greatly to the value to mankind of domesticates such as apples and quinces. In those genera adapted for animal dispersal, the ovary commonly forms a hard core or stone protecting the seeds. Also, in various genera, the seeds produce small amounts of cyanide when crushed, which deters evolution of seed predation as a regular thing. Chewing up a few apple seeds or cracking a handful of peach pits for their kernels is harmless, but eating large numbers of seeds at a time can be lethal.

Domesticated pome fruits, such as apples and pears, and stone fruits, such as peaches, plums, and cherries, originated in regions with cold winters and inherited from their wild ancestors mechanisms for timing winter dormancy and spring bud break and flowering. If there is an unseasonably early warm spell, the trees remain dormant until they have experienced sufficient chilling. The chilling requirement varies greatly between varieties, depending on the frost regime of their homelands. This kind of adaptation is not immutable and can be changed gradually by natural selection or breeding, but it greatly restricts the geographic limits of individual cultivars. A cultivar may grow well in a strange climate, but flowering and fruiting may be suppressed by lack of chilling, or bud break may be triggered before late frost nips the flowers. The geographic consequences are far too complex for detailed consideration here.

***MALUS* — APPLES** (Brown 1975; Roach 1985; Watkins 1976a; Whealy 1989; Zohary and Hopf 1988)

Wild *Malus* species, commonly called crab apples, are small deciduous trees native to northern hemisphere regions with continental climates, i.e., cold winters and hot summers. Their habitats are characteristically edges of woods and other sunny, open places. Before human intervention, *Malus* probably had about two dozen fairly discrete species, partially segregated ecologically and geographically. Wild apple species are obligately cross-pollinated and interfertile, so hybridization has probably always occurred in zones of species contact. Mankind has greatly increased such contact.

Malus domestica (=*Pyrus malus*) — Domestic Apple

The domestic apple has no single wild progenitor; it is a hybrid swarm of thousands of genetically different cultivars. Important genetic contributions were evidently made by at least four wild species: *M. sylvestris*, *M. pumila* (incl. *M. paradisiaca* and *M. praecox*), *M. dasyphylla*, and *M. sieversii*; all four are diploids. They range, singly or together, across virtually all the cool temperate regions of Europe, the Near East, and central Asia. In the Far East, another wild species *M. prunifolia* (incl. *M. asiatica*) has contributed to some cultivars.

Evolution of this compilospecies began prehistorically, probably over a huge area of Eurasia. Long before people began planting apples deliberately, they must have promoted hybridization by seed dispersal and habitat disturbance. Neolithic village sites throughout most of Europe contain remains of apple cores and dried apples, evidently often cut in half and sun dried. Volunteer apple seedlings found open niches on margins of Neolithic grain fields and pastures. Superior variants might have been protected, but constant cross-pollination would have greatly limited effects of casual human selection. There is no evidence of apple domestication anywhere before the Iron Age. Finally, in the 10th century B.C., a Judaean site between Sinai and the Negev had dozens of archaeological apple cores. The location is well outside the range of any wild apples. If the apples were grown locally, as seems likely, they would have needed irrigation. In the 9th century B.C. in *The Odyssey*, Homer wrote of apple trees in the legendary garden of a Phoenician king.

The first record of grafted apple cultivars is from Greece about 300 B.C. The Romans had many different cultivars propagated by grafting, both budding and cleft grafting. It is not known whether grafting was invented in the Mediterranean region or diffused there from the East. It was crucial in allowing propagation of choice highly heterozygous varieties, which could not be rooted from cuttings, as is true of domesticated apples in general. Some wild species used as rootstocks can be propagated by cuttings and suckers, for example *M. pumila* and its derivative the Paradise apple, but during most of the history of apple orchards, seedlings have been used for rootstocks. Commonly seedlings have been allowed to grow large enough to begin fruiting before being top-worked. The rare seedlings that happened to bear especially good fruit became sources of new cultivars.

During the Middle Ages, European peasants and monasteries amassed a lot of apple varieties, which supplied rich genetic variation for selection in Renaissance gardens. From about 1500, European orchards in general were greatly expanded and given much more care. By the 16th and 17th centuries, having diverse and choice fruit trees was a matter of pride among the wealthy and nobility, who sometimes personally participated in selecting and grafting cultivars. Some of the cultivar names in use then have come down to the present, although it is usually speculative how much genetic similarity has been retained. Certainly there have been bud sports even if the same clone has been carried on, and some names have been passed on to seedling progeny.

Pearmain is perhaps the oldest cultivar name still borne by popular modern clones. It became important in England by 1200 A.D., at a time when apple planting and cider making had been much expanded by French influence following the Norman Conquest. Various slightly different Pearmain clones have been introduced to the U.S. since 1800, and others originated here. In the late 19th century, the Pearmain was introduced to Japan, and it has been a parent of important modern Japanese cultivars.

The original Pippin was probably introduced from France to England repeatedly in the early 16th century. One was introduced by Richard Harris, fruiter to Henry VIII. In 1629, Parkinson rated the French Golden Pippin as the greatest and best of apples. About 1700, a Pippin seedling in Yorkshire founded the still popular Ribston clone, and not long after, a seedling on Long Island, New York, founded the Newtown clone, which was for a time the leading commercial apple of California. About 1825 in Kent, England, a Ribston Pippin seedling founded the Cox's Orange clone, another great success.

Another old cultivar, the Gravenstein, probably originated about 1600 in northern Italy. It is a triploid, like several other leading cultivars, and can only be propagated by grafting. By the middle of the 17th century, it had spread northward as far as Denmark. By 1790, it had been introduced to the eastern U.S. from Germany. The Gravenstein was planted, along with various other orchard trees, in northern California by the Russian American Company. The Russian colony was established in 1812 at Fort Ross, its purpose being to supply food crops for their Alaska operations and their fur trade. After the fort was abandoned in 1841, the orchard was left untended, but two Gravenstein trees survived in the mid-20th century. Perhaps the Russians obtained planting material from Mexican settlements in the region with which they are known to have had contacts. Some kinds of apples had been grown in California since the Spanish mission period, although it is doubtful that they were grafted rather than seedlings. Gravensteins are now a major commercial crop in California, largely for canned sauce. The variety may have been introduced repeatedly to the west coast via the Oregon Trail and during the 1849 Gold Rush. During the mid-19th century, many deciduous fruit orchards were planted in the west, including Gravensteins, for a while producing more than the miners and other settlers could eat.

Another widely grown old European cultivar was the Borovitski, also known as Duchess of Oldenburg, and now familiarly as Duchess. Duchess originated in Russia and because of its extreme hardiness and excellent quality for tart pies was widely planted in western Europe. It was introduced from England to North America in the early 19th century, where its hybrid descendants include the Northern Spy and McIntosh.

Eastern North America was the place of origin of a remarkably large share of the world's commercial apple cultivars. The region was a melting pot for apple germ plasm introduced from all the home countries of the European immigrants. Although grafting of named cultivars was practiced from the outset by many colonists, others had orchards that were free-for-all swarms of seedlings. As frontier farming moved westward, five wild North American apple species joined in. The resulting hybrids were mostly poor but some were excellent and were propagated by grafting.

Important modern cultivars that originated as chance seedlings in eastern North America include the Rhode Island Greening and the Baldwin, both triploids, from the early New England colonies; the Jonathan, Northern Spy, and Winesap from the middle Atlantic states about 1800; Grimes Golden and its seedling the Golden or Yellow Delicious, both self-pollinators, from West Virginia about 1800 and 1900, respectively; Rome Beauty from Ohio about 1850; and the all-too-common Delicious from Iowa about 1880. The original Delicious seedling was striped, but subsequent bud sports produced the beautiful deep red color. Unfortunately selection was only skin-deep, and the Delicious is commonly regarded as misnamed. It has, however, been amazingly dominant in the North American market.

Various North American apple cultivars were promptly taken back to Europe where a few of them, notably Golden Delicious, have become commercially important. They were far more successful in Japan where the Jonathan and half a dozen other North American cultivars made up 95% of the crop in the 1950s before new local varieties were bred from them. North American cultivars have also invaded southern hemisphere orchards in Australia, South Africa, and South America; in Chile and Argentina, they became strongly dominant.

Scientific breeding of new apple varieties was begun late in the 19th century in Europe and North America. In North America, the most notable success was the McIntosh, developed in Ontario and released in 1870. In 1898, the New York state experiment station at Geneva crossed the McIntosh with Ben Davis to produce the Cortland. The Cortland was not released until 1915, but has become widely grown. Currently, the most successful new hybrid is the Fuji, a cross between the Delicious and a minor North American cultivar released in 1962 by a Japanese experiment station. The Fuji lacks the pretty red skin of its Delicious parent, but more than compensates by its internal quality — crisp, juicy, and sweet, sometimes with 20% sugar content. Planting was begun in California in 1979, initially as a pollinator for Granny Smith. The demand grew fast, and thousands of hectares of Fuji have been planted in California. About half of the new orchard plantings in Washington state are now Fuji. The Fuji is also being planted in New Zealand.

Some outstanding new cultivars have originated in Australia and New Zealand since introduction of European and North American varieties. The Granny Smith originated as a seedling in New South Wales in 1868. It is now the leading variety in California orchards. Two cultivars of New Zealand origin are now being planted on an increasing scale in California. These are the Gala and Braeburn. The Gala, released in 1965, is a cross between Golden Delicious and Kidd's Orange; Kidd's Orange, released in New Zealand in 1924, is a cross between Cox's Orange and Red Delicious. Braeburn, released in 1952, originated as a chance seedling of uncertain parentage. Braeburn is variable in quality, but at its best may be the finest of all apples.

In Europe, modern breeding, particularly in Britain and Poland, has yielded important new apple rootstock varieties that can be clonally propagated; they combine disease and pest resistance with dwarfing of the tree crown for easier picking.

PYRUS — PEARS (Griggs and Iwakiri 1977; Hedrick 1919; Hutchison 1946; Layne and Quamme 1975; Roach 1985; Simoons 1990; Whealy 1989; Zohary and Hopf 1988)

Pyrus includes about 20 wild species native to most of Europe, the Near East, and temperate Asia. They are nearly all deciduous, but there is one ornamental evergreen from Japan. Natural habitats are sunny rocky slopes, banks of intermittent streams, and open woodlands. The wild species are all obligately cross-pollinated diploids and are capable of interbreeding, but were initially kept fairly discrete by geographic and ecologic segregation. However, species mixing increased as Neolithic farming spread. Wild pears commonly volunteer on the margins of settlements and fields. Also dispersal increased with human use of wild fruit and eventual deliberate planting. Several wild species have long been planted, but are not fully domesticated and retain the edible but small, firm, gritty fruits of the wild species. These include the European snow-pear, *P. nivalis*, one of the favorite species for making perry or pear cider; like apple cider, perry is drunk both fresh and as a wine made by spontaneous fermentation with wild yeast. Other essentially wild species commonly planted are the European *P. austriaca* and *P. salvifolia* and the Far Eastern *P. bretschneideri* and *P. ussuriensis*.

All of these species and various others have contributed genes to the two fully domesticated pears, the European *P. communis* and the Far Eastern *P. pyrifolia*. Like the domestic apple, both are compilospecies, hybrid swarms with hundreds of varieties. It is customary to give these two complexes separate binomials because they were independently domesticated and have remained fairly distinct.

Pyrus communis — Common or European Pear

Pear domestication evidently began only about 2500 years ago. Neolithic and Bronze Age sites in central and southern Europe have yielded many remains of pears, often cut in half for drying, but all of them could have been gathered from wild trees. Evidence of domestication begins only with written Greek and Roman history. About 300 B.C., Theophrastus recorded three named pear cultivars that were propagated by grafting. By about 50 A.D., Romans knew over 40 pear cultivars. Pear cultivation prospered, and varieties diversified in continental Europe and England during the Middle Ages and Renaissance; in some regions, most were crushed for perry cider and wine. By about 1600, the Medici Grand Duke could have his choice of over 200 kinds of pears. Some of these presumably were shipped from distant parts of Europe.

Most of the older cultivars were firm, crisp types, not the soft, buttery types that predominate today. The golden age of pear breeding, when almost all the major modern cultivars originated, was between 1750 and 1850. Belgian and French growers were responsible for selecting, from open-pollinated seedlings, the great majority of the successful cultivars, as shown by names such as Beurre d'Anjou, Beurre Bosc, Doyenne du Comice, Duchesse d'Angouleme. The greatest cultivar of all the Bon Chretien or Bartlett originated in Berkshire, England, in 1770 as a seedling of a French cultivar. Controlled hybridization developed in England and on the continent early in the 19th century. Gregor Mendel experimented with pear hybridization and was given an award for new cultivars.

All these cultivars had to be propagated as scions grafted onto seedling rootstocks. The rootstock can be from another cultivar, a wild pear, or the quince, *Cydonia*. A quince rootstock has a dwarfing effect on the tree, making for easier picking of pears.

Pears, both grafted and as seedlings, were introduced in the 17th century to eastern North America by French, English, and Dutch colonists. Pears, probably seedlings, were introduced by the Spaniards to California in the late 18th century. The pear is less tolerant of extreme continental climate than the apple, but was planted widely in North America and has been planted on a massive scale in the Pacific states for over a century. Yet, unlike the apple, little in the way of new cultivars of *P. communis* has developed since the introductions from Europe. A few seedlings and bud sports have appeared and been propagated on a small scale. Various experiment stations have developed varieties better adapted to local climates. The North American orchards in general remain planted with the Bartlett, Anjou, and other old European pears.

The most notable exceptions are hybrids between the European pear and the Chinese *P. pyrifolia*, which were developed in North America in the late 19th century. Named cultivars of this interspecies hybrid are Kieffer, Garber, and LeConte.

Pyrus pyrifolia — Chinese or Sand Pear

Compared to a common pear, *P. pyrifolia* fruit is russet, crisp, gritty, sweet, and bland. There are many named cultivars, however, with quite different fruits, some extremely large. The pear was probably taken into cultivation in central and northern China about the same time as it was in Europe. Cultivation of pears of fine quality had been extended southward to Kwangtung before 1000 A.D. Marco Polo described pears in markets in Chekiang as white-fleshed, fragrant, and of tremendous size. Westerners commonly mistake Chinese pears for pear and apple hybrids, which do not exist.

The species was introduced to Japan before 800 A.D. During the California Gold Rush, Chinese miners planted *P. pyrifolia* seeds in the Sierra Nevada foothills, and later Chinese and Japanese immigrants introduced other varieties to California, including cultivars propagated by grafting. Mostly, these have been maintained in farmyards and urban gardens of people of Asian descent. Commercial planting for more general use developed slowly in the 20th century in California, Oregon, and Washington. The main commercial varieties include Ya Li and Tsu Li, old Chinese cultivars; Nijisseiki and Chojuro, cultivars originating in Japan as chance seedlings in the 1890s; and Shinseiki, a Nijisseiki and Chojuro hybrid originating in Japan in 1945.

***PRUNUS* — STONE FRUITS** (Bailey and Hough 1975; Fogle 1975; Hedrick 1917, 1919; Hesse 1975; Kester and Asay 1975; Roach 1985; Simoons 1990; Watkins 1976b; Weinberger 1975; Whealy 1989; Zohary and Hopf 1988)

Prunus includes about 150 species of trees and shrubs, deciduous or evergreen according to their climate. They are most abundant in temperate North America, eastern Asia, and Europe, but a few of them are in tropical mountains north of the equator. *Prunus* species generally colonize open and disturbed sites and are thus preadapted to cultivation. They are generally adapted for exploiting scattered, ephemeral opportunities through bird dispersal. Fruits of the wild species, although small, were equally attractive to people. Domestication has resulted in greatly increasing the attractiveness of these fruits.

***Prunus persica* (*Persica vulgaris, Amygdalus persica*) — Peach, Nectarine; *Prunus armeniaca* (=*Armeniaca vulgaris*) — Apricot**

Like most fruit crops in the rose family, these domesticates may be compilospecies with multiple wild ancestry. Wild peaches are reported under various Latin names in the mountains of western China and Tibet. Wild apricots, again under various scientific names, are reported from Manchuria, Siberia, and Korea, across much of China, central and southwestern Asia as far as the Caucasus. The genetic relationships between wild and domesticated peaches and apricots are obscure, and there is no archaeological evidence on time and place of first cultivation.

Their stories, insofar as they are known, begin in ancient China. Archaeological remains of peaches are reported from Chekiang, far east of the supposed native range, dated about 4000 B.C. Peaches are prominently reported in the most ancient Chinese classics. Perhaps more than any other plant, *P. persica* was an object of reverence in traditional Chinese culture. The peach symbolized protection from evil, longevity, and immortality. It figured prominently in myths and legends and was a favorite motif in art and poetry. Mystical properties were attributed to not only the fruit but the wood and the blossoms. Relics of the classical traditions survive in China today in symbolic importance of the species in birthdays, weddings, and New Year festivals. Although much less prominent than the peach, the apricot also was commonly mentioned in Chinese classics and also had traditional aesthetic and magical significance.

In traditional Chinese agriculture, individual farms commonly grew a few trees of several varieties of peaches and apricots. Unlike most apples and pears, these species are self-fertile, and cultivars could be selected and stabilized without grafting by seed propagation. Peaches and apricots in China include many varieties unknown in the West with far greater diversity in flesh colors, fruit size, shape, and culinary uses.

In southwestern China and Turkestan, the smooth-skinned peaches or nectarines are among the commonest cultivars. Nectarines have originated repeatedly from fuzzy peach progenitors by a simple recessive mutation. The reverse mutation, producing a peach from a nectarine, also occurs.

Cultivars were also developed that were adapted to milder climates. The apricot was introduced throughout southern China and even into Tonkin. By 700 A.D., apricots were being grown in Kashmir, as they are today. The peach was also introduced into India and widely planted around mountain villages, but fruit quality is poor.

Both peaches and apricots were introduced to Europe around the beginning of the Christian era. According to Greek and Roman writers, they came over an interior route via Persia and Armenia. During the time of the Roman Empire, they were both well established in the Near East, North Africa, and southern Europe. The Romans evidently introduced cultivation of peaches but not apricots north of the Alps. Peaches survived the early Middle Ages in some French monastery gardens and were introduced to England following the Norman Conquest. Apricot cultivation began in England in the 16th century. For generations, English gardeners have given both peaches and apricots much tender care, growing them under glass or carefully trained on sunny walls for a slim harvest of prized fruit. This is another fine, old institution being lost by modern transport of cheaper imports.

Spread of peach and apricot planting from central Asia into the Mediterranean and thence to northwestern Europe transgressed formidable climatic gradients. Starting in mountains of desert regions, which have some of the most severe winters and hottest, driest summers on earth, the species had to adapt first to the mild Mediterranean and then to the wet Atlantic climates. Being self-fertile, not requiring grafting, and easily carried for almost any distance, the species probably would have spread like wildfire if there had been no climatic restraints.

Introduction of the peach and apricot to the Americas began with seed from the relatively impoverished European gene pool rather than the rich, ancient

Asian stock. During the early 16th century, Spaniards tried planting peaches and apricots in the New World tropics. They had little success until their colonies reached the temperate highlands of Mexico, the Andes, and higher latitudes in Chile and North America.

The Spanish colony at St. Augustine was growing peaches and apricots by 1600, and Spanish missions elsewhere in Florida and what is now Georgia rapidly introduced the peach to the Indians. The alacrity with which the Indians took up this exotic fruit was amazing, especially considering that their agriculture was based on annual crops. The peach was spread from tribe to tribe ahead of European exploration and settlement. Before 1700, the so-called Indian peach was widespread and abundant. It was already present in Carolina and Louisiana (in the original broad sense) when the English and French arrived there. The Louisiana Indians were making peach wine by 1699. The peach had been introduced independently to New England, the middle Atlantic, and Virginia colonies. There also, the Indians quickly began planting it. William Penn wrote that no Indian farm around Philadelphia lacked peaches. By 1750, Indians had spread *P. persica* so extensively that a leading botanical explorer of eastern North America, John Bartram, thought the species might be indigenous. Meanwhile, Spanish settlers and missions had introduced peach cultivation in New Mexico, where it was soon taken up by the Pueblo Indians, including the Hopi of Arizona. By shortly after 1700, peach planting had spread from the Hopi to the Navajo, originally a nonfarming people.

In eastern North America after the Revolution, white settlers moving west planted peach trees by the millions, largely if not mainly for making wine and brandy. As in the case of apples and other frontier orchards, peach orchards were grown from seed and so constituted what was in effect a huge agricultural experiment. Clonal reproduction, usually by budding, spread slowly in the early 19th century. As this practice took hold, innumerable promising seedlings were propagated, and by 1850, North American nurseries were offering about 400 named peach cultivars.

A new era in American peach breeding was begun in 1850 with the introduction of a cultivar that had just been brought to England from China by Charles Fortune, a Scottish plant collecter for the Horticultural Society of London. In eastern North America, this Chinese Cling hybridized with the peaches that had been introduced from Europe earlier. Its descendants include the Elberta, J. H. Hale, Belle of Georgia, and many other named cultivars. Originating as chance seedlings in the mid- and late 19th century, these have been propagated by grafting buds onto seedling rootstocks. As in the case of apples, these mongrel North American peaches have been widely adopted in other temperate regions of the world. They dominate the new peach orchards of Italy, which ship huge quantities of peaches to northern Europe.

Apricots underwent relatively little evolution in North America. Some new varieties have, of course, been developed, but the most important, such as the Blenheim and the Moor Park, were introduced from England long ago. No fresh introduction from Asia comparable to the Chinese Cling peach has been made.

It is remarkable how narrow a genetic base underlies the cultivated peach and apricot crops of most of the world. There are modern breeding programs in many temperate countries, which are constantly developing slightly different new cultivars adapted for their local conditions. However, these generally use the few major cultivars grown in the region rather than tapping the huge store of diversity in the vast Asiatic homeland of the crops. The most important exception is reported to be recent breeding in the Soviet Union and Poland using collections of wild and primitive cultivars obtained from the Caucasus and central Asia.

Prunus dulcis (=*P. amygdalus, Amygdalus communis*) — Almond

Truly wild almonds grow over a wide region of the Mediterranean basin and southwestern Asia as members of *maquis* and oak parkland vegetation. They are concentrated in dry, open sites, such as rocky, south-facing slopes. They are obligately cross-pollinated and hybridize with the domesticate. Almonds are exceptional in the genus *Prunus* in showing no adaptations for bird dispersal. The fruit is not attractive, and the seeds are not well protected by a stony pit. Like many Rosaceae, almond seeds are defended against predators by producing cyanide when crushed. Wild almonds are generally spiny and shrubby.

Domesticated *P. dulcis* is evidently another hybrid species with multiple wild ancestors. The wild species are assigned various scientific names, sometimes binomials, e.g., *P. webbii*, sometimes varieties of *P. dulcis*. The distinction between wild and domesticated populations is blurred by weedy hybrids along field edges and roadsides, which interbreed with both wild and cultivated trees.

The domesticate diverges from the wild almonds in having larger seeds and by a dominant mutation that suppresses the cyanide formation so that the nuts are sweet instead of bitter and toxic. The bitter gene, being recessive, is by definition undetectable in heterozygous state. As a result, some progeny of sweet almond trees produce bitter seeds. Small quantities of cyanide can be metabolized harmlessly, but if a person eats dozens of bitter almonds at a time, they are lethal. The domesticates also diverge from the wild in loss of spines, larger tree size, and, in some varieties, in being capable of self-pollination. Unfortunately, none of these genetic changes are detectible in archaeological material.

Rare finds of broken almond shells are reported from a few Paleolithic, Mesolithic, and early Neolithic sites in the Levant, doubtless from wild trees. Almond remains become more common in late Neolithic and Bronze Age sites in Greece, Cyprus, Anatolia, and the Levant. In some cases, the sites also yield remains of grape and olive. This suggests that planting of nonbitter almonds, which are easily grown from seed without grafting, had begun in the eastern

Mediterranean by at least 3000 B.C. Grafting of special cultivars began during classical Greek and Roman time. In many Mediterranean orchards, seedling almonds continued to be grown up to modern times. Traditionally, the orchardists have to rogue out seedling trees that produce bitter almonds. Before Mendel, the inheritance of recessive genes was a mystery.

Almond cultivation in Europe has remained concentrated in the Mediterranean basin. The trees are hardy farther north, but they flower so early in the year that the blossoms are usually nipped by frost.

Overseas, successful almond orchards were started by the Spaniards first in Chile in the late 16th century, then in California about 1800. These were presumably variable seedling stocks. Starting in mid-19th century, several dozen named cultivars, some seedling landraces and others propagated by grafting, were introduced by California nurserymen. Most of these were from Spain and the Balearic Islands, some from France, and a few from other Mediterranean countries. The vast modern almond orchards of California are not based directly on these original introductions but on chance open-pollinated seedlings derived from them. There are now over 8000 almond groves in California with over 150,000 bearing hectares producing nearly $^2/_3$ of the world crop.

The origin of new cultivars in California can mainly be credited to a private grower A.T. Hatch of Suisun in the Bay Area. In the 1880s, he discovered that the Languedoc cultivar yielded five times as heavily when growing next to chance seedlings as it did when grown in pure stands. In other words, it was partially self-sterile. This led, of course, to interplanting grafted cultivars with seedling pollinators and genetic ferment in consequent seedlings. Hatch then turned to selection of seedlings that were promising nut producers in their own right for propagation by grafting. Some of these had thinner shells than the European cultivars. Some of Hatch's introductions, e.g., the Nonpareil and Ne-Plus-Ultra, became the standard varieties used in California orchards. Subsequent breeding by government experiment stations has contributed relatively little to the stock.

The story of commercial almond planting in Australia followed a similar path, with chance seedlings selected by private growers becoming the basis of commercial orchards.

Traditionally, almonds were considered intolerant of wet soil and were relegated to marginal, dry sites where their yields were relatively low and unreliable. With better techniques of growing, including grafting onto peach rootstocks, which are more tolerant of water, orchards are shifting onto better soils and often being irrigated, resulting in heavy yields.

Prunus domestica **(incl.** ***P. insitita*** **and** ***P. italica*****) — European Plum, Prune, Damson, Gages**

P. domestica is a hexaploid (6x=48). It is usually interpreted as an allopolyploid hybrid between two wild species: the myrobalan, *P. cerasifera* (incl. *P.*

divaricata) and the sloe, *P. spinosa*. The myrobalan is normally diploid (2n=16) and the sloe tetraploid (4x=32). Experimental hybrids between the two have produced nearly sterile F_1 seedlings, but a few F_2 seedlings were obtained, most of which were hexaploid and resembled *P. domestica*. Both of these wild species include variants that are diploid, tetraploid, and hexaploid, so hybridization could have given rise to *P. domestica* type hexaploids in a variety of ways and at different times and places. (Multiple origins are a theoretical possibility in any allopolyploid species.) In any case, the myrobalan and sloe are supposed to have contributed different flesh color and skin color genes, which are recombined in diverse ways in *P. domestica*.

Where and when *P. domestica* originated, beyond the fact that it was in ancient Europe, remain unclear. It and its two putative progenitors are all weedy, commonly volunteering in artificially disturbed places, including roadsides and field edges. European botanists attribute much of the distribution of plums to naturalization following the spread of agriculture. Plum pits found in pre-Neolithic archaeological sites in the Upper Rhine and Danube region resemble both *P. cerasifera* and *P. domestica* of the small-fruited damson type. All these probably represent gathering of fruit from volunteer trees. Possibly there was some human protection and increased dispersal of better variants. Plums are not generally self-pollinated, and as long as reproduction was by seedlings, clearcut domesticates were lacking. The first good evidence of true domestication of plums is from classical Greece and Rome, when various cultivars were being propagated by grafting, along with apples and pears. The Romans evidently introduced these widely, even as far as Britain.

Hundreds of named cultivars of *P. domestica* have been developed in European orchards during the last 2000 years from chance seedlings or bud sports. These were propagated by grafting buds or twigs onto rootstocks started as seedlings or as cuttings from sucker shoots. Rootstocks are chosen according to climate, soil, disease, and pest problems, and for ability to dwarf the tree crown for easier fruit harvesting. Various wild and domesticated *Prunus* species, including *P. domestica* itself, peaches, and cherries, have been used.

P. domestica cultivars were taken overseas early in the European colonial period. Preserved genetically by grafting, some of the old European cultivars are still widely grown. Among these, French varieties are especially important, e.g., Reine Claude, renamed Green Gage after introduction to England in about 1720. The huge commercial prune crop grown for drying in California uses old French cultivars, especially Agen. In contrast to *Malus domestica*, *P. domestica* has produced very few new cultivars from chance seedlings in North America. A notable exception is Mount Royal from Montreal, Quebec. Deliberate breeding of European plums at Canadian and U.S. experiment stations has added a few successful cultivars, most notably the Stanley, a hybrid of Agen and Grand Duke, released at Geneva, New York in 1913. Far more results have been obtained by crossing *P. domestica* with other plum species discussed below.

Prunus americana — North American Plum

North America has about a dozen species of native plums, more than any other continent. All are interfertile diploids (2n=16). *P. americana* is the most wide-ranging and most frequently planted. It is native from eastern Canada and New England south to Florida and west to the Rocky Mountains. The species is common along streambanks, on edges of woods, and along roadsides and field edges, often forming thickets by sucker sprouts from the roots. The fruits are variable in color, size, and edibility. It seems likely that some of this variation was due to prehistoric selection by the Indians, although early accounts contain only hints of this. Plums were recorded as being harvested and dried in quantity by Indians; they were certainly present as welcome volunteers around Indian towns and fields. There are legends, which may very well be true, that early white settlers were given seed of superior plums by Indians. The story is blurred by the presence of several other wild species in eastern North America that hybridize with *P. americana*: *P. munsonii* (wild-goose plum), *P. maritima* (beach plum), *P. angustifolia* (Chickasaw plum), and others; all of which have varieties with edible fruits.

In the 16th century, the native plums of eastern North America were compared favorably to European plums by such explorers as Cartier in the St. Lawrence region, De Soto in the southeast, and Coronado in the western Plains. They were evidently never taken into cultivation by European colonists on the Atlantic coast, where the damson and other European plums were rapidly established. As settlement spread westward into the Appalachians and the Middle West, however, the grafted European plums were evidently temporarily left behind. Frontier and backwoods farmers found the native plums easier to grow without need for grafting and often better adapted to local soil and climate.

For about a hundred years, starting early in the 19th century, *P. americana* and other native plums seemed to be on the verge of successful domestication in farm orchards of the southern and central U.S. This was still in the era of diversified family farms, growing nearly everything they ate usually from seed harvested on their places. *P. americana* could be propagated by root suckers, but mostly seedlings were grown, leading to wonderful genetic diversity for selection. When one bore larger, better than usual fruit, it was commonly given a name, subject to change when introduced to other farms. Select varieties were passed from friend to friend and region to region. By 1900, there were hundreds of named cultivars of *P. americana* and *P. angustifolia*. Commercial nurseries propagated and advertised many of these, and it appeared that important new domesticates were emerging.

Today *P. americana* and the other North American species have almost disappeared from orchards, except among rustic-minded growers of old-fashioned crops. Like much else that belonged on the old general farm, the North American plum succumbed to the expansion of standardized commercial monoculture. North American species have, however, continued to be planted as seedling rootstocks and contributed to some modern commercial cultivars derived from interspecific hybridization, discussed below.

Prunus salicina — Japanese Plum

This diploid (2n=16) species, which has revolutionized North American plum production, was unknown outside the Far East until very recently. Very little is known in western countries about its early history. The species is supposed to be native to China, but Japan is the source of nearly all the cultivars introduced to North America. Various Japanese plum cultivars that are still much grown were introduced to California in the late 19th century. First came the Kelsey in 1870, then several imported by Luther Burbank in the 1880s: Abimdamce, Burbank, and Satsuma (for once an appropriately named cultivar).

Prunus simonii — Apricot Plum

P. simonii is another diploid (2n=16) native to China. It is a cultivar with high quality fruit. Nothing seems to be known about its wild ancestry or early history before it was introduced to France about 1870 by seed sent from China. *P. simonii* joined *P. salicina* in California in the late 19th century. Their hybrid, the Wickson, is widely grown for its large and delicious fruit.

Since about 1900, plum breeders in many parts of North America have concentrated on hybridizing diploid species from different continents. The most successful of the new hybrids all have large genetic contributions from Japanese cultivars. They were developed by private and government breeding programs in many states, including California, North Dakota, Minnesota, Texas, Missouri, Georgia, and Florida. Some of the more important of these hybrid cultivars are Santa Rosa (*P. salicina* and *P. simonii* and *P. americana*), a Luther Burbank concoction that is still a major crop itself and the parent of other major varieties, including Frontier and Ozark Premier; Methley (*P. salicina* and *P. cersifera*, myrobalan); Six Weeks (*P. salicina* and *P. angustifolia*, Chicksaw plum); Bruce (*P. salicina* and *P. munsonii*, the wild-goose plum); Underwood and Waneta (*P. salicina* and *P. americana*).

Together, these part Japanese and pure Japanese diploids have largely replaced the hexaploid European *P. domestica* as the North American fresh and canned fruit crop, leaving the French prunes still dominant for dried fruit.

Prunus avium (=*Cerasus avium*) — Sweet Cherry

P. avium is a diploid species (2n=16) and self-incompatible, i.e., requiring cross-pollination. It is a tall tree native to open, deciduous woodlands over a huge region from northwestern Europe and the Mediterranean to Russia and southwestern Asia. The wild form, or mazzard, has small, red fruits with small stones superbly adapted for bird dispersal.

Domestication involved primarily selection for larger fruits of various colors, from amber to blackish red. Along with this came gradual increase in size of the stones, but unfortunately wild and domesticated varieties have so much overlap in stone size that domestication is seldom evident in archaeological remains. Cherry stones are often present in Neolithic, Bronze Age, Roman, and Medieval sites in much of Europe. They may be from either wild or planted trees.

Historic records report cherry orchards in Greece before 300 B.C. Pliny wrote that there were none in Italy until the 1st century B.C. when the great Roman general Lucullus introduced a superior cherry from Cerasus in Anatolia.

The Romans probably introduced cherry cultivars to northern Europe, even to Britain. The first archaeological remains of cherry stones too large to be from wild trees were found in Roman sites in Switzerland and Germany. Some cherry orchards survived in central Europe during the early Middle Ages in monastery gardens. Cultivation may have died out in Saxon England, but cherry planting was resumed after the Norman Conquest with cultivars brought from France. Orchards expanded greatly both on the continent and in Britain during the Renaissance. By 1600, historical records permit tentative identification of particular cultivars. For example, an amber Spanish cherry may be the progenitor of the modern Royal Anne, also known as Napoleon or Wellington. By the 18th century, many recognizable cultivars were being widely interchanged among European countries, including the Black Tartarian, originally from the Crimea.

P. avium was introduced repeatedly from Europe to both eastern and western North America. In the east, the species was introduced in the 17th century. Throughout the colonial period, orchards were usually seedlings, but by 1757, a New York nursery was offering 20 budded cultivars. During the 19th century, the westward spread of sweet cherry orchards involved old European cultivars that are still grown, e.g., the Royal Anne and the Black Tartarian. These were repeatedly introduced around 1850 to the Pacific coast, which was to become the main North American sweet cherry region. The Royal Anne and Black Tartarian were taken to Oregon from Iowa by wagon train and to California from both New York and France by sailing ship. After 1860, new cultivars were developed in Pacific coast orchards. In Oregon, Seth Lewelling selected seedlings for propagation by budding that he named Bing, Lambert, and Black Republican. The first two now supply most of the North American crop; the third is a major pollinator of the others.

Sweet cherry breeding in government experiment stations began soon after 1900 and continues in various parts of North America, especially British Columbia and Ontario, and in several European countries, especially Germany and Russia. Most research has been done on hybridizing different *P. avium* varieties, but more radical hybridization with other species, including peaches, plums, and apricots has been done also. So far, these have mostly had limited acceptance. Some of the modern cultivars, notably Van, released in British Columbia in 1944, have been widely planted in regions with climates too harsh for the major cultivars.

Prunus cerasus — Sour or Pie Cherry, Morello, Amarelle

P. cerasus is a tetraploid species (4x=32), mostly self-fertile, i.e., not requiring cross-pollination by a different variety and therefore relatively true breeding from seed. It is not known truly wild, although commonly escaping from orchards to colonize fencerows and roadsides.

P. cerasus is generally interpreted as originating by hybridization between cultivated *P. avium* and wild *P. fruticosa* (4x=32), presumably by fertilization of the latter by a freak diploid sperm of the former.

P. fruticosa is native to a huge area of central, eastern, and southeastern Europe and Anatolia. It grows in dry grasslands and other sunny sites as a shrub, forming thickets by root suckers. Its fruits are red, like those of *P. cerasus*, and even more acid.

P. cerasus has not been distinguished from *P. avium* in archaeological, classical, or medieval sources, but sour cherries presumably originated some time after domestication of their sweet cherry ancestor, perhaps in medieval times. Historically, the first clearly identifiable sour cherry cultivar is the Montmorency, grown in France in the 17th century. It was introduced to New England in the colonial period and remains the leading commercial pie cherry in North America, grown mainly in the Great Lakes region. Some old English sour cherry cultivars were also introduced to North America, including Duke, a hybrid *P. avium* and *P. cerasus*. A few named cultivars have been selected in American orchards from bud sports and chance seedlings of the Montmorency. Several North American and European experiment stations have limited sour cherry hybridization programs, including use of germ plasm recently imported from Europe, but the few releases are not major commercial producers.

Various other wild cherry species of Europe, the Far East, and North America produce fruit edible not only by birds but also by people, at least if made into preserves or squeezed for cider. Some of these are commonly planted for home use, e.g., *P. pseudocerasus* (Nanking cherry), *P. tomentosa* (Manchurian bush cherry), *P. besseyi* (western American bush sand cherry). Gardeners have selected seedlings for larger, more palatable fruit, but they are only incipient domesticates.

***RUBUS* — BRAMBLES** (Bailey, L. 1911; Jennings 1976; Ourecky 1975; Roach 1985; Whealy 1989)

There are hundreds of *Rubus* species mostly in the temperate and arctic regions of Europe, Asia, and North America with a few native to tropical mountains, temperate South America, and oceanic islands. The wild species evolved primarily as ecological pioneers of open and disturbed habitats. Like colonizing species in general, they are adapted for wide dispersal, in the case of *Rubus* normally by birds, which are attracted to the berries and pass the still viable seeds. Once a bramble is established from seed, it can generally persist and spread indefinitely by sucker shoots or by rooting where the canes touch the ground. The formidable thorns are presumably an advantage in reserving the fruit for birds as prime dispersers, but have not deterred bears or people from participating.

Human activities are believed to have had a great effect on the abundance and diversity of wild *Rubus* species long before any were taken into cultivation. More important than dispersal, ecological disturbance, especially forest clearance, opened up opportunities for brambles to spread and hybridize. The genus is notorious for its complex interspecific hybrids, a so far almost insoluble problem for orthodox taxonomic treatment. In some groups, e.g., the raspberries, hybridization is usually on the diploid level. In the blackberries, many hybrids are allopolyploids, tetraploid up to dodecaploid (12x). Currently, controlled hybridization is rapidly producing new cultivars, easily propagated vegetatively as clones without grafting, although prolonged clonal propagation often runs into virus problems.

Only the few most widely cultivated *Rubus* species will be considered. Many others have been casually and locally planted.

Rubus idaeus (incl. *R. strigosus*) — Red Raspberry

The wild species has a circumboreal range in subarctic and temperate Eurasia and North America. It is diploid (2n=14), normally cross-pollinated and highly heterozygous and spreads clonally by sucker shoots.

The red raspberry (along with white- and yellow-fruited mutants) was the first bramble taken into cultivation, but the first records are very late. It was present in a few medieval and Renaissance gardens in Europe, but probably as a volunteer or casual planting rather than regularly cultivated. The first clear records of cultivars with larger berries than the wild are in late 18th century Europe and eastern North America. As in the case of other brambles, the late domestication is evidently explained by the abundance of wild and volunteer brambles. In lightly populated regions, e.g., Scandinavia, red raspberries are still so easily gotten along roadsides and in logged forest patches that there is no point in cultivation. The wild berries are delicious.

By 1800, *R. idaeus* had begun to attract increasing attention from horticulturists in Europe and North America, especially in regions where human population had built up enough so the wild berry harvest became inadequate. In 1790, a leading New York nursery could offer only two European and two American red raspberry cultivars. During the 19th century, selection of new *R. idaeus* varieties, both from the wild and in gardens, became widespread in North America and Europe. It became evident that the wide-ranging wild populations were rich sources of genes to develop cultivars suited to diverse local climates and soils. Late in the 19th century, red raspberry breeding was taken up by experiment stations. By 1925, there were several hundred named red raspberry varieties in cultivation. The process of diversification has continued to the present. Various locally successful new varieties have been released since 1950, especially in British Columbia, the northwestern and northeastern U.S., Scotland, and northeastern Europe.

Growing of the red raspberry has remained remarkably diffuse geographically. Except for frozen and other processed berries, the crop is too perishable to be grown for distant markets. Much of the crop is produced by small growers, and if they lack the hands for picking the fruit, it is often offered to customers for picking themselves. It is not the sort of product that engenders agribusiness and monoculture.

Rubus occidentalis — Black Raspberry, Blackcap

R. occidentalis is native to North America from the High Plains to Quebec and Georgia. It is a diploid (2n=14), interfertile with *R. idaeus*, and spreads clonally by arching canes that root at the tips where they touch the ground. The wild fruits are delicious and commonly gathered. Sporadic attempts at domestication began in Ohio in the 1830s, in some cases by Shakers, and by about 1850, some black raspberries were cultivated in New York state. Late in the century, horticulturists and experiment stations in eastern North America named many cultivars, mostly chance seedlings found in the wild or in gardens. By 1925, nearly 100 cultivars were reported in cultivation, and new ones continue to be developed. The crop is mainly marketed locally by small growers and has not become important outside North America.

Rubus neglectus — Purple Raspberry

The purple raspberry is not a species but an F_1 hybrid between the red and black raspberries, later progeny being highly variable. The F_1 hybrid is readily propagated clonally, however, and various named cultivars are grown, mostly in home gardens but with some commercial production in New York state.

Rubus loganobacus — Loganberry

This is a hexaploid (6x=42) that originated in Santa Cruz, California about 1880 in the garden of its discoverer Judge J. H. Logan. It is evidently a hybrid between cultivated *R. idaeus*, the red raspberry, and wild octoploid *R. ursinus*, the California blackberry; the former probably contributed an unreduced (2n) gamete and the latter a reduced (4n) gamete.

Two other complex red raspberry and blackberry hybrids were developed in the 1920s: the Boysenberry in California and the Youngberry in Louisiana; both have 7x=49 chromosome complements. Subsequently, a cross between Youngberry and Loganberry in Oregon gave rise to the Olallaberry. In Scotland, the Loganberry was crossed with the North American black raspberry to produce the Tayberry, which was taken back to North America and is now widely grown.

So far the only bramble cultivars mentioned are derived in whole or in part from two raspberry species. The only blackberry noted so far was *R. ursinus*, one parent of the Loganberry group. There are many other blackberry species native to both Eurasia and North America, and several of these have been taken into cultivation. Blackberry domestication developed more slowly than raspberry growing, especially in Europe, perhaps because of the abundance of hedgerow blackberries. Most blackberry cultivars have originated in North America where private gardeners began selecting superior seedlings from the wild in the mid-18th century. Also, in the mid-19th century, two blackberry species introduced from Europe began hybridizing with the native species. These were the cutleaf blackberry, *R. laciniatus*, and the misnamed Himalaya berry, *R. discolor* (incl. *R. procera*). These hybrids and other interspecific hybrids among North American native blackberries yielded a welter of variants, some of which became successful cultivars. Many are very complex polyploids, far too complex to go into here. Since 1900, experiment stations in several parts of the U.S. have further complicated the story by developing cultivars from controlled hybridization and selection programs. These have been largely aimed at producing finely tuned varieties for different regions. Also, several successful thornless cultivars have been developed. Some efforts have been made to breed more erect canes for mechanical harvesting and pruning. However, blackberry like raspberry growing has remained largely the domain of home gardeners and small commercial growers. The largest commercial operations, mainly for processing, are in Oregon.

FRAGARIA — STRAWBERRIES (Hedrick 1919; Jones, J. 1976; Patiño 1963; Roach 1985; Scott and Lawrence 1975; Wilhelm 1974)

Fragaria includes several dozen species forming a polyploid series from diploids to octoploids. Most species are native to temperate and subarctic Eurasia and North America, but some range southward into North Africa and through tropical American mountains into temperate South America. The fruits are very attractive to birds, and the seeds are evidently dispersed to highly disjunct regions. Presumably, all the species have also attracted humans since very ancient times. Wild strawberries generally spread locally by runners and can easily be transplanted into gardens and propagated as clones. However, only four species are known to have been domesticated, in the sense of having diverged genetically from wild populations. The wild ancestors of the polyploid domesticates are usually dioecious, although hermaphrodite forms occur. Modern cultivars all have hermaphrodite flowers, allowing fruit production from single clones.

Fragaria vesca — Wood Strawberry, Fraise des Bois

F. vesca is a diploid (2n=14) species native to wide regions of temperate Eurasia and North America. This is the commonest wild strawberry in meadows and forest margins of nearly all of Europe, including the Alps. Strawberry seeds, actually achenes, are known from European archaeological sites from Neolithic through Roman and Medieval times, probably from fruit of this species gathered wild and occasionally casually planted in medieval gardens. Clear records of regular cultivation begin in the Renaissance herbals of the

16th century, the plants grown for ornament and medicine as well as food. After about 1530, there is clear distinction between wild and garden strawberries, the latter noted as having somewhat larger fruit. By the early 17th century, several different cultivars were distinguished. Larger-fruited and longer-bearing varieties were developed in France in the 19th century; these are generally called Alpine strawberries. White-fruited variants were selected as being unattractive to birds. *F. vesca* is still widely planted in gardens for home use, but is not the source of the commercial crop. Neither is the following.

Fragaria moschata (=*F. elatior*) — Hautbois or Musk Strawberry

F. moschata is a hexaploid (6x=42) species native to highlands from France and southern Europe eastward through Russia to Siberia. Domestication of this species evidently also began in the 16th century, the first clear record of cultivation being from Wallonia and Germany in the 1570s with many additional French records in the early 17th century.

Domestication of both *F. vesca* and *F. moschata* was arrested in its incipient stage by the arrival in Europe of two New World *Fragaria* species, the ancestors of the modern garden strawberry. The two European native species are still grown on a small scale by devotees of their special qualities.

Fragaria virginiana — Virginia Strawberry

F. virginiana is an octoploid (8x=56) native to woods and meadows over much of North America. The early explorers and colonists of Canada and New England marveled at the flavor, size, and abundance of the strawberries, which were sometimes reported to have been planted by the Indians. Roger Williams wrote in 1643, "this berry is the wonder of all the fruits growing naturally in these parts. It is of itself excellent so that one of the chiefest doctors of England was wont to say, that God could have made, but never did make, a better berry. In some parts where the Indians have planted, I have many times seen as many as would fill a good ship, within few miles compass." Williams also wrote that the Indians mixed the berries with cornmeal in a mortar to make strawberry bread. Various other colonial period accounts extolled *F. virginiana* flavor, fruit size, and abundant yields, sometimes in fields belonging to the Indians. During the 1620s, the species was taken into cultivation in Europe, where hermaphroditic cultivars were selected, some of which are still grown on a small scale, especially for fancy jam. They are all small fruited compared to the Chilean strawberry.

Fragaria chiloensis — Chilean Strawberry

F. chiloensis is another octoploid (2n=8x=56), which probably originated on the Pacific coast of North America. It grows wild in coastal sands from the Aleutians and Alaska to central California. The species was evidently dispersed prehistorically by migratory birds to the mountains of Hawaii and to the beaches and mountains of southern Chile and Argentina. In Chile, the

Araucanians had domesticated *F. chiloensis* before the Spaniards arrived. The Indians propagated the species by planting runners and selected very large-fruited cultivars. They not only used the fresh fruit, but dried it in the sun and also fermented it for wine. The Spanish invaders found the Indian berry crop impressive, producing fruit the size of walnuts and much more delicious than the berries back home. They immediately adopted the species for their own gardens and during the 16th century spread it northward to Santiago, Lima, Cuzco, and Quito. During the 17th century, the Chilean strawberry became generally distributed in the cool highlands of Spanish America, even in Mexico. Curiously, it was not rapidly adopted in Europe. The first recorded introduction to Europe was in 1714 when a French naval officer Frezier brought two plants directly from Chile to the Paris Botanical Garden. Unfortunately, both plants were female, and although the species was propagated by runners and spread widely in experimental plantings, the lack of fruit production was not understood for decades. Finally, by 1740, Breton farmers found that the lacking pollen could be supplied by interplanting *F. chiloensis* with *F. virginiana* or *F. moschata*. From 1750 on, Brittany was the center of commercial planting of *F. chiloensis*, mainly for the London and Paris markets. However, before production of the Chilean species could spread, a new species that was soon to displace it, *F. ananassa*, had appeared.

Fragaria ananassa — Modern Garden Strawberry, Pineapple Strawberry

F. ananassa is another octoploid (8x=56), which originated in Europe about 1750 as a hybrid between the two American octoploid species being cultivated there. From the outset, *F. ananassa* combined the hermaphrodite flowers of domesticated *F. virginiana* with the large fruit of *F. chiloensis*. The first recorded appearance of *F. ananassa* was in Holland, but there are reports from elsewhere in Europe soon after, and the hybrid probably arose repeatedly wherever the two parental species were interplanted. The mode of origin of this distinctive new strawberry was clearly recognized by Antoine Nicolas Duchesne, a 19-year old gardener of Versailles, in a book published in 1766. Duchesne repeated the cross between the two parent species and grew the progeny matching *F. ananassa*. He was ahead of his time, and his explanation was not generally accepted until well into the 20th century.

F. ananassa rapidly became the main strawberry cultivated in Europe and North America. Nurserymen and market gardeners selected many new cultivars, especially in France, Britain, and the U.S., throughout the 19th century. The species has enormous evolutionary flexibility, partly because of its hybrid origin and partly because of its high polyploidy. Its genetic diversity was further enhanced by introduction of germ plasm from the wild California race of *F. chiloensis*, seed of which was sent to Britain by David Douglas about 1830. Soon after, more seed was sent from California to France. Genes from the California race allowed breeders to develop *F. ananassa* cultivars of the so-called everbearing type with greatly extended fruiting seasons.

Breeding of *F. ananassa* was taken up by experiment stations early in the 20th century and has expanded greatly up to the present, with active programs in many countries. *F. ananassa* cultivars have been developed for a remarkable range of climates, including all 50 of the U.S. Controlled hybridization has involved not only *F. ananassa* varieties but also backcrosses to both its parental species. In the 1980s, the University of California, Davis, released new cultivars obtained from crossing *F. ananassa* with a wild Rocky Mountain variety of *F. virginiana*, often called *F. ovalis*. These are day-length neutral, i.e., they flower and fruit under both short and long photoperiods and greatly extend the harvest in regions with long frost-free seasons. Other successful cultivars are very recent releases from programs in various other states and countries. Yields have increased greatly in the last decades; some commercial growers are getting nearly 100 tons/ha/year of superb quality fruit. Availability of fresh strawberries has been greatly expanded geographically, seasonally, and quantitatively. *F. ananassa* is still rapidly diversifying and diverging from its wild progenitors.

Rubiaceae — Coffee Family

The Rubiaceae include about 500 genera and 6000 species, mostly tropical trees and shrubs, especially diverse in the New World. Some members are native to subtropical and temperate regions on different continents, including a few herbaceous genera. Rubiaceae are generally members of forest communities and are not weedy.

Some shrubby genera are grown for their attractive flowers, e.g., *Gardenia*, *Ixora*, *Bouvardia*, *Coprosma*. Economic importance of the family is very limited, considering its size, and is entirely due to the presence of alkaloids, presumably evolved as a defense against herbivory. A few members are medicinal, of which *Cinchona*, the source of quinine, is by far the most important. Native to New World tropical cloud-forests, *Cinchona* became a plantation crop in tropical Asia in modern times. The only other important crop in the family is *Coffea*.

***COFFEA* — COFFEE** (Burkill 1935; Ferwerda 1976; Haarer 1958; Patiño 1969; Purseglove 1968; Rodriguez 1961; Van der Vossen 1985; West and Augelli 1966; Wrigley 1988)

Over a hundred species of *Coffea* have been named, but the genus is badly in need of taxonomic revision, and most of the names may be based on intraspecific variants and interspecific hybrids. *Coffea* as a rule is obligately cross-pollinated by wind and insects; populations are typically quite variable. Most species are diploid (2n=22) and capable of producing hybrids when in contact. Wild *Coffea* species are native to tropical forest regions of Africa,

Madagascar and other western Indian Ocean islands, and southern Asia. They grow as shrubs and understory trees in secondary forests and forest margins, including stream-banks. The cherry-like fruits attract birds and mammals, which disperse the hard seeds. (The familiar coffee bean is a seed divested of ovary and seed-coat; most of the bean is endosperm tissue, the embryo being tiny.)

The alkaloid caffeine is present in many *Coffea* species, not only in the seeds but in the fruit pulp, leaves, and other parts. There are a few suggestions in the ethnographic literature of use of wild coffee as a stimulant. These refer, however, only to chewing of the dried fruit or making tea from the dried leaves, not to brewing the seeds. This is not surprising because the raw seeds have no special flavor or aroma. Discovery of the technology of cleaning, roasting, grinding, and brewing coffee developed outside the range of the wild species.

About 99% of the world's coffee comes from two species with the remainder, which is used in cheap blends, coming from half a dozen other species that will not be discussed here.

Coffea arabica — Arabian Coffee

C. arabica produced virtually all the world's coffee until after 1900 and is still the source of all the high quality beans. The species is evidently an ancient allopolyploid (4x=44), the only nondiploid member of the genus. The species is also exceptional, not only in *Coffea* but among perennial plants in general, in being mainly self-pollinated. Constant selfing tends to produce a pure, inbred line, and true-breeding cultivars are usually propagated entirely by seed. Clonal propagation of *C. arabica* is easily done in research stations, but is rarely done by coffee planters.

C. arabica grows wild in mountain forests at elevations of 1500 to 2000 m in southern Ethiopia and the adjacent Sudan. The region has mild temperatures, a long rainy season, and a 2- to 3-month dry season. Flower buds are initiated during the rains, but stay dormant until after the trees have undergone moisture stress during the dry season and the rains resume; the trees then bloom at the start of the rains, and the fruits mature during the succeeding dry season. *C. arabica* plantations have been most successful where the climate is similar to the homeland and the same periodicity is maintained so the harvest is concentrated in a short dry season. Where there is continuous rain or multiple rainy seasons, the harvest becomes diffuse.

Being evergreen, *C. arabica* needs a deep root system to survive the dry season. The natural habitats include some streambanks and forest openings, but mainly the wild trees grow as an understory beneath taller tree species. In many regions, coffee planters have found that *C. arabica* is healthier and longer-lived if grown under a simulated forest canopy. Usually legume trees, such as *Albizzia*, *Inga*, or *Gliricidia*, are used. These are deciduous during the dry season, which results in reduced transpiration and a protective mulch of leaves; the legumes may also contribute to soil fertility. Elsewhere, notably in Brazil, *C. arabica* is planted without shade. This brings quicker maturity and heavier crops, but the yields begin to decline at an age when shade-grown coffee would be in its early prime.

Coffee brewing and planting evidently began, not in the homeland of *C. arabica*, but in Arabia Felix. Conceivably the species was dispersed to Arabian oases by migratory birds. Perhaps it was taken into gardens initially as an ornamental. Its fragrant flowers closely resemble those of a Near Eastern favorite *Jasminum*, and its foliage and berries are also attractive. It seems likely that the Arabs first used the dried berries for brewing and later hit upon the more complex process of pulping, cleaning, roasting, and grinding the seeds. Sultani coffee, brewed from the dried berry pulp, is a traditional Islamic beverage especially popular in the summer, while coffee brewed from the beans is preferred in winter.

The beginnings of the coffee industry are thus speculative. The first historical records are from Yemen in the 14th century. By the 16th century, Yemen was exporting coffee to other Islamic countries by caravan and ship. Mocha was the main port. The first European ship to load a cargo of coffee there was the Dutch merchantman the *Nassau* in 1616. When the *Nassau* was there, more than 30 Persian, Arab, and Indian ships were in the harbor. Later in the 17th century, ships of the Portuguese, French, and English East Indian companies joined in the Mocha coffee trade. It seems that some of the product was from wild trees in Ethiopia, imported as sun-dried berries by itinerant Arab traders, while some was grown in Yemen. No accounts of Yemeni coffee plantations are available until 18th century reports by European travelers. They described coffee as being grown in the Yemen hinterland on lower mountain slopes, carefully irrigated on terraces, and with a canopy of shade trees. The ripe berries were dried in the sun, and pulp removed from the beans in stone mills. In a culture that eschewed alcohol, rapid acceptance of coffee as a social beverage is easily explained. It is more surprising that the beverage spread successfully from Islam into Christian Europe in competition, not only with traditional wine and beer, but with Mexican cacao, which arrived a little earlier, and with Chinese tea, which arrived about the same time. The first European coffee house (barring Constantinople) opened in Venice in 1615. By 1700, coffee houses had become important social institutions in cosmopolitan cities throughout Europe and even in New York and Boston in America. During the 18th century, coffee had assumed its present place as a pillar of western civilization in general.

Coffee planting remained a monopoly of Yemen until about 1700. There are legends that pilgrims to Mecca had introduced coffee planting to southern India and perhaps elsewhere before 1700. If so, they must have carried freshly picked berries. The seeds have no dormancy and remain viable only if kept moist. The dried beans of commerce are dead even before roasting. Historically documented introductions of the species outside of Yemen all involved careful transport of growing plants, not dried berries or seed. These introductions began about 1700 when Dutch ships carried a few plants from Mocha to the

Malabar Coast, Ceylon, and Java. A coffee seedling begins producing seed within 3 years, so the species was rapidly multiplied in the East Indies. By 1712, coffee grown in Java was being auctioned in Amsterdam. An independent acquisition of *C. arabica* was made in Mocha about 1715 by a French East India company ship. Two plants were successfully established on the island of Bourbon, now Reunion, and from there *C. arabica* was taken to neighboring Indian Ocean islands.

Virtually all the world's *C. arabica* plantations are derived from these two early transfers from Yemen. The Dutch stock was the progenitor of the typical variety (*C. arabica* var. *arabica*), which spread faster and farther. The French stock was the progenitor of Bourbon coffee (*C. arabica* var. *borbonica*), which spread more slowly but eventually joined typical *arabica* in many regions. *Bourbonica* outyields typical *arabica* in some situations. Both produce top quality beans. They differ in a few genes affecting growth form and foliage color, which are recognized by planters. The two varieties have occasionally hybridized to produce new named cultivars, including Mundo Novo, a very important cultivar of Brazilian origin.

From the East Indies, *C. arabica* var. *arabica* was taken to the rest of the tropics via European greenhouses. The already narrow genetic base was reduced to the irreducible minimum; a single tree from Java was shipped to the Amsterdam botanical garden in 1706. Some of its seedlings, established in Surinam by 1718, were the first coffee plants in the New World. During the 1720s, progeny of the Surinam plants were introduced to other parts of the Guianas and to Pará in Brazil. Meanwhile, a seedling of the Amsterdam tree had been presented ceremoniously to Louis XIV in 1714. In the Jardin des Plantes in Paris, it produced seedlings, one of which was successfully introduced to Martinique. During the 1720s, progeny of the Martinique plant were established in French colonies in Guadeloupe and St. Domingue, now Haiti, and also in British Jamaica and Spanish Cuba. Soon after, *C. arabica* was in Mexico, and by 1740, it had been introduced from there to the Laguna region of Luzon by a Franciscan priest. Typical *arabica* coffee was successfully established during the last half of the 18th century, via anastamosing pathways, in various other tropical American regions, including the northern Andes and Central America.

During the first quarter of the 19th century, the Amsterdam lineage *arabica* coffee was introduced to Hawaii, where it became known as Kona coffee. In the late 19th century, typical *arabica* stock from the West Indies was taken back to the Old World. For example, a seedling of Blue Mountain coffee from Jamaica was grown in a greenhouse in the Edinburgh Botanical Garden, and a single seedling was taken from there to Nyasaland by missionaries in 1878. Coffee planting in East Africa was thus initiated not from neighboring Ethiopia, but via the long track Arabia-Java-Holland-America-Scotland. From Nyasaland, typical *arabica* was introduced to Uganda. A redundant introduction of typical *arabica* was made from Guatemala by Catholic missionaries in 1905; from Uganda, planting spread to Burundi. Meanwhile, starting in the 1870s, other missionaries had introduced the Bourbon variety to Bagamoyo and Kilimanjaro in German East Africa. At about the same time, other missionaries independently introduced both typical and Bourbon cultivars to Kenya from Aden and Zanzibar.

Cameroon is the only West African region in which *arabica* planting has been significant. The typical *arabica* was introduced from Jamaica when the region was a German colony in 1913. The crop is still grown in the highlands above 1000 m, mainly by smallholders.

So far, only the initial establishment of *C. arabica* has been traced. Since the trees began bearing within a few years, plantations could spread rapidly, and in some cases, they did. The Dutch East India Company began exporting coffee from Ceylon, India, and Java during the first quarter of the 18th century. The trees were grown by native villagers for a long time before European colonials discovered that *C. arabica* was an ideal crop for estates in the temperate hill lands where Europeans liked to live. European estates, typically of 20 to 40 ha, proliferated in Java, Ceylon, and southern India in the late 19th century.

In the West Indies, European coffee estates were established much earlier. French planters were exporting coffee before 1750 from Martinique, Guadeloupe, and St. Domingue, while British planters were exporting it from Jamaica. Since no native West Indians survived, African slaves did the work. Coffee was usually planted in the hills above the great sugar cane plantations. The Spaniards in the Greater Antilles were not interested in coffee planting, but estates were started in eastern Cuba by refugee French planters after the exslaves took over St. Domingue to form Haiti.

In Haiti, *C. arabica* survived the demise of the colonial estates. Production dropped in the north, which had the worst and longest period of chaos, but coffee planting spread as a peasant crop, and coffee beans replaced sugar as Haiti's main export. Similarly, in the French and British West Indian colonies, few estates survived emancipation, but coffee planting spread as a peasant crop. In Jamaica and other mountainous islands, coffee became a cash crop of pioneer farmers moving into the interior. It thrived on the fertile soil of newly cleared forest lands, but often erosion destroyed plantations on steeper slopes. Many small plantings were successful. *C. arabica* has several advantages as a peasant cash crop. Its genetic uniformity gives high quality progeny when unselected seedlings are grown without need for grafting or breeding. The beans can be processed by simply sun-drying the berries and removing the pulp in a mortar. The raw beans can be stored indefinitely and marketed over primitive transport and trade networks. Smallholders can grow coffee in mixtures with bananas and other crops for sale when prices are up and merely neglect the trees when prices are low, quite unlike the situation of a full-time planter. As a smallholder's crop, coffee became increasingly diffuse in the more rugged West Indian islands. In Jamaica, by the mid-20th century, it was grown all over the remoter parts of the interior by more than 30,000 farmers, most of whom had less than 2/10 ha in coffee and only about 1/10 of whom relied on coffee as their main source of income.

Brazilian estates did not have to cope with emancipation until the 1880s. Brazil had begun exporting increasing quantities of coffee about 1800 and by 1840 had assumed dominance of the world market. Production dropped temporarily after emancipation, but a multitude of new *fazendas* were established by immigrating European *colonos*. Brazil's phenomenal production has been possible only by continual attack by pioneer farmers on virgin forests and soils. From the early plantation regions in Bahia and other tropical eastern states, the main plantation belt has moved southward and westward into Sao Paulo, Minas Gerais, and Paraná, which has a marginal climate with severe frost damage to the plantations in some years. The pattern of advance has commonly been as a hollow ring. New plantations are productive for 15 or 20 years before erosion and soil exhaustion lead to abandonment. Where cover crops and manuring are used, a coffee plantation reaches its peak productivity between about 30 and 50 years of age and can be rejuvenated thereafter by cutting down the old trees and replanting, but this has not been standard practice in Brazil.

Although Brazilian coffee has used the best *C. arabica* germ plasm, it has concentrated on low cost of production rather than quality. Shade trees are dispensed with. The ground is kept bare, so the berries can be swept up off the ground rather than picked into baskets. Instead of repeated pickings of berries as they ripen, the whole crop is stripped off onto the ground at once. The mixture of dirtied berries of different ages often begins to ferment before it has been dried. This is the so-called dry method of processing. (The wet method, by which high quality beans are produced, requires picking the berries at maturity, washing, depulping, controlling fermentation to clean the depulped beans, and rapid drying.) In the trade, Brazilian coffee is rated as hard, harsh, or rank, not mild.

Brazilian coffee estates vary greatly in size with many small to medium holdings and a few great estates. In Sao Paulo alone by the mid-20th century, the estates numbered over 79,000 and occupied several million hectares.

In some years, Brazil has produced more coffee than the world's total consumption for the year. Surpluses and falling prices, of course, lead to uprooting plantations in favor of other crops until coffee prices rebound. When there is a freeze in Paraná, prices shoot up. This is a world-wide problem for coffee growers. When prices are high, everybody plants, but there is a lag of years before new stands come into bearing and glut the market again.

In spite of overwhelming Brazilian dominance of the lower grade world market, coffee planting has thrived in other parts of Latin America by concentration on premium quality. Northwestern South America and Central America have regions with almost ideal physical environments for both the typical and Bourbon varieties of *C. arabica*: mild climates at medium elevations, a rainy-dry season regime similar to the species' homeland, and fertile volcanic soils.

Although *C. arabica* was introduced to the Spanish mainland early in the colonial period, commercial planting developed later than in the West Indies or Brazil. By the mid-19th century, significant exports had begun from the northern Andes and from highland regions of Central America. Soon after, coffee became the main export of Colombia, Costa Rica, El Salvador, Nicaragua, and Guatemala. Mexico also had commercial coffee plantations by the mid-19th century in various volcanic highlands, but the production was mainly absorbed by the domestic market. Initially, Spanish American coffee estates, or *fincas*, were typically small, mostly less than 10 ha, including diversified dooryard gardens and areas in food crops. Usually an overstory of legume trees was planted, and the coffee was closely planted and carefully pruned. In some regions, erosion was a problem, but many *fincas* have been kept productive for generations. Much Spanish American coffee is still grown by small family operated *fincas*, but there has been a recent trend toward consolidation into larger estates.

Processing and marketing have always been more concentrated than coffee growing in Spanish America. Characteristically, a few larger *fincas* operate *beneficios*, i.e., factories for removing the pulp, controlled fermentation to finish cleaning of the beans, and final drying. As time went by, the *beneficios* became more highly mechanized. As Spanish American mountain grown coffee acquired a reputation for quality, outsiders, including British and German investors, acquired *fincas*. German planters were prominent in Guatemala starting about 1870 and soon after expanded across the Mexican border into Chiapas. They found unoccupied forest lands suitable for coffee below the densely populated highlands and employed migrant labor from the highland Maya villages during the picking season. Expropriation of German estates came during World War II. Expansion of highland coffee planting has continued in Spanish America intermittently through the 20th century during periods of high prices.

The last major expansion of commercial *C. arabica* planting took place late in the 19th and early in the 20th century in British and German colonies in the highlands of East Africa and Cameroon.

During most of its history as a plantation crop, *C. arabica* had remained remarkably free from serious diseases in spite of its lack of genetic diversity. Trouble finally began in 1868 when a parasitic fungus *Hemileia vastatrix*, the coffee leaf rust, began attacking plantations in Ceylon. The rust spread rapidly by windborne spores. It was in India by 1869, Java in 1876, and Fiji in 1879. In Fiji, *Hemileia* had reached the eastern edge of monsoonal winds; its spread farther east in the Pacific was blocked by prevailing Trade Winds. Moving westward, *Hemileia* attacked coffee plantations in German East Africa in 1894. It was not reported in Cameroon until 1957. The rust had appeared temporarily in Puerto Rico in 1903, but was eradicated by a vigorous campaign of sanitation. The Atlantic Ocean barrier was finally breached in 1970 when the rust appeared in plantations in Bahia, Brazil. A year later it was in Minas Gerais,

Sao Paulo, and Espirito Santo, having crossed a 50-km wide coffee-free security strip, which had been cleared of *C. arabica* plantations.

The rust can be controlled by spraying with fungicides, but this must be done after nearly every rain. *H. vastatrix* made *C. arabica* uneconomic in many regions, particularly where it was grown at lower elevations. Efforts to transfer genes for rust resistance from wild *Coffea* species, mainly by Dutch breeders in Java, were frustrated by sterility of the hybrids. The narrow genetic base of *C. arabica* plantations limited the possibility of finding resistant variants. Occasionally, a resistant tree was found, e.g., in 1911 in Mysore, India, by British planter Kent, but the resistance of Kent's coffee was not durable. The parasite proved much more flexible genetically than the host. Some 30 physiological races of *H. vastatrix* are now known, and no *C. arabica* cultivar is resistant to all of them. Finally in the 1960s, the United Nations FAO launched a systematic search for wild *C. arabica* germ plasm in Ethiopia, but so far durable resistance has not been found. Thus, the parasite and host provide a classic case history of vulnerability of a genetically homogeneous monoculture. Considering that coffee was planted in large, dense stands in the humid tropics and maintained without rotation for many years, it is surprising that it was not more vulnerable. It is still able to resist the fungus within its optimum highland belt where the climate is similar to its homeland. In Asia, many lowland coffee plantations were converted to *Cinchona*, *Hevea*, or *Thea* plantations. Except in the highlands, planters were able to remain in the coffee business only by switching to another species *C. canephora*.

Coffea canephora (=*C. robusta, C. ugandensis, C. quillon*, etc.) — Robusta Coffee

C. canephora is universally known in the industry as Robusta coffee, although the binomial *C. canephora* has priority in academia. In contrast to *C. arabica*, *C. canephora* is normally cross-pollinated and highly variable genetically. It grows wild in equatorial African forests over a huge region from the Guinea coast and Congo basin to the East African lake region. It grows mainly below 1000 m elevation both along rivers and as an understory tree within the forests. Compared to *C. arabica*, Robusta trees are rather shallow rooted.

In traditional African cultures, Robusta coffee, like *C. arabica*, was of very minor interest. The berries were sometimes gathered, sun-dried, and chewed, according to accounts of explorers such as Speke, Grant, and Burton. A few Robusta trees were planted, usually from cuttings, in dooryard gardens. Interest in the beans for brewing was confined to Arab and later European traders.

About 1900, several large batches of Robusta seedlings were taken from the Belgian Congo and other parts of the native range to Brussels, Kew, and other European botanical gardens. From there, the species was widely distributed in experimental gardens in the tropics of Asia and America. Subsequent advance of Robusta coffee as a plantation crop closely followed the advance of *H. vastatrix*.

Although Robusta beans are used only for cheap blends and instant coffee, the species has become a major crop of the lowland tropics. In addition to rust resistance, it outyields *C. arabica*, can be grown without shade trees and with little pruning, and with less careful harvesting and processing. Robusta now approaches *C. arabica* in total world production. It is extensively planted in much of its native African equatorial forest region, Madagascar, Sri Lanka, Thailand, Indonesia, Brazil, and the lowlands of Colombia.

Programs of genetic research and scientific breeding are currently under way with both *C. arabica* and *C. canephora* in various countries. These have not yet contributed cultivars of commercial importance. The plantation populations of both species are still incipient domesticates with little genetic divergence from the wild progenitors. The most notable exceptions are minor cultivars with conspicuous mutations. For example, Maragogipe, a dominant mutant of typical *C. arabica*, arose in Brazil in 1870; because of its larger beans and special flavor, it has been planted commercially in parts of Latin America. Catura, a dominant mutant of Bourbon, arose in Brazil in 1935 and is occasionally planted because of its compact growth form.

Rutaceae — Citrus Family

The Rutaceae are a family of over 100 genera throughout the world's tropics and temperate regions. The family is remarkable for essential oils, making flowers, fruits, leaves and other parts distinctively fragrant, which has led to human use in ritual and medicine. Commonly cultivated ornamentals include *Calodendron*, *Choisya*, *Murraya*, *Poncirus* (trifoliate orange), and *Fortunella* (kumquat). Kumquats are grown mainly as ornamentals, but the attractive orange fruits can be eaten whole, peel and all.

***CITRUS* — CITRUS** (Burkill 1935; Cameron and Soost 1976; Hutchison 1946; Needham 1986; Patiño 1969; Purseglove 1968; Simoons 1990; Soost and Cameron 1975; Tanaka 1954; Watson 1983)

The genus is native to southeastern Asia from northern India to China and south through Malaysia and the East Indies and Philippines. The peculiar fruit, technically a hesperidium, presumably evolved as adaptations for animal dispersal, perhaps by elephants, swine, or other large terrestrial mammals. The formidable spines generally present on trunks and branches appear to deter climbing mammals. Almost nothing is certain about the natural geography and ecology of truly wild *Citrus*. Both the distribution and genetics of the genus have been drastically modified in Asia by ancient human intervention. The cultivars generally retain strong capacity to escape from plantings and naturalize in open habitats. Bees cross-pollinate wild, feral, and planted trees with consequent hybrid variants.

Over 150 binomials have been bestowed on the genus, but probably only a few of these belong to true species, in the sense of breeding populations, internally coherent and externally discrete. All the important cultivars may be descended from just four original wild species: *C. medica* (citron), *C. grandis* (pummelo), *C. reticulata* (mandarin), and *C. aurantifolia* (lime). Of these, the citron and pummelo reproduce entirely sexually by normal seeds. Seeds produced by the mandarin and lime are partly sexual and partly apomictic, with embryos genetically identical to the mother plant. All the other citrus cultivars (*C. aurantium*, *C. sinensis*, *C. limon*, *C. limonia*, *C. paradisii*, and *C. nobilis*) are believed to be interspecific hybrids between the four basic species and their derivatives. These hybrids occasionally produce normal sexual seeds, but their seeds are usually asexual.

All the hybrid groups evidently originated by accidental hybridization, perhaps repeatedly. Interspecies hybrids conceivably occurred in the wild, but they would have much more likely arisen and survived under domestication. Because of mutation and occasional sexual reproduction, a great many slightly different named cultivars have been selected within these groups. They are commonly propagated by grafting on seedling rootstocks of various kinds of *Citrus* and *Poncirus*.

The early history of cultivated citrus is not yet documented archaeologically, and the written history is commonly blurred by ambiguous folk names. There are occasional helpful descriptions of the fruit, but tracing the historical geography of specific kinds of citrus can only be tentative and sketchy.

Citrus medica — Citron

C. medica may have originated in India and spread prehistorically in cultivation to the Near East and China. Citron fruits have an extremely thick rind, and some have an extremely sour pulp within. The rind, like that of other citrus, is candied as a confection. The main use of citron, however, has always been for medicinal and ritual purposes, not as food. The fruits and other parts of the plant are valued for their distinctive fragrance and appearance. Citron entered written history about 300 B.C. in Greece. Theophrastus wrote a good description of it as an exotic fruit attributed to the Medes or Persians and imported from the Near East. Since at least 100 B.C., citron has been essential in the Jewish Feast of the Tabernacles. By the time of Virgil about 20 B.C., citrons were being grown in Italy. During the Renaissance, citron, along with other kinds of *Citrus*, was grown in northern Europe for the nobility in orangeries, specially designed buildings. Columbus introduced the species to Hispaniola on his second voyage. During the 16th century, it was widely planted on a very small scale in the Caribbean and in Spanish mainland colonies. The species spread with European expansion through the tropics and subtropics in general. However, commercial supplies of candied citron are still mainly obtained from the Mediterranean.

In Asia, citron also had a long tradition as a ceremonial plant. Its spread through China and Southeast Asia was at least partly associated with Buddhism. A mutant variety in which the carpels are not fused and resemble fingers came to be known as Buddha's hand. It was being grown in China by the 4th century A.D. For prized gifts on special occasions, the Chinese traditionally give little potted citron and other citrus trees bearing flowers and fruits.

Citrus grandis (=*C. maxima*) — Pummelo, Pamplemousse, Shaddock

C. grandis, like the citron, is a sexually reproducing species with much genetic variation, especially in tropical Southeast Asia where it may have originated. The fruits of pummelo cultivars vary in size, shape, color, seediness, and amount and kinds of essential oils. Some pummelo cultivars are very acid, some sweet, some bland. Some have fruits the size of a human head with thick puffy rinds.

The pummelo seems to be mentioned in Chinese tribute lists by 500 B.C. and to have been grown commercially as far north as central China by 300 B.C. Like other kinds of citrus, pummelo trees are important in traditional Chinese medicine and are appreciated as much for their fragrance and beauty as for food.

The pummelo was taken eastward into the Pacific islands in prehistoric times. Whistler (1991) lists it as present aboriginally in Polynesian gardens.

Pummelo cultivation evidently spread westward through India and North Africa to Moorish Spain before 1200 A.D. It was probably the citrus that first bore the Spanish name *toronja*. The Spaniards probably introduced the pummelo to the Greater Antilles promptly. Citrus that appears to belong to this species was reported in Puerto Rico and Jamaica before 1600. The Jesuits are credited with introductions of other pummelo cultivars directly to Peru from China and the Philippines about 1600. The pummelo did not thrive in the Peruvian desert climate, but it was widely planted elsewhere in South America during the 17th century. An independent introduction from Asia to Barbados in the 17th century was credited to Captain Shaddock of the British East India Company, whence the name of the Shaddock cultivar.

In more recent times, pummelo varieties have become pantropical in distribution, usually as a minor dooryard garden tree.

Citrus reticulata (incl. *C. unshiu*) — Mandarin, Satsuma Orange, or Tangerine

C. reticulata may also have been domesticated in tropical Southeast Asia or the East Indies. Like the pummelo, the mandarin enters history in Chinese tribute lists about 500 B.C. By 300 B.C., the mandarin was being grown commercially in central China, and by 100 B.C., some growers were said to have a thousand trees in their orchards. By 400 A.D., skillful orchardists were propagating clones by grafting, and an early example of biological control of insect parasites had been found: tree nesting ants were employed to defend the mandarin trees.

The mandarin is the hardiest of all the citrus fruits, but Chinese emperors had to use special greenhouses, the first orangeries, to push mandarin cultivation into northern China. The mandarin has probably always been the most important citrus in China, the one most widely grown commercially and in dooryard gardens. It figures prominently in Chinese cuisine, art, and poetry.

Introduced prehistorically to Japan, the mandarin became the favorite citrus fruit there also, and special cultivars were developed, particularly the satsuma orange.

C. reticulata was curiously late in being adopted outside Asia. Finally in the 19th century, mandarin orchards were established in Spain, Algeria, the West Indies, North and South America, and Australia.

Citrus aurantifolia — Lime

C. aurantifolia is the most strictly tropical citrus fruit. There are apparently truly wild forms in tropical Southeast Asia, the Philippines, and other East Indies with extremely acid fruits. Cultivars have diverse fruits, acid or sweet, large or small, round or oblong, yellow or green, but they all have relatively thin skins. Peoples of tropical Asia appreciate the scent of the lime as much as its flavor. They commonly use it for soap and shampoo. It also has ritual, magical, and medicinal values. Lime juice is administered to elephants to make them sagacious.

Evidently the species was not introduced to temperate Asia or Europe until modern times. It may, however, have been taken westward across North Africa in medieval Islam. In Spanish, different varieties of lime are called *lima* and *limon*, the latter name being shared with the lemon. Early Spanish accounts of introduction of *limones* to the New World are therefore often ambiguous, but evidently refer to limes more often than to lemons.

Columbus introduced what were probably limes from the Canary Islands to Hispaniola in 1493, and during the 16th century, limes were planted widely in the West Indies and in Spanish mainland colonies. The juice was valued as antiscorbutic. The spiny plants made excellent hedges, and the species often escaped from cultivation, colonizing naturally open sites. So-called wild limes later were used by buccaneers to fight scurvy.

Today, extensive commercial lime orchards for export are growing in the Lesser Antilles, especially Dominica and Montserrat, and in Mexico. The trees are now pantropical in dooryard gardens and in small orchards for local use.

Citrus aurantium — Bergamot, Sour or Seville Orange

C. aurantium is a hybrid between *C. grandis*, the pummelo, and *C. reticulata*, the mandarin. It may have arisen repeatedly where the two parents were grown together. The sour orange seems to be recorded in Chinese writings starting about 300 B.C. and in Japan by A.D. 100 A.D. There are clear references in China by 1300 A.D. Sour oranges are traditionally used in China and Japan not as an edible fruit but for marmalade and candied peel. The essential oil is used for soap and perfume, the flower buds to scent tea.

Seeds of the sour orange may have reached Rome about 100 B.C. over the Red Sea trade route from India. Mosaics at Pompeii suggest cultivation in Italy during the 1st century A.D. Cultivation may have been abandoned after the fall of Rome and reintroduced by Islam. Evidently sour oranges were introduced from India to Oman and the Near East in the 10th century A.D., and before the end of the century, they were being grown in North Africa and Spain. Sour orangcs were evidently also introduced to East Africa by the medieval Arabs.

Sour oranges planted by the early Spanish colonists in Hispaniola thrived and spread on their own. By 1550, sour oranges formed thickets infesting savanna pastures. Sour oranges are thoroughly naturalized and often a pest in warmer parts of the Americas, from Georgia and Florida through the West Indies and Mexico to Argentina. Often they are little used. Some go into marmalade and liquers, including Curacao.

Spaniards are credited with introducing sour oranges to the Philippines and Guam. They have naturalized there and also in Fiji and Samoa.

Citrus sinensis — Sweet or Valencia Orange

C. sinensis is also a hybrid between *C. grandis*, the pummelo, and *C. reticulata*, the mandarin, but not from the same parents that produced the sour orange. The pummelo and mandarin have undoubtedly interbred repeatedly, and several different cultivars of *C. sinensis* are known in Southeast Asia, China, and India. Sweet oranges never became nearly as important in China as their mandarin parent. There are possible Chinese references by 200 A.D., but the first Chinese references in which sweet oranges are clearly distinguishable from mandarins date from the 12th century A.D. The earliest record from India was by Ibn Batuta in the 14th century. Some kind of sweet orange may have been grown in the Near East by about 1400. By 1470, it was spreading westward through the Mediterranean, probably initially by Genoese merchants trading to the Levant. A sweet orange variety was already in the Canary Islands in 1493 when Columbus took seeds from there to Hispaniola.

Better varieties of sweet oranges, such as the Valencia, are believed to derive not from this initial introduction via the Levant but rather from direct Portuguese trade and later trade with India and China via the Cape of Good Hope.

The sweet orange taken to the New World by Columbus was probably an inferior variety and soon replaced, but such detail is not decipherable in historical records. Varieties were easily propagated by apomictic seeds, and the Spanish colonists evidently planted them wherever they went. The great chronicler of the Conquest of Mexico Bernal Diaz del Castillo recorded how he planted the first oranges in Mexico in 1518 with seed he had brought from Cuba. He planted the seeds next to an Indian temple at Tonalá in Veracruz. When the seeds germinated, the Indian priests in charge of the idols watered and weeded the young orange trees because they recognized them as new and different from their own plants.

Over and over again, sweet oranges, like the other kinds of citrus introduced by the Spaniards, were reported as abundant within a decade or so of the first colonization. Sometimes they were promptly adopted by the Indians. Often they escaped from cultivation and formed spontaneous groves. By the mid-16th century, thickets of orange trees were considered a plague in pastures in the Antilles. By then, oranges had been widely planted and were abundant on the Spanish mainland from Florida to Panama and Peru. The Portuguese also introduced orange planting promptly to Brazil. Immediate citrus planting by colonists was sometimes prescribed in official orders, suggesting awareness of antiscorbutic value.

From the mid-17th century on, British, French, and other European colonists and navigators joined the Iberians in spreading oranges and other citrus around the world tropics. The seeds, sexual or apomictic, were easily transported, usually without historical documentation. Introductions were often redundant.

In California, for example, sweet oranges were introduced during the Spanish mission period, but the two cultivars that now comprise the state's commercial orchards, the Valencia and the Navel, were introduced repeatedly starting in the 1870s. The Navel originated by mutation in Bahia, Brazil, and arrived in California by at least three pathways: via English orangeries, via Australia, and via a U.S. Department of Agriculture importation from Brazil to Washington, D.C. Unlike the Valencia, which produces apomictic seeds, the seedless Navel has to be propagated as a clone by grafting. Southern California had a great boom in orange, and to a lesser extent lemon, orchards in the late 19th and early 20th century, with formerly extensive cattle ranches and wheat farms broken up into relatively small resident owner orchards. The success depended on organization of cooperative irrigation and marketing. About 6000 growers in California and Arizona market their fruit through the Sunkist cooperative formed in 1909.

It is remarkable how powerfully the spread of commercial orange orchards has been shaped by political boundaries and market locations rather than by physical geographic factors. Most of the world crop is produced in climatically marginal subtropical regions and either needs irrigation or faces frost damage or has to cope with both.

Citrus limon — Lemon

C. limon is a hybrid between *C. medica*, the citron, and *C. aurantifolia*, the lime. Like its lime parent, the lemon can be grown in the tropics, but the lime is generally preferred there. The lemon inherits from its citron parent a thicker skin and longer keeping quality; it has found a niche mainly in subtropical regions.

The place and time of origin of the lemon are obscure; it may have arisen more than once, like other hybrid citrus. It may have spread from northern India to Rome by classical times via the Near East. It apparently died out in Europe with the fall of Rome, like oranges. It was evidently reintroduced to the Mediterranean in medieval times through Islam and spread around the tropics as a minor dooryard garden tree during the European colonial expansion. Like oranges, lemons are grown commercially mainly in climatically marginal subtropical regions, particularly in the Mediterranean and U.S. Florida orchards have been repeatedly destroyed by freezes, a boon to the slightly less marginal California orchards.

Citrus limonia — Canton Lemon or Rangpur Lime

C. limonia is thought to be a hybrid between *C. reticulata*, the mandarin, and *C. aurantifolia*, the lime. It probably originated in Southeast Asia prehistorically and spread to China and India in cultivation. It was in southern China by 1000 A.D. and is now widespread in China, Japan, and the Himalayan foothills. Easily propagated by apomictic seeds, the Rangpur lime is widely popular as a home garden tree in the world's subtropics.

Citrus paradisii — Grapefruit

C. paradisii is a hybrid between *C. grandis*, the pummelo, and *C. sinensis*, the sweet orange. Such hybrids may have occurred more than once, but the modern grapefruit cultivars are believed to have a single origin about 1750 on the island of Barbados, where the parent species were being grown. The original variety was white and with abundant apomictic seeds. It was introduced to Florida by a French naval surgeon in 1809, where it is still grown for juice. The so-called Marsh seedless, which still has some seeds, arose in a Florida orchard as a mutant in 1862. Another mutant in a Florida orchard in 1913 produced the pink Thompson variety, widely planted in Florida and lower Rio Grande, Texas, where it requires irrigation. The deeper red variety (Ruby) arose by mutation in Texas in 1929 and has repeatedly arisen elsewhere. The Marsh seedless and its colored derivatives have been widely planted. In the 1920s, they were the basis for a boom in grapefruit planting in irrigated oases in the hot deserts of southern California and Arizona, commonly as an understory in date groves. They were also taken back to the West Indies, to Central and South America, and to Israel and South Africa. The grapefruit yields superb fruit in the humid tropics. Like the orange, its cultivation in marginal subtropical regions makes sense only because of political and market locational factors.

Citrus nobilis — King Orange or Ortanique

C. nobilis is evidently a hybrid between *C. reticulata*, the mandarin, and *C. sinensis*, the sweet orange, that originated in antiquity in Southeast Asia or southern China. It is grown as a dooryard garden plant there. There are commercial plantings in Jamaica and Florida.

Since the late 19th century, experiment stations in various countries have attempted scientific citrus breeding. Although some new hybrid cultivars have been released, so far all are of minor importance compared to the old cultivars. The most significant contribution of recent breeding has been development of improved rootstocks, not scions.

Solanaceae — Nightshade Family

The Solanaceae include about 75 genera and 2500 species, most diverse in the New World, but cosmopolitan. Many members are ecologic pioneers of naturally open, sunny sites and are fast-growing herbs, vines, and shrubs. The family produces various toxic alkaloids, presumably as a defense against herbivores. The flowers are generally colorful, fragrant, and attractive to insects. The fruits are commonly colorful berries attractive to birds, with many tiny hard seeds that retain viability after passing through an animal. A few species reproduce clonally by tubers. The combination of these characteristics has, of course, attracted human attention. A great many species have been exploited by people since antiquity, and a fair number have been taken into cultivation. In addition to the genera that will be discussed individually, the most notable domesticated genera are as follows:

- Drug plants: *Nicotiana* (tobacco), *Atropa* (belladonna), *Hyoscyamus* (henbane), *Datura* (jimsonweed)
- Ornamentals: *Petunia*, *Salpiglossis*, *Nierembergia*, *Schizanthus*, *Browallia*, *Cestrum*
- Food plants: *Cyphomandra* (tree tomato), *Physalis* (husk tomato)

The three genera dealt with individually include the potato, tomato, and chili pepper.

SOLANUM — POTATO, EGGPLANT, AND OTHER NIGHTSHADES

(Brücher 1985; Brush et al. 1981; Glendenning 1983; Grun 1990; Hawkes 1990; Hawkes and Francisco-O. 1992; Hedrick 1919; Heiser 1969; Laufer 1938; Patiño 1964; Plaisted and Hoopes 1989; Salaman 1949; Simmonds 1976b; Smith, N. 1983b; Tjomsland 1950; Zimmerer 1991)

Solanum is a cosmopolitan genus of about 900 species. The above references all pertain to the potato, which will be discussed at length. Three other Solanum species domesticated for their fruits rather than their tubers deserve brief mention. Heiser (1969) provides excellent fuller accounts. These are as follows:

- *Solanum melongena* — eggplant. Many cultigen varieties have been developed in India evidently from a wild species with small, hard, bitter, spiny fruits. Perhaps *S. melongena* was introduced through medieval Islam across North Africa to Spain and also spread early to the Far East.
- *Solanum muricatum* — pepino. *S. muricatum* is an ancient cultigen in the Andean valleys with large, nearly seedless fruits resembling eggplant. It is propagated by cuttings and apparently originated by hybridization between two wild species.

- *Solanum quitoense* — lulo, naranjilla. This species is evidently a relatively recent but pre-Spanish domesticate in the northern Andes, perhaps also a hybrid between two wild species. The lulo is grown for its delicious, citrus-like juice. It was introduced recently to Central America where it is a commercial crop.

Several other *Solanum* species have been planted in gardens in widely scattered regions for their edible berries, but are only incipient domesticates. One of these is the sometimes deadly black nightshade *S. nigrum*, a cosmopolitan weed, which has nonpoisonous forms that have probably been taken into cultivation repeatedly in different places.

The tuber-bearing *Solanum* species that comprise the subgenus *Potatoe* are all natives of the New World, from southwestern U.S. to Argentina and Chile. Classification of species has been extremely difficult and controversial and has not yet settled down in any generally accepted system. One recent classification (Hawkes 1990) recognizes 88 wild species (plus many interspecific hybrids that have been given Latin binomials) in the subsection of *Solanum* (series *Tuberosa*) that includes all the cultivated potatoes. Other authorities regard this as splitting the species too finely, but there is no doubt that the domesticated potatoes have a phenomenal diversity of close wild relatives. Of the 88 microspecies, three grow in Mexico and the southwestern U.S.; their habitats are pine, fir, or oak woodland at elevations of 2000 to 3900 m. There are none in Central America. The other 85 species are native to the South American Cordilleran system, again at medium to very high elevations, from Venezuela and Colombia south to Argentina. Hawkes credits Peru alone with over 50 species of the *Tuberosa* series.

Loaded with water, starch, and other nutrients, the tubers permit survival and *in situ* proliferation of genetic individuals in environments with extreme seasonality. The exposed shoot systems can resprout vigorously after dying back during dry, cold seasons.

Tubers of the wild potatoes are small and contain bitter alkaloids that make the tubers toxic if eaten in quantity, but they were widely gathered by mountain Indian tribes. Digging the tubers did not necessarily deplete the stands; soil disturbance probably actually aided regeneration from missed tubers. Also, as maize farming spread, some wild potatoes entered the milpas as weeds. Among the numerous wild *Tuberosa* microspecies, human habitat disturbance presumably resulted in much mixing and hybridization. The populations had been segregated ecologically and geographically rather than by internal breeding barriers.

From the vast array of wild microspecies and weedy hybrids, Hawkes suggests that domestication started from a very narrow geographic and genetic source. A single wild diploid *S. leptophyes*, endemic to the Lake Titicaca region and north central Bolivia, may be the primary ancestor of all the cultivated species. It is a low, slender herb that grows in heath, puna, and *Stipa* grasslands of the Altiplano at elevations of 3200 to 4100 m. Hawkes (1990) considers that the most primitive cultigen species, the diploid *S. stenotomum*, was derived from *S. leptophyes* under human selection. From the very beginning, folk propagation of potatoes has presumably been entirely clonal, using tubers, not true seed. However, in addition to somatic mutations, new genotypes would have been frequently available from volunteer seedlings.

Unlike Mexico, which has only small areas on isolated volcanic peaks that are too high for maize cultivation, the Andes have vast areas of arable land above the limits of maize and other basically tropical crops. Use of these areas required development of high altitude crop species. Quinoa, discussed above (chenopod family), became a major Andean cereal grain, and there were several unrelated highland tuber crops, e.g., *Oxalis tuberosa* (oca) and *Ullucus tuberosus* (ulluco), but *Solanum stenotomum* and its later derivatives were the most important and widespread high Andean crops. They made possible not only sedentary agriculture but high civilizations, including one centered at Tiahuanaco on Lake Titicaca, which flourished long before the rise of the Inca Empire.

As cultivation of *S. stenotomum* spread, new cultivars were selected. On the diploid level, *S. phureja* diverged by adaptation to lower elevation, frost-free valleys in northern Peru, Ecuador, Colombia, and Venezuela. Unlike the seasonally dormant tubers of the primitive *S. stenotomum*, *S. phureja* tubers sprout quickly, permitting planting of two or more crops a year. On the other hand, an extremely frost-hardy triploid *S. juzepczukii* (3x=36) originated as a hybrid between *S. stenotomum* and *S. acaule* (4x=48), a wild species of southern alpine meadows up to 4650-m elevation. *S. acaule* is not a member of the *Tuberosa* series; its triploid cultigen derivative is sexually nearly sterile; it is propagated clonally at the upper limits of agriculture in southern Peru and Bolivia.

All the cultigens discussed so far have remained endemic to the Andean highlands. There is one final cultigen species, *S. tuberosum*, which probably also originated in the southern Peruvian-Bolivian highlands, but which became more widespread in prehistoric South America and in historic time became the common potato crop of the outside world.

S. tuberosum (incl. *S. andigena*) is a polyploid (4x=48) derived primarily from *S. stenotomum*, but with germplasm derived from various other diploid species. *S. tuberosum* has been characterized as a genetic sponge because of its capacity for hybridizing with other species, including some outside the series *Tuberosa*. The curious phenomenon of introgression from diploid species into a polyploid involves the high frequency of unreduced gametes produced by various potato species, wild, weedy, and cultigen. When such aberrant 2n gametes from a diploid species unite with normal 2n gametes from a tetraploid, a new tetraploid clone may result.

Some interspecific hybrids involving *S. tuberosum* are triploid or pentaploid clones. Two such cultivars, *S. chaucha* (3x=36) and *S. curtilobum* (5x=60), are hybrids between *S. tuberosum* and other cultigens, *S. stenotomum* and *S. juzepczukii*, respectively. They are grown at extremely high elevations.

A great diversity of potatoes has been maintained where Indian villages still practice subsistence agriculture in the Andean highlands. In Peru, Zimmerer (1991) found that single households plant many clones of several species. A village may have over 100 clones with names recognized throughout the village. Where the traditional peasant open field system is retained, each household plants potatoes in several separate fields at different altitudes and on different kinds of sites. The women who are heads of households select the clones to be planted according to site and culinary uses. There is a strong correlation between elevation, culinary use, and taxonomy as recognized by scientific binomials. The highest fields, about 4000 m and above, are planted with *S. curtilobum*, *S. juzepczukii*, and the other frost-resistant hybrids between wild species and *S. stenotomum* and *S. tuberosum*. These are relatively bitter and are processed by freeze-drying to *chuño* or by soaking in running water. The main crop is grown at elevations between 3500 and 4000 m and is dominated by *S. tuberosum* with many different named clones interplanted in mixed fields. A few clones of the lower yielding but exceptionally flavorful *S. stenotomum* are planted in the same fields. Tubers of both species are stored for several months without processing and eaten daily, simply boiled or baked and seasoned with chili peppers and spices. Other clones of *S. tuberosum* are grown between 2900 and 3800 m. These are watery and less mealy and used in soups (Zimmerer 1991).

Such genetic diversity would seem to provide some insurance against crop failures from weather, disease, and pests. The Indian fields are also only planted to potatoes for a year or two before being rotated to other tuber crops or *quinoa* and also left fallow, so that a 6- or 7-year gap separates successive potato crops.

Under prehistoric Indian cultivation, *S. tuberosum* evolved into two geographically widely separated subspecies, a highland low latitude subspecies in the tropical Andes and a lowland subspecies in cool temperate Chile. Grun (1990) believes the two are different enough that a case can be made for their recognition as distinct species (in which case the Chilean cultigen would be called *S. chiloense*). They are usually treated as subspecies. Unfortunately, the Andean potato has been called *S. tuberosum* subspecies *andigena*, and the Chilean potato has beeen called subspecies *tuberosum*, which is untenable because Linnaeus gave the name *tuberosum* to the Andean cultigen, the only potato known in Europe at the time, and under the international rules the name may not be usurped. The two will be treated here informally as the Andean and Chilean subspecies.

The derivation of the Chilean subspecies is something of a mystery. There is a gap of about 800 km between the northern limit of its cultivation and the southern limit of highland potato cultivation. In the center of this gap are the world's driest desert and some of the world's highest mountains. Hawkes (1990) considered that millenia ago migrating Indians simply carried the Andean potato south and that the Chilean subspecies diverged by natural and artificial selection in isolation. However, Grun (1990) reviews molecular evidence of a more complex derivation. In addition to a large genetic component from the Andean subspecies (or its diploid progenitor, *S. stenotomum*), the Chilean tetraploid has germ plasm considered to be from an unidentified wild diploid. Two species not within the series *Tuberosa* have been suggested as possible sources. These are *S. chacoense* and *S. maglia*, both lowland species of temperate South America.

In an archaeological site dated at 11,000 B.C. on the Chilean mainland near Chiloe, hearths contained potato skins along with other wild plant materials. These were identified by microscopic study of the starch grains as from *S. maglia* (Ugent et al. 1987). Very ancient, presumably wild potatoes have been reported from caves near Chilca, south of Lima, but the dating and identification are highly questionable (Bonavia 1984). More solidly documented archaeological remains were reported from Casma by Ugent et al. (1982) from preceramic levels between 2000 and 1500 B.C. The tubers were small, and the species uncertain. Small tubers dated about 1000 A.D. were also found at Pachacamac near Lima. The sparsity of these finds in the rich archaeological plant remains of the Peruvian coastal oases strongly suggests that potatoes may have been occasionally packed down from the highlands rather than a regular crop on the coast. Also, the potato appears occasionally in the famous ceramics of the Peruvian coastal oases. The potato pots all date from the period between 1 A.D. and the Inca Conquest about 1450 A.D. Potatoes appear on pots of Moche and Chimu styles on the north coast and Nazca on the south coast. Some of these represent freeze-dried *chuño*, which would necessarily have been brought down from very high elevations. The ancient Peruvian potters commonly represented plants at actual sizes; the potato pots are the right size for domesticated Andean *S. tuberosum*.

Written history of the potato began when the Spanish explorers penetrated the Andean high country. The first clear record was in 1537 in the headwaters of the Rio Magdalena in Colombia: in ransacking a highland village, the Spaniards found stores of maize, beans, and what they initially called truffles. By 1550, there were good Spanish accounts of the Indian potato crop from various parts of the Andes, and the Quechua name *papa*, used throughout the Inca Empire, became the common Spanish name. Spanish explorers of the Andes tried eating *papas* and found them good. Spanish colonists soon began exacting tribute from the Indians of part of their potato crop. The forced labor in the high elevation Andean mines such as Potosí, which the Spaniards developed in the 1540s, subsisted mainly on *S. tuberosum*.

The Spanish conquest of Chile was relatively late, and the first European record of the potato there was in 1578 when Sir Francis Drake obtained tubers at 38° S.L. by barter with the Indians. There is, of course, no possibility that any tubers could have survived Drake's 2-year circumnavigation of the world.

During the Spanish colonial period, *S. tuberosum* was taken northward to the isolated Sierra Nevada de Santa Marta of Colombia and to widely separated highland regions of Costa Rica, Guatemala, and Mexico. *S. tuberosum* was introduced to Europe in the 16th century, the exact pathway and time being unknown. There is a famous report that a hospital at Seville was feeding its patients potatoes grown in Spain during the 1570s. There was some uncertainty whether this referred to sweet potatoes rather than potatoes and also whether the tubers might have been shipped from South America rather than grown in Spain. These doubts are discounted by Hawkes and Francisco-O. (1992). The point is not very important, because there is no doubt that *S. tuberosum* had been introduced to Europe by the 1580s. The simplest hypothesis would be that Spanish ships brought tubers from Cartagena or some other northern South American port, along with gold and silver from the Andean mines. It is also possible that Sir Francis Drake's ship brought tubers to England from the sack of Cartagena in 1586 or that some other English ship brought potatoes from a Spanish prize ship. In any case, during the 1580s and 1590s, *S. tuberosum* spread widely in Europe. Carmelite friars were credited with taking it from Spain to Italy, whence it spread through Vatican connections via the Low Countries to Austria by 1588.

The species is abundantly illustrated in the great herbals published in various European countries between the 1590s and 1650s. These and other drawings and early herbarium specimens strongly suggest that the Andean, not the Chilean, subspecies constituted the entire European crop until the 19th century. Meanwhile, in 1621, an independent Spanish introduction brought the same subspecies from South America to the Canary Islands where it is still grown with little genetic change.

The spread through Europe was primarily as a botanical curiosity rather than a crop. Psychological explanations for the reluctance of Europeans to accept the potato have been advanced, but a more practical reason is now generally recognized, namely that yields of tubers were initially very low in Europe. Coming from low latitudes, the Andean subspecies was not adapted to the long days of the European summer growing season. In Spain and other southern regions, it could produce some tubers, but in higher latitudes, it became productive only after evolutionary change, which took a long time. The fact that the species spread through Europe initially as a garden curiosity may have helped in this process, because plants would have been often left to flower and go to seed, not a normal event in the potato field crop. Both human and natural selection were probably involved in developing cultivars productive in high latitudes.

The name Irish potato for these cultivars is apt, because Ireland was the first region outside South America to take the species seriously as a food crop. In Ireland, the species had been a minor garden plant, eaten occasionally as an exotic luxury until about 1650. Then the country went through a long, terrible period of rebellion, British devastation, poverty, and hunger, fully narrated in the classic book by Salaman (1949). By then, some potato clones had been developed that were well adapted to Ireland, producing far more calories per acre than grain crops, and also less likely than grain crops to be carried off by the British. Before the end of the 17th century, the potato had become the mainstay of the peasants' subsistence.

During the first half of the 18th century, the potato became an important food crop in Prussia, Saxony, and other German states. The period between 1750 and 1800 saw the rise of *S. tuberosum* as a field crop in most of Europe: Spain, Portugal, France, England, Scotland, Sweden, and eastern Europe in general. Judging by its folk names, the potato spread from Germany to the Danes and Slavs, but reached Norway and Sweden from Britain. As it had in Ireland, the potato temporarily gave continental Europe a vastly increased supply of cheap food, reducing chronic starvation and allowing the population explosion associated with the Industrial Revolution. In continental Europe, the potato also became a major feed crop for pigs and other livestock, in spite of having to be cooked before being fed. It also provided a cheap raw material for production of alcohol for distilled liquors, e.g., vodka, Schnapps, akvavit, industrial alcohol.

S. tuberosum was taken from England to eastern North America via Bermuda in 1621. Earlier reports of potatoes in eastern North America are evidently based on the sweet potato, *Ipomoea batatas* and on a locally used legume, *Apios americana. S. tuberosum* was again introduced in 1719 to New Hampshire by Protestant colonists from Londonderry. In eastern North America as in most of Europe, potatoes became a field crop only after 1750. The potato may have been introduced to North America independently by the Russian American Company. As they came down the California coast in pursuit of sea otters, the Russians established Fort Ross, north of San Francisco Bay, in 1812, largely as a farming operation to provision their hunters. Among the crops planted were potatoes (Hutchison 1946). Direct introductions from South America to North America only began about 1850.

Meanwhile, *S. tuberosum* had been taken to Asia, perhaps from Europe via the Cape of Good Hope, perhaps directly from South America. It is said to have been in India by 1610, Formosa by 1650, and mainland China by 1700. Thunberg reported it a recently introduced Japanese crop in 1766. The chronology and common names suggest that both the British and Dutch East India Companies were involved, but no good evidence is available. The potato was probably spread widely by European missionaries later in the colonial period. It became a staple only in high mountains, particularly in the Himalaya and southwestern China, but is a familiar minor garden vegetable in temperate East Asia and in tropical hill regions from the Philippines to Indonesia.

Captain Cook brought *S. tuberosum* around Cape Horn into the South Pacific. He left gardens of potatoes and other vegetables at Queen Charlotte Sound in New Zealand in 1769. They were flourishing on Cook's return in 1773, and the Maori showed interest in them. By about 1840, the Maori had adopted potatoes into their own gardens to supplement the previously acquired New World sweet potato, *I. batatas*.

There remains the story of the potato blight, which has been told repeatedly and eloquently and will be merely sketched here. The parasitic fungus *Phytophthora infestans* is presumably native to the New World and had been left behind when the Andean potato was brought to Europe. Little attention has been paid to the geographic distribution of the fungus among wild *Solanum* species or in the South American Indian crop. Like other successful parasites, the fungus is apparently not usually lethal where it and its hosts have long coexisted. The blight was recognized as a problem in South America only after development of commercial potato monoculture, e.g., in Chile and Peru about 1950. In the 1950s, the blight became a problem in the highlands of Mexico where *S. tuberosum* had been introduced during the Spanish colonial period. As commercial potato planting expanded on the great snow-capped volcano of Toluca, *P. infestans* outbreaks began (Fry and Spielman 1991; Robertson 1991). The source was traced to native wild *Solanum* species, which harbor genetically more diverse strains of the fungus than the disease of the commercial crop. Symptoms of the disease are very mild in the wild species and were not detected above 3500-m elevation, although the wild species grow up to 3900 m.

The Mexican studies presumably reveal only the tip of the iceberg. *P. infestans* produces wind-borne sporangia that under suitable weather conditions are believed to be spread for hundreds of kilometers (Fry and Spielman 1991). The fungus is not host specific; it not only infects various *Solanum* species, but is also a common disease of *Lycopersicon*, the tomato genus. If the Mexican example is representative, the potato crop in the Andes may escape the blight at high elevations; the genetic diversity and rotation practices of the Indian crop would also be expected to limit its spread.

The constantly cool, wet climate of the Chilean potato region would have provided optimal conditions for local spread and infection of the crop by *Phytophthora*; selection for resistance would be expected to have operated more strongly there than anywhere else in the aboriginal potato crop. According to Wellman (1972), *P. infestans* was introduced to Europe from Chile in 1835 on potatoes from the island of Chiloe brought as ship's galley stores. Although plausible, this story is unconfirmed. However the fungus arrived, it was widespread in Europe before being identified. Although European farmers were growing various potato clones, all of their original stock was Andean and lacked blight resistance. The destructiveness of the disease in Europe is highly variable. It is reduced by sun and dryness and aggravated by cool rainy weather. After ominous but local failures, catastrophe began in 1845 when the blight destroyed much of the crop in Belgium and the British Isles and famine began in Ireland. The next year, the Irish potato crop was a total failure, and mass starvation began. In the next few years, the population of Ireland decreased by about 3 million, about half of whom had emigrated. The population has never regained preblight levels. The small farmer class was wiped out, and the abandoned farms were combined into large holdings, commonly for pasture rather than gardens.

After the catastrophe of the 1840s, the potato crop was genetically transformed during the last half of the 19th century. With the help of fungicides and new cultivars with enough resistance for practical purposes, the blight became merely a costly nuisance. In both Europe and North America, the period from 1850 to 1900 was the heyday of private potato breeding. Breeders made some crosses between cultivars, but mainly they simply selected mutants or seedlings from open-pollinated lines. In *S. tuberosum*, such seedlings are usually the result of self-pollination.

The Chilean subspecies provided a crucial source of new germplasm during this period. It is generally agreed that the Andean subspecies had been alone in the northern hemisphere until the mid-19th century when it was joined by the Chilean subspecies. Repeated introductions of Chilean clones during this period seem nearly inevitable. Following independence from Spain, Chilean commerce with various European countries had quickly expanded. British and New England whalers, traders, and missionaries were fanning out into the Pacific. Ship traffic around the Horn reached a crescendo, of course, with the California Gold Rush.

The best documented and most consequential potato introduction, however, was not direct from Chile, but relayed via Panama. In 1851, the Reverend Chauncey Goodrich of Utica, New York, obtained clones from the U.S. consul in Panama. Potatoes marketed in Panama could not have been easily brought from the Andes, but California bound ships would normally stop at Chilean ports after rounding the Horn; many also stopped at Panama City to pick up gold seekers who had traversed the Isthmus to avoid the long ocean voyage. Goodrich had an idea that the North American potato crop had become weakened by long vegetative propagation, so he grew large numbers of seedlings seeking to restore vigor. Whatever the limitations of his theory, the result was astonishingly successful. From an original clone that he named Rough Purple Chili, Goodrich selected a seedling that he named Garnet Chili. Other breeders selected seedlings of Garnet Chili, including Early Rose. A seedling of Early Rose was Luther Burbank's Russet, which became the dominant North American cultivar. Descendants of Rough Purple Chili are known to be in the pedigrees of all but two of the 126 North American cultivars whose pedigrees have been traced; the two possible exceptions have incomplete records (Hougas 1956; Plaisted and Hoopes 1989). A more recent introduction from Chile, Villaroela, is in the pedigree of 33 North American cultivars.

In Europe, a Chilean potato thought to have been introduced from Chile in the 1830s was grown by farmers under the name of Daber (Plaisted and Hoopes 1989). Daber was joined in the late 19th century by derivatives of Rough Purple Chili, including Early Rose, introduced from North America. European breeders crossed the Chilean with their old Andean line to produce important new cultivars, e.g., Jubel in Germany, with Early Rose in its pedigree, and Busola in Poland, with both Early Rose and Daber in its pedigree.

Simmonds (1964) found that in the British Isles leaf morphology of potato cultivars changed dramatically after the catastrophic blight; this was shown by study of large numbers of herbarium specimens made at different periods of the crop's history. For 200 years prior to the blight, the specimens all had the Andean morphology with no evolutionary change. In the mid-19th century, the Andean type abruptly disappeared from the record and was replaced by new cultivars with Chilean type leaf morphology or with leaf types that suggest crosses between the two subspecies. It has been suggested that this shift was due to sudden genetic change, under pressure from the blight, in the old British cultivar stocks of Andean derivation rather than to introduction of Chilean stocks, but it is not at all clear why this would have entailed morphological mimicry of the Chilean subspecies. Unfortunately, the pioneer British breeders did not record pedigrees of their selections. It is known, however, that they imported breeding stocks. For example, William Paterson of Dundee, who released the important cultivar Victoria in 1864, had imported Chilean stock (Glendinning 1983). John Nicholls, who released the famous blight-resistant Champion in 1876, obtained stocks from Paterson.

After 1900, potato breeding in both North America and Europe ran into diminishing returns in spite of vastly increased efforts. The small private breeders were replaced by large-scale commercial and government programs making innumerable crosses and growing millions of seedlings for selection. However, for a long time, no new cultivars of lasting value were produced. The first notable product of scientific breeding in the U.S. was Katahdin, released by the Department of Agriculture in 1932. Early Rose and Busola were in its pedigree. In 1970, the northern hemisphere potato crop was still mostly obtained from clones that had originated in the late 19th century.

Concern with the narrow genetic base of the commercial crop and loss of landraces in South America has led to establishment of an international network of so-called gene banks, the main ones located at Lima, St. Petersburg, Braunschweig, Dundee, and Sturgeon Bay, Wisconsin. These maintain huge arrays of cultivars and wild species, both as living cultures of clones and as true seed in frozen storage.

Scientific breeding programs have been largely dominated by searches for resistance to diseases and pests. Introduction of genes from various wild species has been tried repeatedly. For a while during the 1930s, *S. demissum*, a wild hexaploid native to the pine-fir forests high up on the great Mexican volcanoes, was a favorite of breeders. The *S. demissum* and *S. tuberosum* hybrids showed promising resistance to late blight, but it was fugitive. When planted in extensive fields, the new cultivars were soon attacked by new strains of the fungus that had evolved to overcome resistance. However, *S. demissum* has contributed other useful genes to many commercial cultivars. Other wild species now in the pedigrees of many cultivars are *S. chacoense*, *S. maglia*, and *S. acaule*. As noted above, *S. tuberosum* is aptly characterized as a genetic sponge. Programs are also under way to develop new cultivars for special culinary uses and for special environments, particularly the lowland tropics.

All the preceding discussion has emphasized the role of potatoes in high elevation and high latitude agriculture. However, although the crop is most important in those regions, it is also grown in significant quantities throughout almost the whole geographical range of world agriculture, barring the lowland tropics. Potatoes grow very well in subtropical regions during the cool season and also at only moderate elevations in the tropics. Potatoes are grown commercially in every prefecture of Japan, every department of France, every county in California, and in various West Indian islands.

***LYCOPERSICON* — TOMATO** (Jenkins 1948; Luckwill 1943; Rick 1976; Tigchelaar 1986)

About ten species are recognized in the genus, all native to a dry zone on the coast of western South America between Ecuador (including the Galapagos) and Chile. Most of the species produce small, inedible fruits and are not closely related to the cultivated tomato. The species *L. esculentum* is generally broadly defined to include all the common garden varieties of tomato together with the wild Peruvian cherry tomato (*L. esculentum* var. *cerasiforme*, *L. esculentum* ssp. *salendii*), which intergrades via small-fruited cultivars with the large cultivars. All are completely interfertile with each other and with two wild species, *L. pimpinellifolium* (currant tomato), also native to Peru, and *L. cheesmanii* (=*L. esculentum* var. *minor*), endemic to the Galapagos Islands.

These three species are natives of coastal deserts, nearly rainless for long periods between the occasional El Niño episodes when warm ocean water generates heavy rains. Normally, the cold ocean generates stable air masses with persistent overcast. The species grow where condensation of dew at night and from fog drip is the main source of moisture. Away from coastal fog, they may be confined to banks of exotic streams. They are opportunistic in being able to grow indefinitely as perennials if enough water is available and also to grow as ephemeral annuals and survive periods of sunshine and desiccation as seed in the soil. Tomato plants are also remarkably capable of recovering from brief wilting. The wild cherry and currant tomatoes have shiny, bright red berries, evidently attractive to birds. The Galapagos tomato has dull purplish berries that are fed on by the famous giant tortoises and perhaps by iguanas. All three species have indigestible seeds, especially the Galapagos tomatoes, which remain viable after a slow passage through the tortoises' gut.

Only *L. esculentum* has become a domesticated crop, although *L. pimpinellifolium* is casually planted as an ornamental and its tiny red berries are edible.

The historical geography of the cultivated tomato is very curious. *L. esculentum* was evidently never domesticated in its native Peru or anywhere else in South America. This is in spite of the fact that the coastal desert of Peru was the setting for one of the ancient world's most diverse and best documented crop complexes. Other domesticated nightshades — the potato, pepino, and chile peppers — are beautifully documented in Peruvian archaeology, both in accurate ceramic representations and dried remains, but there is no trace of the tomato. Neither is it mentioned anywhere in South America in early Spanish chronicles, rich in information on Indian food plants. The tomato begins to appear in Spanish accounts from South America only after 1600 and then not in Peru or as an Indian crop, but in widely scattered Spanish colonial settlements and under the Spanish name *tomate*, not under any South American Indian names. In fact, to this day, South American Indians seldom grow tomatoes.

The story is quite different in Mexico. Although again archaeological evidence is lacking, there is abundant indirect evidence that the tomato was a diverse and fully domesticated Indian crop before the Spanish Conquest. There is a complication in interpreting early mentions of *tomates* because the Spanish name was taken from the Nahuatl *tomatl*, which is a general term for both *Lycopersicon* and the related husk-tomato (*Physalis*). The kinds were distinguished by a host of prefixes too numerous to discuss here. However, *Physalis* has green fruits eaten cooked and was and still is grown mainly on the high central plateau. A classical source on Mexican botany written for the king of Spain in the 1570s by Francisco Hernandez describes how *tomatl* fruits were ground and mixed with chilis (*Capsicum*) to make an agreeable sauce, which improves the taste of almost all foods and stimulates the appetite. He said tomatoes grew in whatever regions but particularly the hot ones, sometimes spontaneously, sometimes in cultivation. Hernandez evidently used the name *tomatl* generically to include both *Lycopersicon* and *Physalis*.

The Spanish happened to encounter the tomato first in the Aztec empire and adopted the Nahuatl name. This does not prove that the Aztecs were the original domesticators of the species. The crop was and is widely grown by other Mexican Indian groups, including the Maya, Mixtecs, Zapotecs, and Tarascans, each of whom has their own names for the crop, etymologically quite independent of the Nahuatl ones.

We still have no better hypothesis of how *L. esculentum* came to be domesticated than one proposed by Jenkins (1948) long ago. In essence, this suggests that the wild Peruvian cherry tomato was capable of wide bird dispersal, but that outside its native desert, habitats were not open to the species until provided by agricultural disturbance. Jenkins noted that the species, particularly the wild small cherry tomato, has become common over much of the world as a spontaneous weedy volunteer. Jenkins did not explain why South American Indians were not interested in the fruit; perhaps they shunned it because of its poisonous congeners. In any case, it was appreciated by the prehistoric Mexicans, possibly because they already had learned to use the native *Physalis*. They selected recessive mutations for fruit color — tangerine, yellow, pink, white — that would not have survived in the wild. They also selected a simple recessive mutant with multi-chambered fruit as contrasted with the dominant wild type gene for 2-locular fruits in the cherry tomato. Under domestication, more genetically complex changes were selected leading to progressively larger fruits. Mutants with pear-shaped and other fruit shapes were also selected. Jenkins noted that the diversity of tomato cultivars in Mexico is unrivaled except by modern germ plasm collections drawn from the whole world.

Clear evidence of introduction of *L. esculentum* to Europe began appearing in the herbals in the mid-1500s. The first record, in 1544 in Italy by Matthiolus,

was of a yellow-fruited form under the name *pomi d'oro*, a name survivi
today in Italy for tomatoes in general. Matthiolus added a record of t
fruited form. Early European sources noted the fruit was edible, but
through Europe was initially as a garden curiosity, not as a foo
there were rumors that it was dangerous.

In 1572 another Italian, Guilandini, made an interesting statement ... the geographical source of the species. He attributed it to "Temixtitlam," the same peculiar rendition of the name of the Aztec capital used by Cortez in his reports on the Conquest in 1524 to 1526. How Guilandini made this connection is unknown, but there were close communications between the Spanish court and Rome. By 1597, Gerard was growing tomatoes in England from seed received from both Spain and Italy.

The name *tomate* went with the species across the Pacific at an unknown date, perhaps on the Manila galleon. In the 17th century, cultivation under the Mexican name was spreading in the East Indies. Cultivation gradually spread through the warmer regions of Asia.

Acceptance of the tomato as a wholesome food, rather than a love potion, curiosity, or ornamental, was very slow in Europe and North America. McCue (1952) traced this history in detail. Finally, by the late 1700s, tomatoes were being grown and eaten in abundance in Italy and the Iberian Peninsula; they were in common use, raw and cooked, in a great variety of dishes. By about 1800, tomatoes were said to be the commonest of all vegetables in Spain, grown in every village. At that time, the French were tentatively eating tomatoes, but the northern Europeans were afraid of them. The slow adoption in Europe is puzzling because in the New World Spanish colonials had adopted the tomato as food early in the 17th century, not only in Mexico but in the Greater Antilles and widely scattered parts of South America where they had been responsible for its introduction.

In eastern North America, tomatoes were grown only as a curiosity until late in the 18th century. Jefferson grew and perhaps tasted some. Regular use as food began with Italian immigrants to New England and French refugees from Haiti in New Orleans; ketchup was being made in New Orleans in 1779.

Direct introduction from Mexico to what is now the southwestern U.S. would seem likely, but is not recorded in available sources. The earliest record of tomato cultivation in California is at San Diego about 1850. Subsequent development of the huge California commercial crop, now more than 3/4 of the U.S. total, was based on cultivars introduced from eastern North America, which had in turn come from Europe.

L. esculentum evidently became adapted to temperate latitudes during its sojourn in Europe and eastern North America. Selection by growers and seedsmen developed cultivars with earlier and heavier fruit set and larger fruits. The large modern tomato industry in northern Mexican states for export to the U.S. in winter is based on modern cultivars introduced from the north.

Scientific study of tomato genetics and deliberate breeding, as opposed to selection of chance seedlings, began in earnest in the U.S. during the mid-20th century. The tomato is now one of the best known genetically of all species, with hundreds of genes accurately located on chromosome maps. Many hybrids between cultivars and with wild species have been produced, some of which have been accepted by commercial growers and home gardeners. The most newsworthy accomplishment of the University of California breeding was production in the 1950s of cultivars that allowed conversion of harvesting from handpicking to complete mechanization. This involved developing cultivars with short, compact plants, concentrated maturation, and tough, firm fruits, but not necessarily especially delicious. On the other hand, modern breeding in various regions has led to sale of F_1 hybrid seed, combining desirable traits of two parental lines and giving hardier, more disease resistant plants with earlier fruit set, higher yield, and excellent quality. F_1 hybrids are now standard in commercial production in Japan and Israel and are favored by homegrowers in the U.S.

California breeders are now avidly contemplating, especially during droughts, the possibility of combining genes from the wild *L. cheesemannii* of the Galapagos with those of cultivated *L. esculentum*. The Galapagos tomato can grow under irrigation with pure seawater.

CAPSICUM — CHILI PEPPERS (Andrews 1984; Heiser 1976; Patiño 1964; Pickersgill 1988; Pickersgill et al. 1979)

Capsicum includes about 25 wild species, all in the New World, mostly in South America.

The wild species, collectively called bird peppers, are shrubby perennials. They all produce pod-like berries, small, red, and pungent. The pungency is due to capsicin, a volatile aromatic compound that provides the active ingredient in Mace spray. It is concentrated in the placenta (the connective between the fruit and its seeds). The pungency must be tantalizing to birds. Bird dispersal allows the species to colonize widely scattered, open, disturbed habitats, both natural and artificial.

Wild peppers were gathered in prehistoric time by many New World Indians and are still widely sold in Latin American markets. They and their domesticated derivatives are excellent sources of vitamins, especially C, as well as stimulants of the appetite.

The domesticated chilis and their probable wild progenitors are assigned to four species, mainly on the basis of floral characters that have not been modified by human selection. Unlike the wild peppers, which have deciduous fruits, the cultigens retain the mature fruit for picking. Fruits vary in color due to selection of various mutations. Fruit size and degree of pungency are controlled by multiple genes. However, a single recessive mutation blocks all capsacin production and results in a sweet pepper, a type that evidently is unsuccessful outside of cultivation. The domesticates also have been selected for a variety of shapes. Their dazzling diversity is beautifully illustrated in paintings by Andrews (1984). Although many cultivars are strikingly different from their wild progenitors, intermediate cultivars form a genetic continuum that defies formal nomenclature.

Chili cultivars are normally self-pollinated and propagated by seed as pure lines. Whether self-pollination was also characteristic of their wild progenitors is unclear.

C. pubescens is the most distinctive and geographically localized of the cultigen species. Its wild progenitor has not been identified, but there are several close wild relatives in the Andes of Bolivia; the cultigen may be of hybrid origin. It is grown mainly for home consumption at elevations between 1800 and 3000 m in the central Andes. The peppers are sold in local markets and occasionally in nearby cities. A cultivar called *rocoto* has yellow fruits similar to a hen's egg in shape and size.

C. baccatum is native to the same general region but at lower elevations. Wild forms are primarily east of the Andes in southern Peru and Bolivia. Large-fruited cultivars have long, slender, tapering fruits (var. *pendulum*) of various colors. *C. baccatum* cultivars had spread to the Peruvian coastal oases by 2500 B.C., earlier than the introduction of pottery or maize. The cultigen evidently also spread prehistorically northward through Ecuador and eastward across northern Argentina to southeastern Brazil.

C. frutescens (incl. *C. chinense*) is a diverse group and has often been treated as two species, with the wild, weedy, and primitive cultivars called *C. frutescens* and the more advanced cultivars called *C. chinense*. However, the complex forms a continuum (Pickersgill et al. 1979). Wild forms range through tropical lowlands from southern Mexico and the West Indies into Ecuador, Amazonia, and northeastern Brazil. Cultivars taken as a group have the same range and may derive from multiple domestications. Primitive cultivars, e.g., Tabasco, are little different than the wild, except for slightly larger fruits. Advanced cultivars are wonderfully diverse in shape, size, and color. One is *habañero*, perhaps the hottest of all peppers. Cultivars are found in archaeological deposits of Peru, the earliest dated before 8000 B.C. in Guitarrero Cave in the Callejon de Huaylas, an inter-Andean valley (Lynch 1980). Cultivars of this species are not known from the Peruvian coastal oases until about 2000 B.C. This species may also be the source of peppers reported, along with maize and other crops, in the Azapa Valley of northern Chile before 1500 B.C. (Rivera 1991).

C. annuum includes most of the common chili peppers. This is by far the most diverse of all *Capsicum* species containing cultivars of many sizes, shapes, and colors, some hot, e.g., Jalapeño and Cayenne, some sweet, e.g., bell and banana peppers. Conceivably they derive from multiple domestications. The wild form *C. annuum* var. *glabriusculum* (= var. *aviculare*, etc.) has commonly been misidentified in older literature as *C. baccatum*. It ranges from Florida and the Bahamas to Arizona and down through Mexico and Central America to Colombia. Archaeological remains from the Tehuacan Caves in Puebla, Mexico, dated before 5000 B.C. were presumably gathered wild. Cultivation evidently began before 3500 B.C.

These four species were not clearly distinguished taxonomically until very recently. In early historical accounts, a species can sometimes be inferred from its geography. Thus aboriginal peppers of the West Indies and the Orinoco-Amazon region were *C. frutescens*, while those of Mexico were *C. annuum*. Different cultigens overlapped in Central and much of South America. Thus it is necessary to consider the historical record of the species jointly, with only occasional notation of individual species.

C. frutescens, like several New World crops, entered written history in Hispaniola on Columbus' first voyage. The island Arawak name *aji* was to become a *lingua franca* name for *Capsicum* in Spanish America, along with the Nahuatl name *chili*, which the Spaniards picked up in Mexico 30 years later. The Spanish garrison left by Columbus on Hispaniola became better acquainted with *aji* when the Indians bombarded their stockade with calabashes filled with ground pepper and ashes.

On the tropical American mainland, the ubiquity and importance of *Capsicum* among the Indians is clearly recorded in all the classical sources. Oviedo, Las Casas, Hernandez, and others agree that tropical Indians cultivated *Capsicum* everywhere and used it daily in almost everything they ate. Some special Mexican uses involve mixing chilis with cacao for *mole* and with tomatoes for a wet *salsa*. Some sources report *Capsicum* as important in Indian religion, both in ceremonial use and as being banned during ritual abstinence.

Outside the Spanish sphere, Hans Stade documented cultivars of *Capsicum* by the Indians who held him captive in eastern Brazil about 1550. Early accounts repeatedly mention presence of multiple chili cultivars grown together, the greatest diversity evidently being in Mexico and Peru.

Capsicum did not follow maize and other Mexican crops into temperate North America in pre-Spanish time. Even in the Pueblo region of the Southwest, where colorful strings of chilis hung up to dry are a typical sight today, there is no archaeological or early ethnographical record of them. This is a puzzle because dried peppers with viable seeds could easily have been traded for long distances. Also, the wild *C. annuum*, known locally as *chiltepin*, ranges into some of the mountains of southern Arizona where it is avidly gathered by the Papago Indians to be dried for home use and trade (Nabhan 1978). *Chiltepin* is highly prized and scarce, but the Papago did not traditionally plant it with the rest of their crops. Nabhan notes, however, that when they got piped water, some Papago began growing a few pepper bushes in their yards, an example of incipient domestication. The arid climate of northern Mexico and the Southwest would, of course, have hindered spread of chili cultivation, but it is hard to see why it should have been an absolute barrier for Indians who did practice some irrigation.

Introduction of chilis to the Old World and spread in folk cultivation came amazingly rapidly after 1492. Spain had *C. frutescens* from Hispaniola in 1493, and other species were probably brought back to Spain early in the 16th century as exploration of the mainland progressed. With the Conquest of Mexico, a variety of cultivars of *C. annuum*, including sweet forms, were acquired.

Spaniards and Portuguese soon decided that *Capsicum* provided sensational spices, more versatile than East Indian *Piper nigrum*, black or white pepper. In the mid-1500s, Monardes described chili peppers as grown abundantly in every garden in Spain, used with all manners of meats and potages, better than black pepper, and costing nothing but the sowing.

Initial introductions of chili peppers to Africa and Asia are generally credited to the Portuguese, which is quite plausible although details are unrecorded. Various cultivars of both *C. annuum* and *C. frutescens* have long been in folk cultivation in Africa and Asia. The Portuguese presumably acquired both via Spain early in the 15th century and also would have acquired *C. frutescens* directly from Brazilian Indians. In any case, the Portuguese probably brought *Capsicum* to India and the East Indies via the Cape of Good Hope long before 1565, when the Spaniards began bringing Capsicum from the other direction via the Manila galleon. Once introduced to Asia, *Capsicum* rapidly escaped European control. The Indian Ocean region was the hub of the world's greatest network of maritime and overland trade. Arabs, Turks, Persians, Indians, and Chinese were all very active long-distance traders. Dried chili peppers, loaded with viable seed, undoubtedly traveled back and forth every which way. Diffusion was so rapid and extensive that chili cultivation evidently spread back westward through the Ottoman Empire into the Balkans. By the late 1500s, cultivars were coming from opposite directions to meet in central and eastern Europe.

Capsicum became so thoroughly assimilated into the cuisines of various European and Asian regions that it is commonly mistaken for an ancient heritage. It is hard to imagine goulash without paprika, curries without their hottest ingredient, or Szechwan or Thai dishes devoid of chili. Likewise, in the U.S., Texans and Louisianans regard *Capsicum* as part of their birthrights, not a 19th century import.

In contrast to the rapid spread and acceptance of *Capsicum* cultivars outside their ancient range, their distributions within Latin America have remained remarkably stable. There have been a few recent introductions of cultivars to new regions, for example, of *C. baccatum* to Costa Rica, of *C. pubescens* to a few places in the highlands of Mexico and Central America, and of *C. annuum* to Brazil. In general, however, Latin American cultivars retain their pre-Columbian ranges.

The world's *Capsicum* cultivars are essentially a heritage from the tropical American Indians. Only a little scientific breeding has been carried out, concentrated on the sweet types.

Sterculiaceae — Cacao Family

The Sterculiaceae are pantropical and subtropical with about 50 genera and 800 species. Several genera are commonly planted for their spectacular flowers: *Brachychiton*, *Dombeya*, and *Fremontia*. Two genera of rain forest trees have seeds that contain caffeine and other alkaloids that act as stimulants in humans. These are *Cola* of West Africa and *Theobroma* of tropical America. Only *Theobroma* will be discussed here.

***THEOBROMA* — CACAO** (Barlow 1949; Cope 1976; Cuatrecasas 1964; Hunter 1990, 1991; Patiño 1963; Purseglove 1968; Stone 1984; United Nations 1964; Urquhart 1955)

Theobroma includes about 20 species of small trees native to lowland and lower montane rain forests of Central and South America. Most are understory species in the shady, sheltered, constantly warm and humid environment beneath the forest canopy. Others grow in less sheltered riparian habitats. Inflorescences are not borne on young branches, but are cauliflorous, i.e., developed from latent buds on trunk and old branches. The wild species are generally self-incompatible, i.e., require cross-pollination with another genotype. The tiny flowers are pollinated by midges. The melon-like fruits develop year round. They are orange or red when ripe and attract animals, which tear open the shell to reach the sweet, juicy pulp. Monkeys and parrots, for example, remove the seeds, eat the mucilaginous outer seed coat, and reject the hard-shelled seed within. The embryo is rich in fats and other foods that give it a strong start for germination in dim light, but the seeds are bitter and loaded with alkaloids, especially theobromine. The seeds are nondormant and only briefly viable.

Relationships between *Theobroma* and the aboriginal American peoples followed two quite distinct patterns. For South American Indians, *Theobroma* species were wild trees exploited only as a source of a minor fruit. Usually, they simply sucked the pulp off the seeds, but sometimes they squeezed the juice for fermenting to wine, using the same apparatus they had devised for squeezing the toxic juice from bitter manioc. The seeds were discarded as inedible. Nowhere in South America do early Spanish accounts of Indian plant uses, which are often quite detailed, mention planting of *Theobroma*.

By contrast, in parts of Mexico and Central America, early Spanish accounts, starting with Columbus on the Nicaraguan coast, depict *Theobroma* seeds as articles of commerce obtained from a cultivated crop. Indians in this region had discovered how to produce what we call chocolate, a corruption of the Mexican (Nahuatl) name *xocoatl*. This is a complex mixture of organic substances, some naturally present in the seed, including the fats or cocoa butter. The flavor and aroma of chocolate are not present naturally, but result from complex manipulation, including a sequence of anaerobic and aerobic fermentation involving various microorganisms and enzymatic changes associated with death of the embryo. After fermentation, the seeds were dried rapidly in the sun, toasted, ground, and mixed with maize, chili, and other ingredients to make a drink or a food. The chocolate could be sweetened with honey, but it was also used for *mole*, a fiery hot sauce in which turkey meat was cooked. (The turkey was an ancient domesticate in Mexico.)

Two *Theobroma* species were cultivated in Mexico and Central America.

Theobroma bicolor — Patashte (Nahuatl *patlachtl)*

The species has been found wild in primary lowland forests in Chiapas and adjacent Guatemala. The pioneer medico-botanical work of Francisco Hernandez, written about 1575, describes it clearly as cultivated in Mexico. The seeds provide a relatively low quality chocolate, but have continued to be planted widely in dooryard gardens for home use. *Patashte* trees do not require shade, unlike the better Mexican cacao. During the Spanish colonial era, *patashte* along with *cacao* was introduced widely in tropical America, usually under its Mexican name. The seeds are sometimes mixed with those of the following species, but are of very minor commercial importance.

Theobroma cacao — Cacao (in the strict sense)

The name, from the Mexican (Nahuatl) *cacahuatl* became cocoa in English.

The definitive taxonomic work (Cuatrecasas 1964) recognizes two wild subspecies:

1. *T. cacao* subsp. *cacao*, the Criollo lines, is native to forests of southern Mexico (Vera Cruz, Chiapas) and east to Belize. The fruits are elongated and pointed and are ribbed and rough; cotyledons are whitish.
2. *T. cacao* subsp. *sphaerocarpum* (=*leiocarpum*), the Calabacillo lines, is native to South American rain forests from the Amazon headwaters in Peru and Ecuador east to the Orinoco basin and the Guianas. The fruits are rounded and nearly smooth; cotyledons are purplish.

The two subspecies presumably diverged in isolation following ancient long-range dispersal by migratory birds. A tantalizing alternative has sometimes been suggested, namely that cacao was not native to the Maya area but was introduced from Amazonia by ancient traders. Among several reasons for discounting this possibility is the brief viability of the seeds. Cacao seeds are viable for only a couple of weeks while fresh and moist; they do not survive drying.

The first subspecies *T. cacao* subsp. *cacao* is the ancestor of the cacao cultivated in Mexico and Central America at the time of the Spanish Conquest. This group of cultivars, the so-called "Criollo" types, is very diverse and presumably evolved under long selection by the Maya and neighboring peoples. The Indians valued the seeds highly and used them in trade almost like currency. The common people probably consumed them only on rare ceremonial occasions. Evidence that chocolate was prized by the Maya nobility during the Classic period was recently found by archaeologists excavating a spectacular ruin at Rio Azul, near Tikal, in the Peten. A tomb dated at about 450 A.D. contained a superb polychrome pot bearing the glyph for cacao and containing residue that has been identified chemically as chocolate (Adams 1990).

Moctezuma, the last Aztec emperor, was inordinately fond of chocolate, and he and his court consumed astounding quantities. His tribute list (Barlow 1949) recorded annual levies of hundreds of *cargas* or loads, each supposedly of 24,000 seeds (3×20^3 in the Aztec duodecimal system). This tribute was carried to Tenochtitlan, now Mexico City, from what is now Vera Cruz, Chiapas, and Guerrero. After the Conquest, the Spaniards took possession of cacao growing there and in other parts of southern Mexico and from Guatemala south to Costa Rica, all in territory that had been frequented by pre-Conquest Mexican traders. The Spaniards reported no Indian planting or use of cacao seeds from anywhere south of Costa Rica.

Cacao was promptly sent to the Spanish court and spread then among the royalty and nobility of Europe. The Aztec recipes were altered by adding Old World products — cane sugar and cinnamon — as well as the New World vanilla to make sweet drinks and confections. Cacao was esteemed not only for its taste but as a stimulant and aphrodisiac. The rapid increase in European demand led to extensive planting of cacao in Spanish territories of the West Indies and South America. The earliest, made during the 16th century, were in Jamaica, Trinidad, and Venezuela. By 1650, Spaniards had expanded cacao planting to all the Greater Antilles and to Colombia and Ecuador, while the French had begun some planting in Martinique. In the 1650s, the British took possession of Jamaican cacao; previously they had only what they could loot from Spanish production. The Dutch on Curacao, much too dry for cacao planting, became traders in cacao from Spanish territories very early.

Seeds obtained from wild *Theobroma* species in the Amazon and Orinoco basins became a minor component of the cacao supply during the 17th century. Missions among the forest Indians enlisted their neophytes in extractive cacao gathering, and lay entrepreneurs soon joined in the trade. The product was comparatively poor, and none of the species other than *T. cacao* became successful in cultivation. However, cacao planting in South America involved both the Mexican Criollo cultivars and the local wild Calabacillo type, i.e., both *T. cacao* subsp. *cacao* and *T. cacao* subsp. *sphaerocarpum*. Being cross-pollinated and completely interfertile, the two subspecies produced hybrid progeny, some of which became extremely important new cultivars. The best hybrids yield cacao of nearly as high quality as Criollo on trees that are less demanding environmentally and higher yielding than Criollo. The hybrid cultivars are known collectively as Forasteros. These proved acceptable for less expensive products such as milk chocolate, breakfast cocoa, and so on. Trinitarios are a subset of the Forasteros selected later, around 1800, in Trinidad. They resulted from further hybridization between Forastero and Criollo cultivars and are intermediate in quality, yield, and environmental tolerances between the two.

The Portuguese were remarkably uninterested in cacao in the early colonial period. Brazil produced only a little extractive cacao from wild trees until planting finally began about 1750, when Forastero cultivars were introduced to Bahia. After 1880, emancipated slaves leaving the cane plantations joined in the cacao planting on small holdings. By the 1930s, Brazil had become the largest American cacao producer, with Bahia growing nearly the whole crop. The state has over 10,000 cacao planters, mostly with farms of less than 20 ha.

The New World retained strong domination of world cacao growing until early in the 20th century. By then, world demand was skyrocketing while Latin American production was flat. Moreover, within Latin America, domestic consumption was taking more and more of the crop, and some once important exporting countries — Mexico, Guatemala, and Colombia — became net importers.

The Old World tropics have many regions suitable for cacao planting, and certain of these are now important producers, but the introduction of the crop overseas was erratic and generally very late, perhaps largely due to brief viability of the seeds. Nevertheless, by 1670, Spaniards reportedly had introduced Criollo cacao to the Philippines. The Trade Wind passage from Acapulco, which is in an old cacao growing region, to Manila would have been fast enough to effect this. In the 20th century, the Philippines are anomalous in retaining the fine Criollo types; the government has discouraged replacement with the higher yielding Forasteros except in regions too dry for Criollo, which is widely scattered through the archipelago. Perhaps the Philippines were the source for Criollo plantations begun in the late 19th century by German colonists in New Guinea, Neu Pommern (now New Britain), Samoa, and the Caroline Islands.

During the 19th and early 20th centuries, Forastero cultivars, including Trinitarios, were spread through British, French, and Dutch colonies in the tropical Indo-Pacific Ocean region by a network of introductions too complex to pursue and perhaps usually by transport of living seedlings or cuttings. Cacao has been grown commercially in Reunion, Madagascar, Ceylon, Indonesia, the Solomons, New Hebrides, and Fiji but always on a very small scale. The whole region produced only about 2% of the world's cacao until very recently when Malaysian plantations were sharply expanded.

West Africa is a quite different story. Although planting did not begin until the 19th century, West Africa since 1930 has produced 2/3 of the world's cacao. The crop is almost entirely from Forastero types, of which the self-compatible Amelonado is strongly dominant. It had been introduced to islands in the Gulf of Guinea — Principe, Sao Tomé, and Fernando Po — by the 1820s, almost certainly from Bahia, Brazil. The crop was not taken to the mainland for half a century. There is evidence that cacao was grown at mission stations in the Gold Coast, now Ghana, before 1880. Except on Sao Tomé, where there were cacao estates under Dutch control early in the 19th century, the crop was never an enterprise of European estates. Rather it spread among African farmers on village lands and small private holdings. Planting is often in selectively thinned forest or in clearings with bananas for temporary shade and legume trees for permanent shade. Cacao is grown all through the West African equatorial forest belt from Sierra Leone to Angola, but the leading producers are in the center of this belt, from the Ivory Coast east to the Cameroons. Government experiment stations in some of these countries have introduced high quality cultivars of the Trinitario type and clones developed by the United Fruit Company in Central America, but the less delicate, higher yielding Forasteros remain strongly dominant.

Africa usually produces about 1 million tons of dried cacao beans a year and the rest of the world about half a million tons. Some of the best farms produce a ton/ha, but most get much less. The total area planted to cacao is probably about 5 million ha, which is a lot of land but only a small percentage of the tropical rain forest climatic region. There are vast areas of country that are climatically and edaphically suitable for cacao where none is grown and probably never will be. Even for people able and willing to survive in a rain forest environment, cacao is not an easy way to make a living. Although there are seasonal peaks, the fruits ripen year around and must be promptly picked and processed. Extracting and fermenting the seeds, putting them to dry in the sun, covering them during each rainshower — all require intensive hand labor. Clonal propagation is usually prohibitively expensive, and seedlings are genetically variable and yields unpredictable. Even on the best cultivars, yields are commonly disastrously depressed by pests and diseases. Some parasites specific to *Theobroma* have been escaped by plantations overseas. However, there is no escaping black-pod disease, caused by *Phytophthora palmivora*, which has many hosts and is cosmopolitain in distribution. Small farms have, as a rule, coped with these difficulties better than large plantation monocultures. A striking example is Costa Rica where cacao has long been successfully grown in Indian gardens and on small colonial-type estates. The United Fruit Company had huge banana plantations in the Caribbean lowlands of Costa Rica, which were destroyed by a soil fungus, Panama disease, in the 1920s. The company tried to replace bananas with cacao and built great expensive factories at Limon to ferment and dry the seeds. The cacao monoculture was an easy target for black-pod and other diseases, and the company abandoned the plantations, converting some land to oil palm plantations and distributing much of the rest among pensioned former employees, the same West Indian blacks who had grown the bananas. They continued to grow and process cacao on small, labor-intensive farms, although not very profitably.

Worldwide, it is estimated that only about 20% of the cacao crop is produced on large plantations. Cacao is often cited by planners and theoreticians as an example of a crop that could be integrated into sustainable agroforestry to replace the degraded tropical monocultures.

Starting in the 1930s, advanced scientific breeding programs have been carried on in Trinidad and several other countries. It was found that cacao could be readily propagated clonally by cuttings and by budding and other forms of grafting. Elite clones selected for yield, disease resistance, and so on were made available, but did not catch on among planters. Cacao is one of a very few crops in which a trend from heterogeneous seedlings to standardized clones has been reversed. Seedling orchards proved to be cheaper to establish, less vulnerable to pests and diseases, and also yielded a crop with traditional flavors that manufacturers preferred to the crop from clonal orchards. Scientific breeding has so far had little effect on *T. cacao*.

Vitaceae — Grape Family

The Vitaceae include a dozen genera and several hundred species mostly woody vines (lianas) native to tropical and temperate regions of both northern and southern hemispheres. Genera grown as ornamentals include *Parthenocissus* (Boston ivy, Virginia creeper) and *Cissus*. The only crop plants are the grapes.

VITIS — GRAPES (Bailey 1911; Einset and Pratt 1975; Hutchison 1946; Olmo 1976; Patiño 1969; Pitte 1983; Renfrew 1973; Roach 1985; Sauer, C. 1971; Simoons 1990; West and Augelli 1966; Whealy 1989; Zohary and Hopf 1988)

Vitis includes about 60 species native mainly to North and Central America and to East and Southeast Asia. All the species are diploid, and nearly all have the same chromosome complement (2n=38) and are capable of interbreeding and producing fertile hybrids. In nature, the species are kept discrete mainly by ecological and geographical segregation. Three species native to southeastern North America are genetically isolated from the rest by a different chromosome complement (2n=40); this is the subgenus *Muscadinia*, which includes *V. rotundifolia*, the muscadine grape. Interspecific hybridization has become important in evolution of cultivated grapes only in recent times. For most of the history of viticulture and wine making (enology), a single species was involved.

Vitis vinifera — Eurasian Grape

Unlike many ancient, complex crop species *V. vinifera* is considered by the experts to have been derived from a single, fairly homogeneous wild progenitor *V. sylvestris*. The thousands of different clones of wine, raisin, and table grapes that were developed in the Near East and Europe arose under cultivation by human selection of mutations and their recombinations; their propagation was easily done clonally by root cuttings, without the need for grafting until parasite problems developed in modern times, as noted below.

The native range of *V. sylvestris* extends from Iberia and the northwest African coasts eastward through Europe and southwestern Asia to the Himalayas. The bulk of the range is in Mediterranean summer-dry climate, but wild populations range into adjacent mild climates in the Rhine and Danube valleys and in western Asia. *V. sylvestris* contributed several important traits to its domesticated descendants: dense clusters of berries with attractive red/purple/blue anthocyanin coloration; high sugar content, up to 25%; mildly acidic; thin-skinned; ripening evenly and retained on the vine when ripe; turning into raisins in proper climate; and few and small seeds. The fruits clearly evolved primarily as adaptation for bird dispersal. The natural habitats were discontinuous open sites along streambanks and other forest margins. Human gathering of the fruit and making of wine (by spontaneous fermentation of the juice with wild yeasts naturally present on the berries) evidently began long before domestication and probably independently throughout the range of *V. sylvestris*.

The clearest and most consistent difference between wild *V. sylvestris* and domesticated *V. vinifera* is in reproductive biology, a character unrecorded in the archaeological record. All wild grapes are normally dioecious with equal numbers of male and female plants in a population. Almost all cultivars have bisexual flowers and are capable of self-pollination to set fruit. The change was accomplished by a single dominant mutation, which prehistoric farmers realized could eliminate cultivation of fruitless male plants for pollination. Many cultivated clones are heterozygous for this mutation, retaining the primitive recessive genes; thus, when seedlings are grown, some are unisexual, which has promoted intercrossing of cultivars.

Following domestication, there were gradual and rather subtle changes in seed morphology, as the species was no longer dependent on seed dispersal by birds. Carbonized seeds of the wild type have been found in many Neolithic and Bronze Age sites in France, Belgium, Germany, Switzerland, Italy, and the Balkans, some perhaps from primitive cultivation. The first clear evidence of the transition from *V. sylvestris* to *V. vinifera* seeds comes from a series of finds in Macedonia dated between 4500 and 2000 B.C., suggesting the establishment of viticulture in Greece during the early Bronze Age. Even earlier archaeological evidence of Bronze Age cultivation comes from the Near East. From about 3200 B.C. onward, Jericho and various other Jordan Valley region sites have yielded abundant remains of grapes, including seeds, dried berries, and wood. The seeds are of the wild type, but the location is slightly outside the natural range of *V. sylvestris*, so viticulture is indicated. By 2000 B.C., production of grape wine is documented from the Aegean to Mesopotamia by wine storage vessels, art, and references in clay tablets. By 2000 B.C., there is also evidence of grape planting, raisins, and wine in Egypt, far outside the range of wild *V. sylvestris*. Egypt is climatically unfavorable for viticulture, and it is probable that grapes were grown there only in the Nile delta as a luxury crop, while raisins and wine were imported from the Levant or beyond. Wine has been an important part of commerce and colonization in the Mediterranean ever since the rise of seafaring states, particularly the Phoenicians and Greeks. Wine, like olive oil, was a high-value commodity that could be produced generation after generation on dry, rocky hillsides and carried in casks or ceramic amphorae wherever ships could go. Labeling with vintage, grape variety, and winery names began in classical antiquity. The Greeks and Romans were great connoisseurs of wine and paid extremely high prices for choice vintages. Wine of course became much more than a commercial element in western civilization. It has been embedded in religious ceremonies, from the lusty cults of Dionysius and Bacchus to the austere Judaeo-Christian rituals. As for the social role of grape wine, it is hard to visualize how different daily life in southern Europe would have been without it. Most of the spread of viticulture in Europe was completed during classical Greco-Roman time, when innumerable varieties were already grown. The Romans expanded grape cultivation in Iberia and Gaul, on the Rhine, even to Britain perhaps.

The most successful expansion was within the natural climatic limits of wild *V. sylvestris*. The crop thrives in the mild, moist winters and warm, dry summers of the south and is pushed northward mainly on sunny, south slopes. During early medieval times, a few monasteries in England and Ireland grew some grapes, probably on sheltered, sunny walls, but it is unlikely that they produced enough wine for their needs. In the late Middle Ages, following the Norman Conquest, the French bishops and nobility made a determined effort to establish vineyards in England. Some of these were modestly successful, especially during unusually warm and dry years, but most were abandoned during the 14th century, partly due to the economic crisis following the Black Death. During the Renaissance and after, some diehard viticulture has persisted in England, but has always been something of a tour-de-force.

As would be expected, little is known about the spread of particular cultivars during ancient and medieval time. According to legend, the Petit Sirah was introduced to the Rhone Valley by the Romans. A great many other of the superb old wine grape varieties were probably selected in medieval Europe, especially in France. The Chenin Blanc, for example, appears to have been definitely recorded in Anjou about 850 A.D. Hundreds of other cultivars have unknown origins.

In Asia, the earliest evidence of cultivation of *V. vinifera* and of wine making is from Iran and Baluchistan, dated before 2000 B.C. It is uncertain whether domestication there and in the Near East was independent or came from a common origin, perhaps in Anatolia or the Caucasus. *V. vinifera* cultivation spread very slowly into India and China. There were several other grape species native to eastern and southeastern Asia, some of which have edible fruits. These have often been gathered wild, occasionally planted and used for wine, but none were domesticated. By about 100 B.C., Indian and Chinese sources began mentioning grape cultivation and wine making, presumably with arrival of *V. vinifera* from Southwest Asia. Grapes, raisins, and wine became important in China late in the 1st millenium A.D. during the T'ang period, when Chinese authority was extended far westward over Persian and Turkic peoples. Raisins and wine were among the goods exacted as tribute from those regions. Also, new varieties of grapes were introduced to northern China, particularly Shansi, where viticulture has been practiced ever since. Marco Polo reported that in the late 13th century, the capital of Shansi Province had many excellent vines supplying plenty of wine that was shipped throughout China. Grapes have remained a minor but not trivial crop in Shansi and some other parts of northern China. Several varieties are grown including hybrids between *V. vinifera* and local wild *Vitis*. North Chinese viticulture involves some unusual and laborious techniques of winter protection of the vines by unearthing, pruning, and burying them in bundles in pits. Holes are provided for ventilation in warm weather; during extreme cold, the holes are plugged. The Chinese also developed unusual methods of storing the fruit in cold cellars for nearly year-round supply.

Japan acquired *V. vinifera* cultivars from China during medieval times, but grape growing and wine making were carried on in only a few localities and on a very small scale until late in the 19th century. Then, during the Meiji era when Japan began seeking occidental technology, Japanese delegations were sent first to California and then to France to study viticulture and wine making. Choice French cultivars were introduced, and in 1893, a fine, large winery and cellars on the French model were built in the Kofu region of southern Honshu. In 1900, its wine won medals at a Paris exposition. However, grape wine has never become popular in Japan, and what little is drunk is nearly all imported. On the other hand, after World War II, table grape growing expanded in all the major islands, generally in sheltered, dry inland basins. By 1980, table grapes were being grown on over 80,000 Japanese farms, usually in tiny, intensively cared for plots, and sold at extremely high prices. Often tourists come out from the cities and pay to be allowed to picnic under the grape arbors and pick the bunches themselves.

In the west, the Spaniards and Portuguese began spreading *V. vinifera* far outside its old range as soon as they began venturing overseas, first to the Canaries, Madeira, and other Atlantic Islands. Columbus introduced grape cuttings to Hispaniola in 1494 where they survived but were not very fruitful. During the 16th century, the Spaniards attempted viticulture, and sometimes wine-making, in the West Indies and on the mainland. The Portuguese in Brazil and later the French in the Caribbean did the same. Generally, the harvests were poor, not only because of the unsuitable physical environment but because of losses of fruit to parrots and the ubiquitous ants. The first true success during the Spanish colonial period was in Peru where grapes thrived in the irrigated oases of the coastal desert. By about 1550, vineyards were established at Lima, and before 1600, more extensive vineyards were planted in oases of southern Peru. Peru became the supplier of wine to other Spanish colonies on the Pacific coasts of South America and Central America. Drake captured a ship loaded with wine bound from Peru to Nicaragua. Early Spanish colonists in Chile found even better grape-growing country around Santiago.

Meanwhile, *V. vinifera* was planted widely in Mexico immediately after the Spanish Conquest on direct orders from Cortez. The vines were successful enough in drier parts of Mexico that the Spanish crown attempted to suppress the industry to eliminate inroads into the market for wine from Spain. Viticulture and wine and brandy making survived in scattered remote areas of Mexico, such as the spring-fed oasis of Parras de la Fuente in the Coahuila desert, whose wines and brandies supplied the northern mining centers during the late colonial period. Also the frontier missions in the northwest had to make their own sacramental wine, with a little left over for the table. By the late 18th century, the missions had introduced grape cultivars to Baja and Alta California. As in Chile, the vines flourished upon re-entering a Mediterranean type climate, a tiny embryo of what was to come.

Meanwhile, a much more complex chapter in evolution of grape cultivars had begun in eastern North America. Eastern North America has about a dozen native grape species that have been taken into cultivation, at least casually, and used in breeding cultivars. Aboriginally, the Indians exploited the fresh fruit and dried quantities of it for raisins; there is no record of their making wine. Being ecological pioneers adapted to open habitats, especially riverbanks, the grapes commonly grew well and yielded profusely around Indian clearings and villages. Being skilled farmers, the Indians might have begun domestication of the grapes, but the historical record has only a few hints that they did. In the Chesapeake Bay region, for example, Verrazzano observed in the 1520s that competing vegetation had been cleared around some grape vines so they would be more productive. Another 16th century account of the same region, written by a Jesuit, reported that a Powhatan Indian vineyard was as well laid out as those in Spain, and the vinestocks were laden with very large, fine, white grapes. This sounds like a description of the Scuppernong, generally regarded as a cultivar derivative of the wild muscadine *V. rotundifolia*. The Scuppernong was cultivated by the early English colonists of Virginia and became a favorite wine grape of southeastern North America.

As a rule, however, the colonists were slow to try planting the native grapes. All nationalities, not only the Vikings, were impressed by the abundance and quality of the wild grapes. Many early colonists made wine from them by the hogshead: the Spaniards in Florida; the French in the Great Lakes; even the English in New England. But when it came to planting, they usually tried cuttings of *V. vinifera* brought from Europe. Persistent and serious attempts were made throughout the colonial period to introduce cultivars from various European regions, for example by French Huguenots in the Carolinas and Georgia in the late 17th century. New importations of choice cuttings from Portugal and Madeira were tried in Georgia in the mid-18th century. These and innumerable other attempts failed. It was eventually realized that *V. vinifera* cultivars suffered from the severe climate of eastern North America and also from the indigenous plant diseases and insect parasites. Viticulture succeeded in the region only after new cultivars, nearly all hybrids between *V. vinifera* and the native species, were developed. Some of these that are still very important commercially originated between about 1800 and 1850 as chance seedlings selected and propagated by farmers and amateur gardeners. Examples are the Catawba, found on a North Carolina farm; the Delaware, found in a New Jersey garden; and the Concord, a volunteer in a house yard in Massachusetts. All of these are of the slip-skin type, like the native grapes, but in other characteristics, they resemble *V. vinifera*. Expert opinion is that all appear to be hybrids between European cultivars and the wild *V. labrusca*, the fox grape, native from New England to the Great Lakes and south to the Carolinas and Tennessee.

Since 1850, grape breeding in eastern North America has become more complicated, with deliberate hybridization involving several native grape species, first by amateurs and later by experiment stations, continuing to the present. It is this array of European and American hybrid cultivars that became the foundation of viticulture and wine making in eastern North America and more recently in the Pacific Northwest. New York is able to grow a few hundred hectares of pure *V. vinifera* on Long Island in vineyards warmed in winter by the Gulf Stream; these bring premium prices to upgrade wines made with American and European hybrids.

North American grapes and their hybrids with *V. vinifera* were introduced to Europe with quite some unexpected consequences. The plants taken to Europe during the mid-19th century evidently carried with them an insect parasite *Phylloxera* and various parasitic fungi, which were devastating to the pure *V. vinifera*. Starting in 1860, French vineyards began dying from the attacks of *Phylloxera*, and soon the whole of European viticulture was threatened. *V. vinifera* was saved by grafting scions of its cultivars onto American grape rootstocks, instead of simply root cuttings as had been the ancient method of clonal propagation. Three species, *V. rupestris*, *V. riparia*, and *V. berlandieri*, provided *Phylloxera* resistant rootstocks. French breeders also developed hybrid rootstocks, some involving up to six American species and *V. vinifera*. Massive introductions of wild North American germ plasm were brought to France mainly from the midwest. In order to escape the trouble and expense of grafting, breeders began programs of hybridization intended to combine the fruit quality of the European with the pest resistance of the American grapes into a so-called direct producer variety. After over a hundred years of such breeding, the direct producer varieties are still inferior to the pure European varieties in quality. Some of them have become commercially successful, however, in regions unsuited to the preferred *V. vinifera* cultivars. They occupy over a 1/4 of the area of French vineyards. They are also important in expanding grape cultivation and passable wine production in other regions with climates marginal for *V. vinifera*, both in Europe and the midwestern U.S. Most Japanese vineyards are now planted to hybrid European and American cultivars. Pure *V. vinifera* scions still produce nearly all the premium quality European wine and table grapes. The line between wine and table grapes has always been vague, various cultivars being fine for both purposes.

Europe has continued to produce most of the world's grape crop, but overseas production has expanded greatly since the mid-19th century, especially in temperate regions with warm, dry summers suitable for the premium varieties. California has become the most important region outside Europe using pure *V. vinifera* scions. The original Spanish Mission variety, which produced an extremely sweet wine, was joined since the 1850s by a host of other cultivars brought directly from Europe. A Hungarian aristocrat Agostyon

Haraszthy began introducing and planting grapes on a large scale at San Diego in 1851 and at Sonoma in 1856. He toured the principal grape growing regions of Europe in 1861 with a commission from the governor of California and purchased planting material for hundreds of cultivars; he arranged for importing others from the Near East. Haraszthy was mainly interested in fine wine grapes, but one of his introductions, the Muscat of Alexandria, became a leading table and raisin variety, as well as making cheap, sweet wine. A cultivar originating in Anatolia, the Sultanina, otherwise known as Kishamish and Thompson Seedless, became the leading California table and raisin grape during much of the 20th century before better seedless varieties were developed locally. An old seedless cultivar, the Black Corinth or Zante Currant, is still grown commercially in California for special, small raisins, which are used like true currants (species of *Ribes*).

Viticulture boomed in many parts of California during the late 19th century. Grapes were easily started from cuttings, began bearing heavily within a few years, and could be made into a high-value, nonperishable wine or brandy. Most of the wine was ordinary at best, but gradually some respectable wineries emerged.

California wineries made considerable progress during the early 20th century until national prohibition struck in 1918. Wine consumption in the U.S. doubled during the 15 years before repeal due to a loophole permitting legal home wine making in batches up to 200 gallons and also due to abundantly available cheap bootlegged wine in unlabeled bottles. While the professional wineries closed, grape shipments boomed and acreage expanded. The premium cultivars were abandoned in favor of those that gave the greatest tonnage of fruit and withstood bulk shipment.

Since repeal, California vineyards and wineries have gradually emerged from limbo. Old vineyards were replanted to better varieties, and much new acreage has been added, especially in central and northern California. The University of California's Department of Viticulture and Enology at Davis helped with research and training of graduates for jobs in the industry, both in large corporate operations and in the explosive proliferation of little wineries. The living germplasm collection at Davis includes thousands of *V. vinifera* cultivars and a worldwide representation of wild species. Yields of classical cultivars, such as Cabernet Sauvignon and the Chardonnays, have been multiplied without loss of quality by simple selection of mutants within clones. Crosses between classical cultivars have produced new so-called University varieties that are being used for a large share of new commercial plantings. New University table grape varieties, some hybrids between the European and American species, are in commercial production; some are polyploids with exceptionally large berries. In short, California is no longer a poor imitation of the great European grape tradition.

Similar advances have gone on recently worldwide, most dramatically in the southwestern U.S., northern Mexico, temperate South America, Australia, and New Zealand. Also, the traditional European core area has not lagged in advances. Wine making has developed an international network of exchanges of germ plasm, technology, and personnel. The industry has emerged entirely from the pattern of secret recipes of the monks and varieties grown on a single castle's hillside. Wine making became a science more than an art more slowly than brewing did. Although Pasteur learned the basic microbiology of both, wine fermentation was harder to bring under scientific control. Unlike grain, the grapes come into the winery with complex, flourishing populations of wild yeasts and bacteria already fermenting the juice. Pure yeast cultures have only recently become standard in wine making. Also, although the main components of both beer and wine are relatively simple chemically, the character and quality of wine depends on a kaleidoscopic array of complex organic molecules present in only trace quantities. Some of the compounds important in the odor and flavor of wine are detectible by a wine expert in concentrations of less than 1 part per billion. The 20th century advances in organic chemistry, particularly in chromatography, finally made possible scientific monitoring and control of the changes in wines while they are being made. Fortunately, the result has not been standardization and reduction of quality for mass production but the opposite.

Monocots

Agavaceae — Century-Plant Family

The Agaraceae include about 20 genera native to the tropics and subtropics in general. The Agavaceae differ from the related lily and amaryllid families in having strongly fibrous leaves, usually clustered in a rosette on a short perennial trunk. In some, the trunk is massive and tall, e.g., the joshua-tree and other *Yucca* species and in the dragon-tree and other *Dracaena* species.

Several genera yield long, hard fibers and have been taken into folk cultivation for making rope, nets, sacking, and many other uses. *Furcraea* is cultivated commercially for cordage fiber in its native Central and South America and has been introduced overseas, notably in Mauritius. Many other genera are widely planted as ornamentals, including the New Zealand flax, *Phormium*, and the African cast-iron plant, *Sansevieria*. As a crop, the important genus is *Agave*.

***AGAVE* — CENTURY-PLANTS, MAGUEY** (Bahre and Bradbury 1980; Castetter et al. 1938; Gonçalvez de Lima 1956; Hodgson et al. 1989; Lock 1969; West and Augelli 1966)

Maguey was a West Indian name picked up by early Spanish explorers. It became the *lingua franca* term on the continent for *Agave* and *Fucraea* species in general.

Several hundred species have been described, and more are still being discovered. Most are native to Mexico with some in the southern U.S. and in the West Indies and Central America. Agaves are sun-loving xerophytes of cliffs, rocky ravines, and deserts. Some species survive with less than 15-cm annual rainfall and in rainless years. After rains, they quickly develop extensive shallow root systems, which die back during drought. They store water in succulent leaves with a very thick cuticle and sharply limit transpiration losses by a crassulacean acid metabolism (CAM). CAM drastically cuts transpiration by temporal separation of gas exchange and photosynthesis; CO_2 is taken in through open stomates at night when humidity is high and stored in organic acids, which are used in photosynthesis during the daytime when stomates are closed. They slowly accumulate enough photosynthates to put forth the huge inflorescence, not after a century but usually after a decade, more or less. The inflorescence, tree-like in many species, attracts swarms of hummingbirds, bats, and insects as pollinators. The fruits attract larger birds, which presumably disperse the seeds. Agaves generally also reproduce vegetatively by basal shoots forming new sucker plants around the old one, which dies after flowering. Some species reproduce by bulbils and plantlets, which develop in place of flowers in the inflorescence.

A great many wild *Agave* species have been exploited since ancient times by various New World tribes, and several species were taken into cultivation in Mexico before Columbus. There is prehistoric evidence of cultivation in Arizona by the Hohokam.

Agaves provided *mescal*, a staple food of nearly all the hunters and gatherers in desert and semi-arid country from Texas to southern California and much of Mexico. *Mescal* was made by roasting piles of *Agave* trunks in large barbecue pits. The *mescaleros* had usually removed the inflorescence buds from plants about to bloom and left them to live and accumulate photosynthate for a year or so before harvesting the trunks. Roasting produced a sticky, molasses-flavored mass, which was very nutritious and nonperishable. For many tribes, it was the main food much of the year. Mixed with water, *mescal* was also used to make wine by fermentation with wild microbes. This was done by a few North American tribes, the Apache, Zuñi, and Pima-Papago, and by many Mexican tribes. Initially, use for food was much more important than for alcoholic beverages. In colonial Mexico, however, technology for distillation of wine to brandy was introduced from Spain and also from the Philippines. *Agave* brandy or *aguardiente* usurped the name *mescal*. Distillation of *mescal* from wild *Agaves* remains an important rural industry, usually illegal, particularly in Sonora. Legal, tax-paying distilleries producing advertised brands of *mescal* brandy developed a large domestic market during the 19th century. They are located in several warm, dry regions, the most famous being at Tequila in Jalisco. Among *mescals*, *tequila* has become a name like cognac among grape brandies. By law, the name can be applied only to brandy made mainly from *A. tequilana*. Exports have boomed since 1940. The species is planted on thousands of hectares, sometimes with maize and other crops between the rows of *Agaves*.

Pulque, a quite different beverage, is produced in a cooler region of the central plateau of Mexico, mainly from the huge *A. salmiana* and closely related species. Again, the inflorescence bud is excised from a plant about to flower, but in this case, an open wound is maintained from which sap is collected, often thrice daily, for fermentation. The sap flows in amazing quantities for months; an individual plant sometimes yields over a thousand liters before giving up. The fresh sap, or *aguamiel*, is a sweet, refreshing drink. The vast bulk of the *aguamiel* is taken to local wineries for fermentation, which takes about a week. The wine, *pulque*, is highly perishable and is not aged but consumed promptly, usually nearby.

Pulque is abundantly recorded in Aztec hieroglyphic codices. According to them, they learned to cultivate *Agaves* and make wine, called *octli* in

Nahuatl, during the 13th century. Presumably the practices were developed in very ancient times by the earlier inhabitants of the central plateau region and adopted by the late arriving Aztecs. *Octli* in the Aztec empire was tightly controlled as a drink for serious ceremonial use. It was drunk ritually by the priesthood and given to defeated warriors before they were sacrificed. It was drunk by the general populace only during great religious festivals.

After the Spanish Conquest, in spite of various efforts to control its abuse, *pulque* became the popular intoxicating beverage of the general population, especially in mining towns and cities of the central plateau. *Pulquerias* became the saloons or pubs of the working class.

Since before the Conquest, *A. salmiana* and the other *pulque Agaves* have been partly wild, partly planted in central Mexico. The distinction is often vague because plants are easily established along paths and field edges by planting without subsequent cultivation. Some *Agave* hedges have reproduced *in situ* for centuries and formed a sort of terrace by trapping soil washing down slopes. Much *pulque* is still made from wild or casually planted *Agaves*, but true plantation monoculture became increasingly important during the 19th century when many *Agave* haciendas were established. Some of these were expropriated after the 1910 Revolution, but many tens of thousands of hectares are devoted to *pulque Agave* plantations today. Much of the area is in a few states adjacent to the federal district, but significant plantations are widely scattered in states farther north. A federal bureau, the Patronato del Maguey, was established to promote modernization of *pulque* production, including pasteurization and bottling. Unlike *tequila, pulque* has not become appreciated beyond its homeland.

Wherever they grew, wild *Agaves* were traditionally important fiber plants for American Indian peoples. The fibers were extracted from the leaves in various ways: by scraping away the other tissues of fresh leaves, by retting, by heating them out of dried leaves, or by salvaging fibers from leaves baked for *mescal*. Many tribes used them for all sorts of twine and rope, including tackle to build pyramids. They were important for nets, bags, and hammocks. Fibers of various *Agaves* were also woven into textiles, not as soft as cotton but in some cases remarkably fine, comparable to linen. The tribute list of the last Aztec emperor Moctezuma shows that bundles of *Agave* cloth were exacted as annual tribute from some of the central plateau provinces climatically unsuited to produce cotton (Barlow 1949).

Two kinds of *Agave, henequen* and *sisal,* have become important commercial fiber crops. They have been named as species, *A. fourcroydes* and *A. sisalana,* respectively, although they are really clones. Both are sterile pentaploids that were developed as cultivars prehistorically in Yucatan by the Maya. (*Henequen,* like *maguey*, was a name the early Spanish explorers picked up from West Indian natives and applied to different species on the mainland.) The leaves of both contain fibers over a meter long, which make excellent twine, rope, and sacking.

Henequen became a commercial plantation crop in northwestern Yucatan in the mid-19th century when cattle haciendas began planting it for export. In the late 19th century, demand for *henequen* boomed with use as binder twine in wheat harvesting. Until 1920, Yucatan led the world in exports of *Agave* fiber, mainly *henequen*. The crop has since dwindled to insignificance due to rising foreign competition, weak markets, and conversion of the big haciendas to *ejidos* occupied by tens of thousands of small farmers. *Henequen* was peculiarly adapted to the thin, stony limestone soils of Yucatan. Although grown on a small scale in many countries, *henequen* proved much less successful away from home than *sisal*. Also, *sisal* is generally preferred because it lacks *henequen's* spiny leaf margins.

In 1836, *sisal* was introduced to southern Florida, where it became naturalized on some of the keys. Florida became the source of *sisal* introductions to Africa and South America later in the century, after Mexico embargoes export of plants to try to maintain its monopoly.

The first important introduction was to German East Africa (now Tanzania). Bulbils of *A. sislana* were sent there in 1893 by a Florida nursery. By 1898, these had multiplied sufficiently to plant 40 ha near Pangani. Plantations spread steadily, first along the coast, later inland as railway construction proceeded. By the start of World War I, the colony was exporting 20,000 tons of *sisal* a year. Expansion of plantations continued under British rule, and for decades, *sisal* was the leading commercial crop and main export of Tanganyika. East African *sisal* estates usually had 1000 to 2000 ha of *Agaves*, sufficient to keep busy one or two of the special machines that extracted the fiber. About 90% of the weight of the harvested leaves was discarded as waste, which made long transport uneconomic.

During World War II, *sisal* prices rose with increased demand and cutoff of the competing Manila hemp with Japanese occupation of the Philippines. During the early 1950s, this single clone occupied a quarter of Tanganyika's farm land and comprised $^2/_3$ of its exports by value. Production peaked at nearly 300,000 tons in the early 1960s. By then, the price of *sisal* was falling fast because of overproduction. As has happened often with tropical commercial crops, too many countries jumped on a bandwagon. By the early 1960s, Tanganyika was competing for a flat *sisal* demand with growing exports from Kenya, Mozambique, Natal, and especially Brazil.

Brazil began *sisal* planting in 1945, mainly by low income small farmers in Paraiba and adjacent areas of the dry northeast. Instead of the power machine decorticators used in East Africa, the Brazilian farmers extract the fiber by hand. Brazilian production was overtaking Tanganyika's by the mid-1960s. After Tanzanian independence, the *sisal* estates were broken up for more diversified small farming, and *sisal* declined to a quite minor role.

Sisal planting has suffered worldwide by competition from synthetic fibers. *Sisal* remains a common, cheap cordage fiber and has become a source of special, strong papers, e.g., coffee filters and tea bags.

Araceae — Aroid Family

Araceae are a family of over 100 genera, mainly tropical but with a few temperate zone members, e.g., jack-in-the-pulpit. The family is easily recognized by its peculiar inflorescence type: many tiny flowers on a spike-like spadix and a sheet-like spathe, as in the calla. Most members are rather succulent, perennial herbs; some are epiphytes with aerial roots, e.g., *Philodendron, Monstera.* A few are grown commercially as ornamentals, e.g., *Caladium, Anthurium.* Aroids have a special defense against herbivory: their tissues contain bundles of needle-like calcium oxalate crystals, which puncture mouth and throat tissues. However, humans learned to overcome this by simply cooking the plants.

COLOCASIA **AND RELATIVES — TAROS** (Bellwood 1980; Kirch 1991; Massal and Barrau 1956; Merlin 1982; Patiño 1964; Plucknett 1976; Powell 1976, 1982; Purseglove 1972; Simoons 1990; Yen 1991; Yen and Wheeler 1968)

Four species in as many genera were prehistorically domesticated as food crops, primarily for their starchy tubers, with the leaves also eaten like spinach as a potherb. All four species are normally propagated vegetatively as clones, although volunteer seedlings occasionally occur. In folk cultivation, many named clones are distinguished with varying ecology and food uses, but the differences between cultivars and their wild progenitor are rather subtle, and in each genus, all are usually lumped under a single Latin binomial. Three of these are Old World natives: *Colocasia esculenta, Alocasia macrorrhiza, Cyrtosperma chamissonis;* the other, *Xanthosoma sagittifolia,* is native to tropical America. All are sometimes loosely called taro, but the name properly belongs to *Colocasia,* being given by the Polynesians in Tahiti and spread thence by European usage. Aroid tubers provide a heavy yield of carbohydrate, but when used as a staple, they need to be combined with fish or other good protein sources.

Colocasia esculenta grows wild in freshwater marshes and along streams over a huge range from Southeast Asia through the East Indies to tropical Australia where the aborigines gathered the wild tubers until modern times.

It has been repeatedly suggested that taro may have been one of the most ancient of domesticated crops. It has been deduced from indirect evidence that taro was a more ancient domesticate than rice in such diverse places as Formosa, the Philippines, Assam, and Timor.

Techniques of flooded field taro growing have been proposed as precursors to the more elaborate rice paddy management. Unfortunately, the lack of undecomposed plant evidence in archaeological sites leaves these as hypothetical possibilities. There are, however, archaeological ditches in the highland valleys of New Guinea very similar to modern taro patterns and have been ^{14}C dated at 4500 B.C. This implies that the Australoid aborigines of Melanesia initiated crop growing independently. In historical times, taro has been one of the staple crops of New Guinea and other Melanesian islands, including the Solomons, New Caledonia, New Hebrides, and Fiji. Many different clones are grown; these vary in ecology, some being grown in newly cleared rain forest without irrigation; they also differ in uses and cooking quality.

Taro is also an essential staple of many Polynesian islands, especially the better watered ones. The archaeological record suggests that the precursors of the Polynesians came from the Asian mainland as already a Neolithic people. They were evidently skilled seafarers. Their distinctive Lapita pottery was left in coastal and offshore island sites through much of Melanesia between 1500 and 1000 B.C. They had limited interbreeding with the Melanesians, but might have acquired taro from them. By 1000 B.C., the Polynesians had begun moving eastward from Melanesia into uninhabited Pacific islands, such as Tonga and Samoa. During the 1st millennium A.D., they settled the rest of Polynesia, including New Zealand, Easter Island, and Hawaii. The pioneer voyagers presumably carried only a few taro clones, but many have been brought from Tahiti to Hawaii during the period of the repeated contacts in the 12th and 13th centuries A.D. At the time of Captain Cook's arrival, the Hawaiians had hundreds of named taro clones, and the plant was embedded in their myths and folklore. A great deal of labor and engineering skill had been put into terracing for fields flooded with slowly flowing water. Taro was eaten in many ways, much of it as poi, which is prepared by pounding the tubers to a pulp, which is briefly fermented to make a nonalcoholic, slightly acid food.

Modern Hawaii does not have nearly enough taro. With growing population and urbanization and with destruction of many of the old flooded fields for sugar plantations, taro has become an expensive luxury. Many of the clones have been lost. Efforts are underway to conserve clones in gene banks. There is still one large commercial region in Hanalei Valley on Kauai, a spectacular landscape of beautiful flooded fields below rugged mountain slopes.

Like rice, taro was introduced to Madagascar prehistorically by East Indian seafarers.

Meanwhile, about the time the Polynesians were beginning their voyages into the central Pacific, taro was spreading northward in China. It was recorded in written records in central China at about the beginning of the Christian era, and by 500 A.D., it was widely grown, especially in the south. In northern China and Japan, taro has never compared with rice as a staple. However, some excellent clones have been selected in China and Japan, including some triploid ones which have been widely introduced overseas in recent times. Since early in the 20th century, the U.S. Department of Agriculture has been introducing taro clones to the southern U.S. One of the best of these, known as dasheen, was imported from the West Indies, but is believed to have been of Chinese origin. Some taro is grown commercially in the Imperial Valley of California for the ethnic Oriental and Pacific island markets.

In Africa, taro was first recorded in Egypt about 500 B.C., having been brought from India or Arabia. It probably was repeatedly introduced to East Africa during medieval times. By unknown routes, taro was introduced across Africa to the Guinea coast before the arrival of Europeans.

Colocasia may have been taken from West Africa to tropical America repeatedly during the colonial period. The earliest available clear records are from the Guianas where it was used for provisions for slave plantations in the 18th century. Any earlier New World reports cannot be disentangled from references to the native *Xanthosoma*. *Colocasia* has become a minor crop in the humid tropics of the New World in general but never, so far as is known, in flooded fields.

Alocasia macrorrhiza, giant taro, and *Cyrtosperma chamissonis,* giant swamp taro, are natives of Southeast Asia and the East Indies, the same general region as *Colocasia*. They both have enormous leaves, often twice as tall as a person, and produce huge starchy tubers. They are easier to grow than true taro, but are generally considered inferior. They evidently spread as prehistoric crops through Melanesia and Polynesia where they are generally less important than taro. *Alocasia* is often only semi-cultivated and used as emergency food. *Cyrtosperma* is a staple on land unsuitable for true taro. In the Solomons, it is grown in coastal marshes by inhabitants of artificial islands. In Fiji, the huge bulbous tubers of giant swamp taro are conspicuous in the great Suva market, but they are offered on the sidewalk outside the covered central market where taro is sold.

Cyrtosperma was the traditional mainstay of a tiny part of the world's people: the inhabitants of the atolls of Micronesia. In many islets, ditches, or pits were dug in the coral sand so the *Cyrtosperma* could obtain moisture from the lens of freshwater above the seawater. The huge tubers developed slowly; sometimes baskets were carefully woven around each and kept filled with compost.

Xanthosoma sagittifolia, American taro, was a minor crop of the Greater Antilles before Columbus; it was reported from Hispaniola and Puerto Rico by some of the earliest Spanish chroniclers under the Arawak name of *yautia*. It was also grown by Caribs in the Lesser Antilles and by tribes on the mainland coast of the Caribbean in South and Central America.

Xanthosoma was introduced to West Africa in the mid-19th century and at about the same time was introduced by missionaries to the South Pacific. It yields about 20 tons of tubers/ha, about the same as true taro, but does not require irrigation. Both crops are now nearly ubiquitous in the humid tropics.

Arecaceae (Palmae) — Palm Family

Arecaceae are a very ancient angiosperm family, recognizable in the fossil record since the Upper Cretaceous. The 200 or so living genera are collectively extremely widespread in a great range of tropical and subtropical habitats.

Most wild palms are exploited locally for a variety of useful products including timber, thatch and cordage, sweet sap for toddy, and edible fruits and seeds. Many kinds of palms have been planted in gardens, but only a few have become plantation crops, three of the most important of which will be discussed, along with one still in folk cultivation.

PHOENIX — DATE PALMS (Aschmann 1957; Lee 1963; Mason 1915, 1923; Nixon and Carpenter 1978; Oudejans 1976; Patiño 1969; Zohary and Hopf 1988; Zohary and Spiegel-Roy 1975)

About 15 wild species of date palms are native to tropics and subtropics of the Old World from the Canary Islands through Africa and the Near East to southern Asia and the East Indies. Generally they grow in open, sunny sites with plenty of water, as along streams, swamp margins, and in desert oases. *Phoenix* palms are dioecious: male and female inflorescences are borne on separate trees. Thus, a single seed introduction is not enough to establish a new disjunct colony. However, the species did successfully populate geographically fragmented habitats, evidently by bird dispersal. Dates are typical stone fruits with the seed protected by a hard shell and surrounded by fruit pulp attractive to birds. Like most palms, the wild species have multiple folk uses: in addition to producing edible dates, the sap is tapped for making sugar and wine or toddy; the pith is a source of sago starch; the leaves are used for matting; and the trunks for timber. Various species are planted widely as ornamentals. Only one species has become cultivated as a fruit crop.

Phoenix dactylifera — Near Eastern Date Palm

The species is native to arid regions of North Africa and southwest Asia. Untended, apparently wild groves are reportedly still growing around springs and seepage areas, fresh or brackish, in desert areas of Jordan and the Iran-Iraq border region. The fruits are small and not very edible. The species probably had a much broader natural range in oases westward across the Sahara and eastward in other deserts on the Persian Gulf. However, it is no longer possible to draw a line between wild date palms and those whose heredity and geography have been modified by human selection and dispersal. Likewise, it is impossible to tell whether early archaeological finds of dates were from groves that were natural or human influenced.

The archaeological record of dates presently available is curiously scant and begins surprisingly late. No finds of dates are known from Near Eastern archaeological sites for several thousand years after domestication of cereals and legumes. A few presumably wild date pits have been found in Egypt and southwestern Asia, dated to the 6th and 5th millennia. During the 4th millennium what are possibly cultivated dates are reported from several Near Eastern sites, but considering the durability of date pits in a desert, the evidence suggests they were not yet important. During the Bronze Age from about 3000

B.C. on, date growing seems to have become widely important in the desert regions of the Near East, including Egypt. Written history of the ancient Near Eastern city states shows dates were a major crop.

The late domestication of the date palm may be explained by peculiarities of its ecology and biology. Linnaeus gave the genus the name of the legendary firebird. The tree has been characterized as growing with its head in fire and its feet in water. For pollination and fruiting, it needs rainless summers with fierce heat; it also needs constant and abundant water supply for the root systems. It would not fit into the Neolithic Near Eastern pattern of upland crops grown during the winter rains. Neither was it suited to primitive Near Eastern irrigated agriculture utilizing the seasonal flooding of the Nile and the Tigris and Euphrates. Expansion of date groves from their native oases must have had to await advanced technology and engineering that could provide controlled year around irrigation with an annual water input of about 2 or 3 m. In certain locations, instead of irrigation, palm gardens could be planted by digging holes and trenches to reach natural water tables. In any case, it is not surprising that planting was delayed until the transition from stone to metal tools had begun in the 3rd millennium. In the high civilizations of antiquity, e.g., in Babylonia, Assyria, and Egypt, written records and artistic depictions of date gardens abound.

When people began planting dates, the wild populations must have offered much genetic diversity for selection. The species presumably underwent dramatic range changes during and after the Pleistocene; in particular, the Sahara is known to have had repeated Holocene pluvial periods much less arid than the present. It has been suggested that *P. dactylifera* acquired much of its genetic diversity by hybridization with several other species in both Africa and southwest Asia. However, evolutionary change in the species due to human selection has been very slow. When an individual female palm turns out to bear superior fruit, her seedlings generally revert to the ordinary level. Even today, when seedling date palms are grown, their yield is generally low and of poor quality. Also, in seedling date groves, about half the trees are males. It was discovered in antiquity that even with wind pollination many of these could be dispensed with to make room for more bearing trees. By 2300 B.C., cuneiform tablets record hand-pollination, permitting reduction of male trees to 1 per 25 or 50 females.

The most important step in date cultivation came with the discovery of vegetative reproduction. Unlike some other *Phoenix* species, *P. dactylifera* produces branch shoots or offsets at the base of the trunk, which eventually develop their own root systems. In nature, this results in a clump of daughter palms around an old mother tree. Date planting in the ancient Near East became truly productive when gardeners learned how to use offshoots for propagation. It is not entered into lightly, like scattering seeds, because the rooted offshoot ready to plant may weigh 25 kilos or more and can be separated from its mother only by hard work with metal tools. However, such clonal propagation became standard among serious date growers in antiquity, with many superior clones being given folk names. Some remained within a single garden or single family or oasis, but often offshoots were sold and spread through a region. All the oases of the Libyan Desert, for example, grew hundreds of thousands of Saidy dates, one clone of particularly fine fruit. Such diffusion may very well have had to wait domestication of the camel, believed to have been much more recent than most Near Eastern animal domestication. The sparse archaeological record on the domesticated dromedary suggests that it was introduced to Egypt in Early Dynastic time about 3000 B.C. A close interdependence of cultivated dates, camels, and desert nomads developed. Up to modern times, the Bedouin have carried dates, usually about 150 kilos per camel, from interior oases to city markets and seaports for export. Generally the Bedouin prefer so-called dry or bread dates, self-curing on the tree in full desert climate, and solid when ripe for ease in packing and transport.

Unlike the Saidy, many prized, named clones produce soft dates, which are perishable unless properly processed. They need to be picked at the proper stage and cured by sun-drying to reduce the moisture and reach high enough sugar concentration to prevent spoilage. Curing is an art, especially in regions without extremely high temperatures and extremely low humidity. After curing, soft dates were traditionally packed tightly in a sticky mass in baskets or bags woven from the palm fronds. Such packages have long been widely traded, not only by caravan but by ship. They were one of the main cargoes carried by the fleets of Arab sailing ships in their annual voyages to the east African coast.

In humid coastal areas, such as the Nile Delta, some prized date clones, such as the Samany, are grown for crisp, half-ripe fruit eaten fresh.

P. dactylifera became important in ceremonies of Judaism, Christianity, and Islam. Dates also became traditional food for pilgrims to Mecca. The Moors introduced the species to southern Spain in medieval times. Seedling groves have continued to grow there at Elche, valued as much for fronds to sell for Palm Sunday as for the low quality fruit.

Whether from Spain or North Africa, dates were evidently taken along as ship's provisions on early Spanish voyages to the New World. Where Spanish sailors left the seeds along beaches or other open sites, volunteer groves occasionally developed. By 1550, such groves were reported in Hispaniola, Mexico, and Panama. Later during the Spanish colonial period, date groves were reported at many other places in tropical America, presumably grown from seed of Old World dates imported for consumption. Sometimes, given enough hot sunny weather, they produced edible fruit. Dates developed into a regular crop in only two regions of Spanish America: the coastal desert of Peru and the central desert of Baja California. In Peru in the mid-17th century, date groves were reported to be producing well at the oases of Trujillo in the north and Ica in the south. In 1942 near Ica, I photographed a large stand of apparently untended date palms, of all ages and sizes, strung out along a dry wash. In 1966, extensive new plantings of palms had been made in the same region, between Ica and Pisco, in large pits dug to approach the water table. Small numbers of date palms are grown in various of the irrigated valley oases of Peru.

In Baja California, date palms were planted at some of the Spanish missions in the mid-18th century. The San Ignacio mission produced about 50 kilos of dates in 1765. Production there and at neighboring oases increased exponentially until the mid-19th century when the expansion of the groves was completed. Today they are the main crop of four major oases and a dozen minor ones in the central desert. Cultivation is very casual and based entirely on seedlings rather than offshoots; male palms are left to grow. Dried dates are offered at the roadside at very low prices; they are small and dry and not very sweet.

Spanish missions in Alta, California also had seedling date palms, with occasional descendants still growing in the region, but the coastal climate is not suitable for serious date growing. Seedling date palms, usually unproductive, have also been widely grown in the Gulf States from Florida to Texas.

The modern commercial U.S. date crop did not develop from seedlings introduced via Spain and Latin America. It is based on selected clones introduced directly from North Africa and the Near East. The U.S. Department of Agriculture initiated such introductions near the end of the 19th century and by 1910 had several high quality varieties established in experimental gardens in the Salt River Valley of Arizona and the Coachella Valley of California. The Coachella Valley already had palm oases, with the native *Washingtonia* growing around many springs. These experimental plants impressed prospective private date growers, who imported large numbers of offshoots from Algeria, Egypt, and Iraq between 1911 and 1922. Groves were started at many places in the Arizona desert, but proved commercially inviable because of frequent summer rains. The Coachella Valley proved hot and dry enough for dates to be highly productive. By the late 1950s, the valley was harvesting over 20,000 tons a year from 2000 ha, more than was being imported. By 1970, following nationalization of French-owned date groves in Algeria, California became a major date exporter. Like much other California agriculture, date growing has become increasingly mechanized, and small private holdings have been replaced by agri-business. Instead of climbing the trees with ladders for pollination, leaf pruning, thinning, and picking, workers are hoisted by hydraulic cherry-pickers, and instead of hand-pollination, great special pollen-blowing machines are used. Exotic irrigation water is now imported by canals. The groves have to compete for water and land with expensive and expanding desert resort and residential developments. As they are displaced, some of the grand old palms are trucked elsewhere for landscaping, often into coastal climates where they will no longer provide a fruit crop. Enough groves remain in the Coachella Valley to draw tourists to an annual date festival with Arabian dances and camel racing.

Genetically the California commercial date crop is based simply on clones transported from the Old World. The leading commercial variety, the Deglet Noor, is essentially unchanged from the original stock collected at the Algerian oasis of Biskra. Other less common premium quality varieties are also old clones with Arabic names. Some new varieties originating as accidental seedlings in California have been named and are being propagated, but their value is uncertain. Deliberate breeding of dates has scarcely begun and is not expected to give quick results. One prospect recently suggested is selection of male palms for superior pollination performance. This would involve not only quantity of fruit set but also its quality. The date fruit, like the maize grain, exhibits xenia, i.e., its character is affected by genetics of the pollen parent. Theoretically, certain male palms might have a favorable effect on fruit ripening and quality and could be propagated clonally, like the females are, to replace the unselected males now used.

In Australia, a few date palms have been planted since the late 19th century at many places in tropical and subtropical Queensland. Most of these are in areas with summer rain, and only a few are in the more suitable desert interior. Perhaps water is too valuable there for successful commercial production of a crop with such high irrigation requirements.

COCOS **— COCONUT** (Bruman 1944; Buckley and Harries 1984; Burkill 1935; Harries 1978; Patiño 1969; Purseglove 1976; Sauer 1965, 1967, 1983; Scott, R. 1961)

Cocos is now considered to include only a single living species; other species formerly included were reassigned to other only distantly related genera. Wild and cultivated coconuts do not form discrete, genetically distinct populations, and both are customarily given the same species name *C. nucifera*.

The coconut evidently evolved in the Indo-Pacific Ocean region where Tertiary and Quaternary fossils have been found. This remains the region of greatest genetic diversity of the species and its parasites, including many totally dependent insect species. The coconut, unlike the date, reproduces only by seed and is monoecious, so a single seed suffices for establishing a new colony. The palm is tightly and superbly adapted to a very specific habitat, the outer edge of tropical beach thickets at the upper limit of wave reach, precisely the place where drifting nuts are left stranded. On low atoll islets subject to washover, the palms occupy a wider belt. The species has no mechanism for natural dispersal beyond reach of the waves. Coconuts are known to be capable of remaining viable while floating in the sea for over 6 months, no maximum limit being established. Harries found that coconuts are highly polymorphic, particularly in the Southeast Asia-western Pacific region, with extreme varieties producing small-seeded, very thick-husked fruits and large-seeded, thin-husked fruits. All are well adapted for drift dispersal, but the thicker-husked forms may float longer and germinate more slowly. The larger-seeded coconuts presumably have an advantage in competitive, local establishment because of larger food supply and quicker germination. This polymorphism may have evolved with the drastic Pleistocene sea-level changes between glacial and interglacial periods. Falling sea levels must have exposed coral reefs and wave-built beaches to subaerial erosion. Conversely, during deglaciation when the sea level rose so fast that coral growth could not keep up, coral sand supplies for beach building would have been shut off, and many atolls became unvegetated submarine banks until the sea level stabilized long enough for reefs to reach the surface and beaches to be built from reefal debris. For *C. nucifera,* this would have meant repeated episodes of local extinction, recolonization, and increasing competition with vacillating natural selection promoting genetic diversity. During Holocene time, the range of sea-dispersed wild coconut palms probably spanned the tropical Indian and Pacific Oceans from East Africa to the Pacific coast of Panama and Costa Rica. Populations were established only where rainfall was sufficient to maintain a freshwater lens floating above the sea water under the beaches. The palms can survive occasional flooding with sea water, but need plenty of fresh water for normal growth and reproduction.

Wild coconuts must have been much exploited by prehistoric coastal peoples. The endosperm, liquid in a green nut and solid in a ripe one, provides both a refreshing drink and a rich foodstuff. Casual planting probably was begun independently in many places. Even the Australian Aborigines, archetypal hunters and gatherers, occasionally planted coconuts rather than simply consuming them; on Cape York peninsula, stranded nuts were sometimes moved up the beach above the tidemark, and the trees were then regarded as property of the planters and their descendants (Buckley and Harries 1984).

In Southeast Asia and the East Indies, coconut palms became standard components of the ancient dooryard garden complex. In addition to the fresh drink from the green nuts, coconut inflorescences were tapped for toddy, which was fermented for alcoholic *tuba*. The endosperm was prepared in many ways, not usually eaten raw, but cooked and incorporated in many complex recipes. By long, intimate acquaintance, peoples of Southeast Asia and the Indo-Pacific islands learned to classify and utilize the nuts according to precise stages of development. Later this practice spread with colonization of the South Pacific islands by agricultural peoples. Presence of wild coconuts undoubtedly greatly aided this colonization, particularly on waterless atolls. The Polynesians almost certainly were responsible for expanding the range of the species to Hawaii in the 12th century. The Hawaiian Islands get drift seeds from the North rather than the South Pacific.

It is unclear whether prehistoric planting as a dooryard garden species had any appreciable effect on coconut evolution. Harries suggested that larger-seeded, thinner-husked varieties would have been favored for planting and developed into a cultivar, but this is speculative. Conceivably such nuts would have been preferentially consumed, and rejects left on the ground to germinate.

At least in some cases, Pacific islanders are reported to have taken germinating coconuts for planting unselectively. Also, the long generation time probably prevented much evolution under folk cultivation.

C. nucifera was absent from the Atlantic-Caribbean region until introduced by the Portuguese. Vasco da Gama's expedition brought coconuts back from the Indian Ocean in 1498, and during the 16th century, the Portuguese initiated coconut planting on the West African coast, the Cabo Verde archipelago, and Brazil. The earliest recorded introduction to the West Indies was from the Cape Verde Islands to Puerto Rico in 1582. By then, *C. nucifera* had evidently been introduced from both Panama and the Philippines to Acapulco and the Pacific coast of Mexico. By the 1670s, coconut plantations on the coast of Colima were producing wine with supposedly marvelous qualities, most likely by technology brought from the Philippines. The early planting in western Mexico was an isolated case at the time.

In the West Indies and on the mainland of tropical America, coconut planting spread slowly and was of slight economic importance through the 18th century. In contrast to their rapid adoption of sugar cane and various other introduced crops, the American Indians were curiously uninterested in *Cocos*. There is no record of pre-European exploitation, let alone planting, of the wild coconuts on the Pacific coast of Central America. The earliest record of Indian interest in coconuts was in the late 17th century among the Cuna on the opposite side of the Panama isthmus from the wild groves. This was after the species had been introduced there by the Spaniards. Coconuts were planted in small numbers soon after 1600 in many Spanish settlements on the coasts of Central and South America, the original sources being unspecified. Likewise a few coconuts from unspecified sources were planted in the 17th century in some of the Antilles, notably Cuba, Jamaica, and Guadeloupe. At the end of the 18th century, Captain Bligh brought a few Tahitian coconuts to Saint Vincent and Jamaica. In San Andres and Providencia, where coconuts are now almost a monoculture, planting was not recorded until 1820.

Large-scale planting of coconuts for commercial export of oil and dried copra began abruptly in many regions during the mid-19th century. Some of these were in places such as Ceylon and Malaysia where small-scale garden planting of coconuts was an ancient practice.

Outside the region of traditional folk planting, commercial coconut plantations were begun about the same time by European colonials who had been operating slave plantations of other crops, such as sugar cane and cotton. For example, after the British Empire abolished slavery in 1835, plantation owners in such far-flung colonies as the Cayman Islands and the Seychelles planted coconuts as a low labor crop. In the Seychelles, the commercial plantations, which remain the islands' main resource today, were established with seed from the wild palms fringing the seashore. This native variety bears relatively thick-husked and small-seeded nuts, but attempts to introduce better varieties have repeatedly failed. In 1905, for example, 15,000 Ceylonese coconuts were started, but most soon died, partly due to susceptibility to native beetles, and a single unhealthy individual survived after 50 years.

Wild *C. nucifera* was taken into cultivation independently during the 19th century on various other Indian Ocean islands, including the Agalega Islands, the Chagos Archipelago, and Cocos-Keeling Atoll. All these were uninhabited until British and French colonists arrived to exploit the wild seashore coconuts, which were soon used to establish plantations.

Since the mid-19th century, several hundred million coconut palms have been grown in plantations, large and small. The great majority were planted in the East Indies, including the Philippines, and other Indo-Pacific regions where the species grew wild. The planted palms now vastly outnumber the wild ones. Although most of the production has continued to be consumed locally, copra became the main export of Ceylon, the Philippines, and many other Indian and Pacific Ocean islands. Coconut endosperm contains a high proportion of unsaturated fatty acids, and the oil is slow to become rancid. The nut meat is converted to copra by simply drying in the sun or in ovens heated by burning the husks. It could be kept for a long time awaiting arrival of a schooner and could be carried without spoiling to distant markets. The oil, extracted in the consuming country, is in demand for cooking, margarine, high quality soaps, and for various industrial applications. Premium coconut oil and grated endosperm are used in confectionery. The oil is a solid fat at ordinary air temperatures, but melts in the mouth at body temperature. For about a hundred years, until overtaken by soybeans about 1950 and later by *Elaeis*, coconuts were the leading commercial source of vegetable fats and oils. *C. nucifera* is planted on millions of hectares throughout the world's humid tropics. Most of the healthy trees have their peak nut production when they are between 30 and 60 years old, but continue to be fruitful until they are over 100. However, many plantations have been wiped out prematurely by disease, especially in the Caribbean region where they presumably had a narrow genetic base. The old varieties have commonly been replaced with so-called Malay dwarf varieties. Scientific breeding programs have been initiated, but they are too recent to have any general effect on the producing populations.

***ELAEIS* — OIL PALM** (Burkill 1935; Dalziel 1948; Hardon 1976; Hartley 1977; Patiño 1969)

Elaeis includes two wild species — *E. oleifera* of Central and South America and *E. guineensis* of West Africa. Both are ecological pioneers of streambanks and swamps in tropical rain forests and adjacent seasonal forests. Like the coconut and various other palms, *Elaeis* endows its seeds with a large oily endosperm providing abundant energy for germination of the embryo. However, it was not the oily seeds but a more unusual characteristic of *Elaeis* that initiated human exploitation, namely its oily fruits. The mature ovary is differentiated into an attractive red skin, a thick, soft, and palatable mesocarp, which contains about 50% oil, and a hard inner shell enclosing the seed. The oblong fruits average 3 to 4 cm in length. The basic adaptation, attracting birds or other animals yet protecting the seed, is of course shared with the date palm and many unrelated plants. However, instead of sugars, *Elaeis* offers the metabolically more costly oils. Evidently this paid off for both species by bird dispersal through geographically broad but very fragmented habitats. Like *Cocos, Elaeis* is monoecious, so dispersal of a single seed is enough to start a new colony.

Aboriginal use of *E. oleifera* in the American tropics was very minor, and the species has never become a plantation crop. It is occasionally planted experimentally and is considered a possible source of useful genes for future breeding of *E. guineensis*.

West African farming peoples entered into a symbiotic relationship with *E. guineensis* in prehistoric times. The palms were not usually planted, but came up in forest clearings made for yams, taro, and other staple crops. When the clearings were abandoned, some of the palms survived in the forest regrowth until the next cycle of shifting cultivation. Eventually oil palms overtopped the forest canopy and continued to produce harvestable fruit in extensive semi-wild groves. Groves planted both accidentally and deliberately also developed in dooryard gardens. These became important resources for the Guinea coast and Congo villages that had created them. The male inflorescences were tapped for toddy or palm wine. The female inflorescences matured into huge compact bunches of fruits, edible fresh or more commonly used for cooking oil. West Africans traditionally extracted the oil from the fruit pulp and discarded the hard-shelled kernels, although minor use was made of the oily seeds by cracking the kernels. Folk taxonomy recognizes in addition to the normal hard-shelled form a thin-shelled form, which could be cracked by the teeth so the seed could be eaten. This so-called *tenera* form became of interest to modern breeders; it is heterozygous for a mutant gene that produces a seedless, abortive fruit if homozygous.

During the medieval period of Arab trading and slaving, human seed dispersal probably initiated isolated groves of *Elaeis* on the coasts of East Africa and Madagascar.

Early European activities along the Guinea coast concentrated on slaves, ivory, and gold, not palm oil. However, *E. guineensis* fruits were presumably commonly loaded as provisions on slave ships. The seeds were easily carried, and casual unrecorded overseas introductions probably took place repeatedly. By 1700, some of the palms had been planted in Jamaica, and the fruits were being used by Africans. The species evidently was introduced to Brazil with the early slave trade; it is traditionally important in African creole cuisine in Brazil, especially in Bahia. The species had been introduced to Martinique before 1763 when famous French botanist Jacquin named it *E. guineensis*. By the mid-19th century, it had been introduced to a few other places in the West Indies and to European colonies in the Indian Ocean, including Mauritius,

Reunion, and Java, but only as an ornamental or botanical garden specimen, not as a crop.

Exploitation of palm oil as a commercial export crop started early in the 19th century. When the British began suppressing the slave trade, West African chiefdoms gradually began supplying palm oil instead of captives to European trading ships. Exports were mainly from the vast Niger delta and the adjacent Bights of Benin and Biafra. Initially the traders did not go ashore because of fearful mortality from diseases, but lay offshore in river mouths to load oil brought by native boats. When British colonization began in the late 19th century, the region was known as Oil Rivers. The pulp oil found a ready market in Europe and later in North American for margarine, soap, candles, and industrial metal plating. By the start of World War I, about 100,000 tons of pulp oil and 250,000 tons of palm kernels were being exported annually. Kernel exports exceeded pulp oil because most of the latter was consumed locally. Nigeria was by far the leading exporter, but significant quantities were shipped from other British, French, Belgian, and Portuguese West African colonies.

Commercial planting of the species finally began in 1911, not in its homeland but in the Dutch East Indies. The species passed through an extremely narrow genetic bottleneck in its passage from Africa to Malaysia. According to orthodox theory, this loss of genetic diversity should have made the progeny slow to evolve and vulnerable to disease, but so far the vast Asian plantations are thriving. There have been attempts to broaden the genetic base by fresh introductions but without notable success in displacing the initial stock. This consisted of four seedlings planted in 1848 in the botanical gardens at Buitenzorg, now Bogor, Java. All may have the same parent tree growing in a botanical garden in Mauritius. Seeds from the Buitenzorg trees were widely distributed in the East Indies in the late 19th century, not as a crop but for ornamental plantings. Avenue palms on tobacco estates in the Deli region of Sumatra caught the attention of a Belgian who had seen native exploitation of the species in the Congo. He and a German established the first large scale oil palm plantations of their estates in the Deli region in 1911. By 1912, Deli seed was used for plantations in the Selangor district of British Malaya. Within 20 years, over 50,000 ha of oil palms had been planted in Sumatra and Malaya. Fortuitously, since it was originally disseminated as an ornamental, the Deli form is more productive than the average native West African groves. However, it is not outside the range of variation in those groves and is of the normal hard-shelled wild type. There have, of course, been many efforts by planters and experiment stations in Southeast Asia to improve the Deli variety by simple selection and to hybridize it with fresh germ plasm brought from Africa, but it has remained essentially unchanged as the basis for commercial plantations in the East Indies and Malaya. In both regions, the story was complicated by Japanese occupation and later political upheavals. However, with some interruptions, oil palm plantations have continued to expand. In Sumatra in the 1970s, large injections of capital to rehabilitate and expand the plantations were made by the World Bank and Asian Development Bank. Oil palms have been planted on large areas that had been degraded to *Imperata* grass savanna by shifting cultivation. In Malaya after World War II, many former rubber plantations were converted to oil palms. Expansion into new areas has accelerated since 1960. On the Malay peninsula, oil palm plantations have expanded from about 55,000 ha in 1960 to an estimated 600,000 ha in 1980. On the island of Borneo, the new Malaysian state of Sabah became an important oil palm planter after 1960 with about 70,000 ha planted in 15 years. By the mid-1970s, Malaysia, including Sabah, was shipping about 1 billion tons of palm oil per year, including palm kernels, well over half of the world's exports. More recently *Elaeis* planting has begun in New Britain.

Meanwhile, West Africa has continued to rely mainly on palm oil from volunteer groves and has only slowly and tentatively developed plantations. The earliest, abortive attempt at planting by European colonists was by Germans in the Camerouns just before they lost that colony in World War I. The first European colonials to make significant oil palm plantations were Belgians in the Congo in the 1920s. Planting was initiated by colonists who had seen the beginnings of oil palm plantations in Sumatra and was originally on private estates. The colonial government soon became involved and established what was to become a great oil palm experiment station. Attention was focused on the rare thin-shelled *tenera* form as exceptionally productive of oil. Its hybrid genotype was elucidated and methods worked out to produce *tenera* seed for planting. The program included upgrading village groves and furthering planting by villagers. Modern mills were built for crushing the fruits and seeds, and local manufacture of margarine and soap was begun. By 1958, the Belgian Congo had about 150,000 ha of European-owned oil palm plantations and 100,000 ha owned by natives. In 1960, with independence and the subsequent civil war, expansion of plantations was suspended, but Zaire remained the leading West African exporter of palm oil until 1970 when expanding population and internal consumption ended export.

Very little oil palm planting was done in British and French colonies in West Africa, which included the vast majority of wild and volunteer village groves. The colonial authorities generally disapproved concessions for plantations by Europeans, out of respect for village land rights, even in lightly populated regions. Late in the colonial period, which ended about 1960, some start was made on oil palm research and fostering small-scale village planting. Since independence, increasing population has changed the former palm oil surplus to a deficit in most of the countries. In several of them, during the 1960s and 1970s, government research stations were established, and selected seed was provided to foster plantations. The most successful program was evidently in the Ivory Coast; by the early 1970s, there were about 40,000 ha of large plantations and 25,000 ha of village plantations. By then, Dahomey and Camerouns each had about 20,000 ha of oil palm plantations.

In the New World tropics, commercial planting of *E. guineensis* developed in the 1930s, largely as an attempt by banana plantations to diversify following catastrophic disease problems. Much of the early planting material came via the United Fruit Company's experimental gardens at Lancetilla, Honduras, but there were other independent introductions to South and Central America. Although seed from diverse geographic sources has been experimented with, plantations have commonly relied mainly on the hard-shelled Deli form. By the early 1960s, 10,000 ha had been planted in Honduras and Costa Rica. By 1972, a United Fruit oil palm plantation at Golfito, Costa Rica, was producing oil for margarine that was marketed widely in Central America.

Although *E. guineensis* did not become a regularly cultivated crop anywhere in the world until after 1900, it soon overtook *Cocos nucifera* in oil production and has become the world's most rapidly expanding tropical plantation crop. In spite of serious modern efforts at breeding, the vast modern plantation populations are scarcely changed genetically from the stock used for the first plantations in the Dutch East Indies and the Belgian Congo. The plantations are the progeny of only a few individual wild trees. In other worlds, the crop is only an incipient domesticate.

BACTRIS (INCL. *GUILIELMA*) — PEACH PALM AND RELATIVES (Clement 1988; Johannessen 1966; Patiño 1963; Prance 1984; Seibert 1950; U.S. National Research Council 1975)

Bactris is a genus of over 200 species widely distributed through the humid tropics of the New World. Some species grow as understory palms in rain forests; others occupy riparian habitats and margins of brackish and freshwater swamps. The trees are monoecious, insect-pollinated, and generally bird dispersed. The fruits have fleshy outer tissues and a hard inner shell protecting the seed. The trunks are usually densely covered with spines that deter climbing fruit predators.

Bactris gasipaes (=*Guilielma speciosa, G. utilis*) — Pejibaye, Peach Palm

The pejibaye is a cultigen not known in truly natural habitats. Three closely related wild species, with smaller fruits, are native to riparian forests of the Amazon headwaters in Brazil, Peru, and Bolivia (*B. microcarpa, B. insignis, B. ciliata*). The cultigen had a huge range at the time of the Spanish Conquest in dooryard gardens (and occasionally in large groves) belonging to a host of Indian tribes in northern South America and southern Central America. The pejibaye has at least ten etymologically unrelated names in different Indian language families, suggesting multiple domestications rather than spread as a crop. The Central American name pejibaye was spread during the Spanish Conquest as a *lingua franca* name.

Spaniards first encountered the pejibaye near the northern border of its range on the Caribbean coasts of Costa Rica and Panama. A possible reference to the species there was a report by Columbus of Indian weapons made of black palm wood and of Indian wine made from a spiny palm. Unmistakable accounts begin in the 1540s when this region was contested by rival Spanish expeditions based in Panama and Nicaragua. Each force cut down thousands of the pejibaye palms belonging to the local Indians, partly to build palisades with the spiny trunks, partly to feed on the delicious apical meristems, the so-called palmito or hearts of palm. The latter depredation was committed by Indian allies of the Nicaraguan party. Unfamiliar with pejibaye cultivation, they presumably thought them wild and free for the taking. This so outraged the owners of the palm groves that they overwhelmed and massacred the intruders. An inquiry into this debacle elicited the testimony that the local Indians valued their pejibaye trees above everything else but their women and children.

There is abundant evidence that the pejibaye was extremely important in traditional South American Indian economies from Colombia to the Guianas and in the upper and middle Amazon basin but not the lower Amazon. The main product was the prolific fruit crop, individual trees sometimes producing two crops a year of over 100 kilos in each harvest. Fruits are usually bright orange or red. They vary in size from about 2 to 6 cm in diameter. The flesh is not sweet but oily in the primitive races, starchy in the more highly selected ones. The fruit is also a fair source of protein and some vitamins and minerals. The ripe fruits can be left on the tree for some weeks without spoilage. The fact that most Indians did not select mutant forms with spineless trunks may be due to preferring trees that stored fruit safe from climbing predators. When the palms grew too tall to harvest the fruit from the ground with poles, the Indians used ladders and ingenious scaling devices.

Pejibaye fruits are delicious, resembling chestnuts when roasted or boiled. During the months of the harvest, many Indian tribes traditionally ate little else and were said by Spaniards to be especially sleek and sharp. For some Indian groups, the pejibaye harvest was the basis of their calendar reckoning and the focus of festivals and rituals. Alcoholic drink was widely made from the fruit, the microbiology of the fermentation process being obscure. At least in some cases, the fruit was masticated along with maize meal before fermentation. Some tribes also had ways of preserving the fruit by fermenting in pits and by drying. After about 75 years, trees are about 15-m tall, hard to harvest, and declining in productivity. They then may be felled, providing the delicacy of the apical bud and valuable timber. The green wood is relatively easy to work, but hardens into something resembling steel, extremely strong and flexible, and a beautiful black color, which was used by the Indians to make blow guns, lances, spears, and musical instruments.

Pejibaye palms, like date palms, are propagated by both seeds and vegetatively by basal offshoots. Seeds are slow to germinate, but once started, a seedling begins to bear fruit quickly, often within 5 years. Offshoots are even more precocious and were often preferred by Indian farmers, while black and white farmers are more likely to use seedlings.

Although Spaniards and other outsiders generally found cooked pejibaye fruits good eating, they seldom took over planting of the palms, and the crop has declined in much of its former area with the breakup of Indian cultures. Costa Rica is the outstanding exception. Although its original Indian custodians are almost extinct, the pejibaye remains a favorite food of the country. Pejibaye palms are common in dooryard gardens of humid lowland valleys on both the Caribbean and Pacific slopes. The cooked fruits are sold at railroad stations and on city streets and are universally appreciated as solid, savory food. Planting of pejibayes in Costa Rica for sale as a cash crop began in a small way early in the 20th century. Seed was obtained from old Indian groves and from markets, with casual selection for quality. Planting expanded after World War II, and there are several dozen commercial plantings now, mostly no more than a few hectares in size. The largest commercial plantings in Costa Rica, totaling about 2,000 ha, are not for the fruit but for the hearts of palm, which is canned for export. In the world market for this delicacy, the pejibaye has to compete with various other palms, including wild species extensively exploited in Brazil.

There has been no great geographic spread of pejibaye planting in modern times. Post-Columbian expansion may be responsible for planting in Nicaragua, Trinidad, and eastern Brazil. Agricultural experts have recognized the potential value of the crop for the Old World tropics, but so far only experimental plantings are known.

Bromeliaceae — Bromeliad Family

Bromeliaceae are a family with about 60 genera in the New World tropics and subtropics. Most bromeliads are epiphytes, but there are terrestrial species in deserts and arid habitats, including *Ananas*. Bromeliads have strong xerophytic adaptations remarkably convergent with those that have evolved in unrelated families such as the cacti. Among these are succulent, water-storage tissues that make the plants resistant to desiccation, spiny defenses against herbivores, and crassulacean acid metabolism (CAM), discussed under *Agave*. Most bromeliads, including *Ananas*, are tank plants; rain and dew are funneled to reservoirs at the leaf bases, supplementing the supply of both water and mineral nutrients from the root system.

Some bromeliads, including *Ananas*, have brilliant flower colors attractive to hummingbirds. Many genera are grown as ornamentals, often as house plants. A few are planted as spiny hedges in the tropics. The only species that has been domesticated as a crop is the pineapple.

ANANAS — PINEAPPLE (Patiño 1963; Pickersgill 1976; Purseglove 1972)

A. comosus is a cultigen, genetically and geographically distinct from the wild *Ananas* species, all of which are native to eastern South America. The most likely wild progenitors are believed to be *A. bracteatus* and *A. fritz-muelleri*; both have ranges straddling the Tropic of Capricorn in the Paraná-Paraguay basin and adjacent Brazil. Both probably contributed genes to the cultigen; they are interfertile with it and each other. Both wild species have smaller fruits than the cultigen, and their fruits are full of thousands of hard seeds. Unlike the fruits of other wild *Ananas* species, which are dry, the fruits of *A. bracteatus* and *A. fritz-muelleri* are edible when ripe and attractive to animal dispersers. The fleshy pineapple fruit is a unique structure formed by the fusion of many individual flowers and inflorescence branches into a single mass with the main stem of the plant at the core. The true fruits, i.e., the ripened ovaries, are lost in this mass. The plants are not killed by removal of the ripe fruit; new plants develop from basal offshoots; also, the leafy crown from the top of the fruit can take root and grow into another plant.

A. comosus evolved to essentially its modern noble character under prehistorical Indian selection for fruit size and quality. Early in the domestication process, mutants were evidently selected for self-incompatibility and parthenogenetic fruit development; unlike its wild relatives, the cultigen sets seed only after cross-pollination by a genetically different clone and develops seedless fruit without such pollination. Seedless pineapples were obviously a great improvement in the Indians' view, particularly since vegetative propagation gives a crop quicker than seed. Even for dispersal of the crop to new areas, vegetative propagation was probably used. The offshoots and fruit crown remain alive for weeks or months until being rooted.

Unlike some other clonally propagated cultigens, *A. comosus* did not become sexually sterile. Clones have to be grown in isolation, or they intercross. I have been surprised by good looking but seedy pineapples in the West Indies.

From its original homeland, prehistoric Indian cultivation evidently carried the pineapple northward through the Amazon and Orinoco basins to the Caribbean and thence westward through valleys of the northern Andes to Central America and southern Mexico. The species, and even single clones, proved amazingly tolerant of variation in rainfall. Although *A. comosus* is a true xerophyte and can be grown under less than 50 cm of rainfall a year, with good drainage the crop can be grown with 10 times that much rain. Except for drainage, it also proved remarkably indifferent to soil variables.

The pineapple entered written history in 1493, when Columbus encountered it on Guadeloupe in the West Indies in Carib cultivation. Columbus met it again in Panama where he found the Indians growing it in large quantities and making wine with it. The pineapple was noted by various early Spanish explorers as cultivated by the Indians, for example by Grijalva on the Bay of Campeche and by Orellana in central Amazonia. The Spaniards did not bother with the local Indian names for it; it immediately became *piña*, the fruit resembling a pine cone. A Brazilian Indian name, *nanas*, was adopted by the Portuguese and went with the plant to the East Indies and Asia.

Spaniards and other Europeans were not slow to appreciate the pineapple. They raved about its flavor, odor, and beauty, sometimes at length. During the 16th and 17th centuries, *A. comosus* was widely introduced overseas, initially by Portuguese and Dutch ships sailing around Africa to the East. By 1700, it was established in St. Helena, on the Guinea coast, at the Cape of Good Hope, in Madagascar, Reunion, Mauritius, Java, in India, Nepal, Burma, Thailand, Malaya, Formosa, China, and the Philippines. Often it was rapidly taken over as a crop by the native farmers. Once in a while, a fruiting pineapple plant grown in a tub had been shipped to Europe as a royal gift. Shortly after 1700, greenhouse cultivation of pineapples was begun in England and on the continent as an extreme luxury. The fruit became so familiar to the elite that when Joseph Banks arrived in Brazil on Captain Cook's first voyage and reported on tropical fruits there, he treated it as a well-known fruit, unlike bananas, which he described at length as a completely strange fruit.

Several distinct clones are grown commercially in various parts of the world today, with the greatest diversity in the West Indies and tropical America. By far the most widespread and important clone is Cayenne. It was introduced from French Guiana to France about 1820. It spread rapidly in greenhouse cultivation in Holland, Belgium, England, and the Azores. After 1850, Cayenne was introduced from European greenhouses to field cultivation in many tropical and subtropical regions, including Florida, Jamaica, South Africa, Ceylon, and Australia. When Hawaii began pineapple planting in the decade 1885 to 1895, the Cayenne clone was imported from a dozen different countries. Since 1920, Hawaii has sent planting material of Cayenne to countries all around the world.

Starting early in the 20th century and continuing for over 50 years, the Cayenne clone was the basis of huge Hawaiian plantations and canneries. The uniformity and predictability of a vegetatively propagated crop were advantageous in large scale operations. The monoculture was also greatly helped by a lucky peculiarity of the pineapple plant, namely its unique response to hormone spraying. It is the only crop whose flower initiation and consequent fruit development can be triggered by spraying with ordinary auxins, an obvious advantage in scheduling harvesting and canning. On the high islands, particularly Oahu and Maui, pineapple plantations were relegated to areas unsuitable for sugar cane. The low island of Lanai, in a rainshadow behind Maui's mountains, was a rather unproductive livestock ranch until bought by Dole for a great pineapple plantation. By mid-century, the plantations had become highly mechanized and capital intensive and owned by a few companies, one of which could plant 10,000 ha and ship 400,000 tons of pineapple a year. By about 1970, however, it was evident that Hawaiian pineapple planting was anachronistic because of enormous inflation of land and labor costs. About half of the former plantation area has already been converted to residential and other uses, while the companies have started new commercial plantations in the Philippines, Taiwan, and Thailand.

It seems odd that great agribusinesses, with multimillion dollar advertising budgets, are dependent on a crop essentially unchanged genetically from prehistoric Indian cultivars. The Cayenne clone has produced a few slightly different subclones, but they have not replaced the original. Modern breeding has been attempted with *Ananas* by hybridization and selection of seedling progeny but so far without practical results.

Musaceae — Banana Family

The Musaceae include six genera of mostly large or giant herbs native to humid tropics and subtropics of all continents and various islands. The species are generally pioneers of sunny, disturbed sites capable of vegetative spread as perennial clones. Many species have brilliant flowers attractive to birds and produce berries attractive to other birds, which disperse the hard seeds.

Strelitzia, Heliconia, and *Ravenala* are commonly cultivated as ornamentals. *Ensete* is an important staple crop in the highlands of Ethiopia; soft tissues of the banana-like trunk and other parts are sources of starch for bread and cooked as a vegetable; the crop is propagated vegetatively and is not clearly different genetically from its wild progenitor. The most economically important genus is *Musa*.

MUSA — BANANA, PLANTAIN, MANILA HEMP (Patiño 1969; Purseglove 1972; Simmonds 1966, 1976c; Stover and Simmonds 1987; Wilson, C. 1947)

Musa includes about 40 species of treelike perennial herbs native to naturally open habitats in Southeast Asia, the East Indies, the tropical Pacific, and tropical Australia. The wild species are diploids dispersed by seed and spread *in situ* as clones. One of these, *M. textilis*, is Manila hemp or *abacá*. *Abacá* has been cultivated commercially in the Philippines and Borneo for the long fiber in the leaf sheath used for cordage. The species has not evolved notably under human selection.

Edible banana cultivars, by contrast, are strikingly different from their wild progenitors. Two wild diploid species, *M. maclayi* and *M. acuminata*, were independently domesticated in antiquity. In both cases, mutations were selected that convert fruits full of hard, stony seeds and little edible pulp to parthenocarpic seedless fruits full of edible pulp. The cultigens were easily propagated as clones by sucker shoots at the base of a mature, fruiting plant.

M. maclayi, a wild species of New Guinea, the Solomons, and Queensland, is believed to be the progenitor of the cultivated *fe'i* bananas of the Pacific islands. They are diploids with 2n=20 chromosomes. Like sugar cane and taro, *fe'i* clones were taken eastward into the Pacific by prehistoric Melanesians and Polynesians. *Fe'i* bananas have declined in importance in modern times. Clones have persisted in the rugged interior of Tahiti after cultivation was abandoned and are still harvested by banana hunters. The starchy fruit is cooked.

M. acuminata, native to the Malay Peninsula and adjacent regions, is believed to be the primary wild progenitor of all the other kinds of cultivated bananas. It is a diploid with 2n=22 chromosomes. As in the *fe'i*, prehistoric human selection produced seedless cultivars on the diploid level (AA genomes). These have fine, sweet fruit, but are less productive than their triploid derivatives. They are still widely grown as a minor garden crop, but are important only in New Guinea.

Triploid mutants (AAA genomes), quicker growing and with larger fruit, probably arose repeatedly in the ancient banana gardens in the Malaysian rain forest regions.

A complex array of other edible bananas originated by hybridization between *M. acuminata* and *M. balbisiana,* a wild diploid (2n=22) species (BB genomes) native to seasonally wet and dry monsoon regions over a huge area from India to the tropical Pacific. Hybridization probably took place repeatedly as cultivation of edible AA-type bananas spread into territory of the wild BB type. Primitive AA cultivars, although female sterile and seedless, produce good pollen. Some of the resulting hybrids are diploid (AB), but most are triploids (AAB or ABB). All are highly sterile and seedless. The B genome contributed broader climatic tolerance. However, all bananas are intolerant of dry or waterlogged soil. Ideal sites are well drained with very frequent rain. If there is less than 5 cm of rain in any month, the plants are severely stressed unless irrigated.

Cultivars with a single *M. balbisiana* genome (both AB and AAB) are generally sweet; they include the so-called ladyfinger and apple or *manzana* types. Clones with ABB genomes are starchy, cooking bananas widely called plantains. The names plantain and banana, however, are inconsistently applied in different regions and are ambiguous unless defined.

The Linnaean names *M. sapientum* and *M. paradisiaca,* often used in older literature, are both based on AAB-type clones; neither can be validly applied to important commercial clones, with are AAA. Only the AAB-type clones followed the *fe'i* banana into the Pacific islands in prehistoric times.

All three triploid types (AAA, AAB, and ABB) were introduced to Africa prehistorically. Some presumably came via Madagascar where bananas arrived, along with rice, with the Malgache immigrants from Sumatra; others probably came from India. In any case, bananas were important crops in Africa from Zanzibar and the highlands of East Africa across the continent to the Guinea coast before arrival of the Europeans.

Classical Greek writers knew about bananas by hearsay; Alexander the Great's army met them in India. Bananas may have spread westward into

Mesopotamia before the rise of Islam. During the 10th century A.D., they spread with Islam through Egypt and into Spain, especially in Granada and Málaga.

Bananas were introduced to the Canary Islands early in the 15th century, perhaps both by the Portuguese from West Africa and from Spain.

The first known cultivations in the New World were in Hispaniola and Puerto Rico by some of the first Spanish settlers. Bananas were abundant in Hispaniola by 1520, reportedly brought from both Granada and the Canaries. Back in the rainy tropics, they flourished and immediately became a much more important food than they had been in Spain. During the early 16th century, one of the first things Spanish colonists did when founding new settlements was to make substantial banana plantings. Before 1530, bananas were a significant food crop on both the Caribbean and Pacific sides of Panama. By 1550, they were being grown on both sides of Mexico, in Costa Rica, Ecuador, and again under heavy irrigation in Peru. By then, the Portuguese had begun growing bananas on the tropical coast of Brazil.

Initially linked to Spanish and Portuguese colonies, bananas soon spread inland among escaped African slaves and among Indian tribes. The first Spanish explorers of the Amazon basin found no bananas among the Indians in the mid-16th century, but Amazon Indians had the crop soon after 1600.

It is evident from 16th century accounts that the early Spanish banana introductions to the New World involved several clones, including both AAA- and AAB-types. A Spanish expedition evidently brought additional clones, probably AAB, from Tahiti to Lima in 1775.

Starting in the late 18th century and increasing during the 19th century, British and French expeditions collected various banana clones and dispersed them through botanical gardens to their Caribbean colonies, whence some of them were dispersed to Latin America. Two closely related groups of AAA-type clones, the Cavendish and the Gros Michel, eventually became virtually the entire base of commercial banana planting. Cavendish bananas are a group of clones differing by a few mutations that affect stature and pigmentation of the plants. The group was grown only in Southeast Asia, south China, and the East Indies until the 19th century when repeated introductions overseas began. Probably the first was by the Philbert expedition, which brought plants from Indochina to France in 1820, and very probably left plants of the clone in Cayenne and the Canary Islands en route. In 1826, Telfair introduced the dwarf Cavendish from southern China to Mauritius, whence it was sent to England for greenhouse cultivation. Cavendish was the family name of the Duke of Devonshire, in whose greenhouse the clone flowered in 1836. Suckers from England were taken to Samoa by a missionary, and the clone was taken from there to Tonga, Fiji, Tahiti, and Hawaii by 1855. The dwarf Cavendish went on through the tropics and subtropics to become by far the most widespread of all banana clones. It became the basis of nearly all subtropical commercial plantations, typically on family farms, not only in the Canaries but in Israel, South Africa, Australia, and Brazil. In the New World tropics, the Cavendish group had to wait a long time before taking over dominance of commercial plantations from the Gros Michel.

To consumers, Cavendish and Gros Michel are about the same, but for growers and shippers, there are important differences. Initially, commercial planters in tropical America chose the Gros Michel mainly because it was easier to ship. When green, it could take rougher handling and could be shipped as whole bunches, i.e., the whole inflorescence as a unit, without wrapping or cutting off hands for packaging. For a long time, it was the whole source of export bananas from the New World tropics.

The Gros Michel was originally grown, along with other clones, in gardens of Burma, Thailand, Malaya, Indonesia, and Ceylon. It is believed to have been introduced to the West Indies about 1825 through the botanical garden of Saint Pierre, Martinique, which had many Asiatic plants. It was introduced to Jamaica about 1835 and by 1875 was widespread in the Caribbean region. Gros Michel was taken from the Americas to the Pacific islands: from Jamaica to Fiji in 1891, from Nicaragua to Hawaii in 1903.

All by itself, the Gros Michel clone was the biological basis for the development of the great commercial banana plantations of tropical America, which developed in the late 19th century for export to North America. A casual export trade in bananas had begun in the middle of the century, with schooners bound for Atlantic and Gulf ports picking up deck cargoes of bananas from peasant growers in various parts of the Caribbean. A crucial change in the trade began in 1874 when Minor C. Keith, a young New York entrepreneur, began acquiring extensive landholdings for banana plantations in the Caribbean lowlands of Costa Rica. Keith was building the first railroad in Central America outside Panama and saw banana exports as a source of revenue for his shoestring finances. He relied heavily on West Indian, particularly Jamaican, labor force, since the plantations were established by clearing rain forest in a region with few native or Spanish inhabitants. Plantations were to profoundly affect the ethnic patterns of other Central American countries, with enclaves of English-speaking black Protestants. While Keith was expanding banana planting elsewhere on the Caribbean mainland, Boston entrepreneurs had begun export production in Jamaica, Cuba, and Hispaniola. Except in Jamaica, the Greater Antilles plantations were eventually converted to sugar cane but not before the Boston Fruit Company pioneered in developing fast banana transport by specially built steamships and by its own railroad car fleet distributing in the U.S. In 1899, Keith's interests and the Boston Fruit Company were merged into the United Fruit Company. This combined huge potential for production in vast tropical forest regions with capacity for rapid overseas transport and distribution. Soon after 1900, the United Fruit Company began refrigerating its ships and developing a radio network, the beginning of the Great White Fleet, the first regularly scheduled steamship service in Central America, and Tropical Radio, the first radio network. Its plantations were expanded into other Central American countries, first on the Caribbean, and after opening of the Panama Canal in 1914, on the Pacific side. United Fruit, commonly known simply as La Compañia, built ports and towns complete with schools, hospitals, and utilities. Being so powerful and so foreign, it was not generally beloved. Its dominion of the so-called banana republics is now past history, however. United Fruit was gradually overtaken by competitors, especially Standard Fruit, a subsidiary of Castle & Cook (Dole), and Del Monte, who were more flexible in adapting to changing conditions. One of the crucial changes was the arrival of a parasitic fungus to which the Gros Michel was susceptible.

Panama disease is caused by soil fungus, *Fusarium oxysporum* believed to be native to the East Indies where it is not a serious problem because many clones of bananas are grown in scattered locations. The fungus was first identified in Australian banana plantations in 1876 and in Panama in 1893, presumably introduced with imported banana plants. By 1903, it was present in Jamaica, and it has spread slowly but inexorably through the American tropics. On susceptible clones, there is no known control, and once present in a plantation, the fungus persists indefinitely in the soil. The Cavendish clones are not susceptible to Panama disease, but rather than switch from the Gros Michel, the United Fruit Company for a long time moved the banana plantations to virgin land and attempted to grow other crops, such as African oil palm, on the old plantations, or simply abandoned the land. This was largely responsible for eventual major geographic changes in their plantations, including abandonment of whole early Caribbean plantations and the shift to the Pacific coast.

The Standard Fruit Company, based initially on the Caribbean coast of Honduras, was quicker to shift from Gros Michel to Cavendish bananas, partly because they had more limited options. Export of the Cavendish fruit demanded innovations in shipping, notably cutting the hands off the stem and packing them in cardboard boxes. In the long run, this was an advantage in permitting mechanization of shipping and also in speeding distribution, reducing cost of handling in the importing country where labor was more expensive than on the plantation. Also Cavendish clones, including the Valery, commonly proved more productive than Gros Michel.

Along with shifts in the kinds grown, there came changes in size and concentration of plantations. The banana companies have learned to reduce their vulnerability to hurricanes, labor strikes, taxation, political pressure, and other problems by contracting with small growers. Since 1950, Ecuador has become the world's leading banana exporter, mainly produce of numerous independent growers contracted to shippers.

Meanwhile, in Jamaica equally drastic changes took place. Until World War II, the island was a United Fruit stronghold. When exports resumed after the war, the old pattern was broken. Instead of Gros Michel, a Cavendish clone, locally called Lacatan, was the main crop. Produced by independent growers organized into cooperatives with automatic insurance against hurricane losses, all fruit was shipped to Britain. In the 1950s, a similar pattern was established in the Britain Windward Islands.

Since the 1930s, the French West Indies have exported considerable quantities of bananas to France, and there are other trades, e.g., from Brazil to Britain. Altogether, the New World exports make up about 70% of the international trade. Currently, the fastest growing trade is from the Philippines to Japan, developed largely by Castle & Cook (Dole).

All this discussion of commercial production should not obscure the fact that about 85% of the world's bananas are grown by small farmers and consumed locally. In the tropics in general, starchy, cooking bananas are a staple food. They are boiled, baked, fried in oil, and incorporated in complex recipes. Probably about half of the world's total banana crop is grown and consumed in the humid tropics of Africa.

Orchidaceae — Orchid Family

The Orchidaceae are the largest of all plant families with about 700 genera and 20,000 species native to all tropical and temperate regions. The family specializes in exploiting scattered, rare habitats, which a species colonizes by means of producing astronomical quantities of almost microscopic seeds, borne by the wind. A single seed pod may contain over a million seeds. The family also specializes in peculiar and complex adaptations to specific insect pollinators involving spectacular flowers, commonly long-lasting as an adaptation to rarity of pollinator visits. In tropical forests, orchids are characteristically epiphytic, and cultivation of the popular ornamental orchids is generally not in ordinary soil but in microhabitats resembling a tree branch. The only genus cultivated for utilitarian purposes is *Vanilla*.

VANILLA — VANILLA (Bruman 1948; Dequaire 1966; Purseglove 1972)

Vanilla is a genus of over 100 species native to both eastern and western hemispheres including remote Pacific islands. It includes perennial vining herbs, both terrestrial and epiphytic, with adventitious roots clasping tree trunks. The fragrant, attractive flowers last only a day and are structurally incapable of self-pollination without assistance from insects or other animals.

V. pompona is occasionally cultivated in the West Indies and *V. tahitensis* in Tahiti and Hawaii, but a single species *V. planifolia* is the source of nearly all commercial vanilla.

Vanilla planifolia (=*V. fragrans*)

V. planifolia is a terrestrial vine native to seasonally wet and dry tropical forests from Vera Cruz and Colima, Mexico south through Central America and all across northern South America. It requires light shade, a tree trunk to climb, and good humus-rich, well-drained soil. The vines are xerophytic, i.e., shallow-rooted, succulent, and resistant to desiccation. Flowering and fruiting are seasonal.

There is no evidence of aboriginal cultivation of the species or even of interest in gathering the wild pods except at the northern fringes of its range where it had minor use in the Aztec Empire. At the time of the Spanish Conquest, vanilla was sometimes used by the Aztec nobility to flavor chocolate beverages. However, unlike cacao, vanilla was not recorded in tribute lists of the Aztec Empire; cacao was evidently used in far greater quantities and in many different ways. Presumably, small quantities of wild vanilla pods were traded up to the Mexican plateau from the Pacific and Gulf Coast lowlands.

Vanilla flavoring is a curious product because it is not naturally produced by the wild plant. The flavor, vanillin, is a simple organic molecule ($C_8H_8O_3$) produced by fermentation after an immature pod is harvested and subjected to several months of complicated manipulation during curing. Conceivably, the technique was only discovered once.

In any case, the Spanish conquerors of Mexico learned about both chocolate and vanilla from the Aztecs and soon introduced them to Europe. In the 1580s, consumption of hot chocolate as a beverage, flavored with vanilla and sugar, became a craze in colonial Mexico and Guatemala and exports to Europe began before 1600. Chocolate rapidly became popular throughout Europe during the 17th century and demand for vanilla greatly exceeded the supply. In 1676, the English pirate William Dampier reported that vanilla exports to Spain were coming mainly from the Bay of Campeche and the coast south of Vera Cruz. He found wild vanilla in the forests of Bocas del Toro, northwestern Panama, apparently unexploited. The product was valuable enough, so Dampier tried to harvest and cure the pods but could not find out how to do it. He regretted not being able to go to Bocas del Toro yearly in the dry season to cure and load a ship with vanilla pods. Soon after 1700, wild vanilla was being exported from the forests of Oaxaca, the Mosquito Coast, Colombia, and Venezuela.

About 1750, the first record of planting vanilla was among the Totonac Indians of Vera Cruz, at the margin of the natural range. Previously, the Totonacs had been gathering wild vanilla in the hills and selling the pods to buyers in the town of Papantla, who did the curing. By about 1760, the Indians had begun plantations. Propagation is simple; plants are multiplied clonally from pieces of the vine with its adventitious roots. Rooted cuttings start to bear in ±3 years. The contribution of the Totonacs was integrating vanilla planting with shifting cultivation of maize. As a milpa was abandoned, vanilla was planted in the young forest regrowth and flourished for about 10 years before the shade became too dense. By about 1800, nearly all the vanilla consumed in Europe came from Mexico and over 80% of it from the Totonac region.

Meanwhile, *V. planifolia* had been introduced in greenhouse cultivation in Europe, probably repeatedly. The overseas transfer was easily made because cuttings survive desiccation for months. However, lack of pollinators prevented pod development until Belgian botanist Charles Morren published a method of artificial hand pollination in 1838. The technique was introduced to Vera Cruz about 1840 by French colonists and was soon adopted by the Totonac with a consequent fivefold increase in the fruit set and harvest. However, artificial pollination was a mixed blessing. It allowed establishment of vanilla plantations overseas, outside the range of natural insect pollinators. Although Mexico has continued to produce premium quality vanilla, most of the world crop has been produced elsewhere since the mid-19th century. By 1800, *V. planifolia* was available from European greenhouses for introduction to tropical colonies. In 1819, the Dutch introduced it to the Buitenzorg Botanical Gardens in Java. Plantations were established in Java in the 1840s for Dutch markets.

Far more important was early 19th century introduction to the French Indian Ocean colony of Reunion, whence cultivation spread via Mauritius to the Seychelles and to Madagascar and the Comoros. The western Indian Ocean now leads the world in vanilla cultivation. Vanilla plantations in all these islands are individual family enterprises and are extremely labor intensive. The vines are grown in artificial woodlands that provide light shade and litter for mulch. Mostly *Casuarina equisetifolia*, native to Australia, and *Gliricidia sepium*, native to Central America, are used. Both have rough bark and symbiotic nitrogen-fixing bacteria associated with their roots. Trees are carefully pruned, and the vanilla vines are carefully trained. Flowers are hand-pollinated daily during early mornings by women and children. Pod maturation takes months and curing additional months, with much skillful tending. Each of the Seychelles producers that I visited cured her harvest on the clean, safe, flat roof of her house. The total annual crop of pods from most islands is measured in tens of tons; Madagascar, the world's leading producer, has a crop of some hundreds of tons of pods. However, the value to the local economies is important. Fortunately, the natural vanilla still commands a good price. Although vanillin is chemically identical, whether obtained from the orchid or from coal tar or waste from paper mills, the natural extract evidently contains other plant products that enhance its odor and flavor.

Poaceae (Gramineae) — Grass Family

This family was placed early in classical taxonomic systems because of a now abandoned assumption that the simple floral structures represented a primitive evolutionary stage. Grasses are now recognized as being a late lineage in monocot phylogeny. Much of their evolutionary diversification presumably took place in the Tertiary as grasslands, which were simultaneously the cause and result of frequent natural fires, spread at the expense of forests.

Within the family, the bamboo tribe of arboreal grasses, sometimes insect pollinated, is considered to be relatively primitive. Bamboos are extremely useful to mankind, particularly in the Far East, for manufacture of innumerable artifacts. They also produce large quantities of edible grain but only at intervals of many years, so the grain is taken as a bonus (or a curse, if it brings forth an irruption of rats) and is not relied upon as a regular crop. Although some bamboos are planted sporadically, reproduction of the populations is not generally dependent on human aid.

The various tribes of herbaceous grasses are of far greater importance. Harlan (1982) wrote that mankind is sustained more by the grasses than by plants of any other family. About $^{2}/_{3}$ of the calories and half of the protein in human nutrition come from cereal grains. Also, grasses are paramount in feeding the livestock that supply most of our meat and dairy products.

By definition, cereal grains are the caryopses of grasses, i.e., single-seeded fruits in which the ovary wall and the seed coat are fused around the embryo and endosperm. People and other animals may eat the caryopses whole, but the main nutrition comes from the endosperm. Endosperm is a triploid tissue, inheriting two chromosome sets from the mother plant and one from the pollen parent. In that respect, grass seeds are like angiosperm seeds in general. The special value of the grasses as seed crops depends on how quickly, abundantly, and easily the crop can be grown. This is largely due to special ecological adaptations inherited from the wild ancestors. The progenitors of our staple cereal crops all evolved as ecological pioneers in naturally open habitats. They were preadapted to grow in the full sun and mineral soil of tilled fields. Although the great majority of wild grass species are perennials, as a rule the ancestors of domesticated cereals are annuals, which complete their life cycles in a single growing season. They allocate a remarkably large share of their photosynthate to seed production instead of vegetative growth, a crucial legacy for the staple grain crops.

Noncereal grass crops, e.g., sugar cane and pasture grasses, are also descended from pioneer species of streambanks or other naturally open habitats, but they are perennials capable of vigorous vegetative reproduction. The pasture grasses inherit adaptations for survival under grazing and trampling by hoofed animals that probably coevolved with large ungulates in the Tertiary.

ORYZA **— RICE** (Bender 1975; Burkill 1935; Chang 1976, 1989; Harlan 1975; Morinaga 1960; Patiño 1969; Rutger and Brandon 1981; Simoons 1990; Vishnu-Mittre 1977; Watson 1983)

The genus of domesticated rice includes about 20 wild species, natives of tropical and subtropical regions all around the world. The species are not well adapted for long-range dispersal, and the migrational histories behind these huge disjunctions are obscure.

Wild *Oryza* species generally grow as emergent aquatics in seasonally or permanently flooded sites. Like other marsh and swamp plants, they require special mechanisms to supply air to root systems growing in anaerobic soil. The leaf epidermis of *Oryza* is nonwettable, so a continuous film of air extends from emergent to submerged foliage. As oxygen is used by the root systems, air from leaves flows down through special ducts in the plant body while the CO_2 produced by root respiration is given off in solution to the surrounding water (Raskin and Kende 1985). For plants capable of solving the gas exchange problem, growing with their foliage in air and their roots in water permits enormous productivity.

The grain of all the wild *Oryza* species is excellent food; it was almost certainly gathered since very ancient times. Wild rice is still gathered by Australian aborigines and native peoples of other continents. Like wild grasses in general, wild species of *Oryza* have shattering inflorescences, i.e., when the grain is ripe, it is not retained on the inflorescence but scattered. Mutations for nonshattering inflorescences occasionally occur, but are eliminated by natural selection. This situation would probably not change by human exploitation of natural stands of wild grain. Nonshattering mutants would be more likely to be taken away by human harvesters and eaten, while the normal shattering forms escaped harvest and reproduced *in situ*. If people were not satisfied with the yield of natural stands and began planting seed elsewhere, the nonshattering forms would be favored and spread. The switch from shattering and self-sowing to nonshattering and human dispersal is a key step in evolution of cultigens, not only in rice but in grain crops in general. It involves morphological differences detectable in archaeological material. Human dispersal and planting of rice, or any crop, implies some sort of tillage, partly to remove competing vegetation. In undisturbed natural vegetation, species distribution and abundance are not as a rule limited directly by the physical environment. Rather species are confined to a fraction of the physically suitable area, the fraction where they grow better than their competitors. Ancient people must have been fully aware of this and practiced weeding and tillage accordingly. Once tillage and planting were begun, another plant characteristic was exposed to new selection pressure. Wild plant seed remains dormant in the soil until dormancy is broken by an environmental trigger; some seed remains dormant for many years. With human harvesting and planting, nondormant mutants are strongly selected because they are most likely to contribute to the next crop.

In *Oryza*, evolution of nonshattering, nondormant grain took place prehistorically in two regions: West Africa and Southeast Asia. The West African domesticated rice *O. glaberrima*, was of only local importance and is of mainly antiquarian interest.

Oryza sativa — Asiatic Rice

It is customary to divide domesticated Asiatic rice and its close wild relatives into three species: *O. sativa*, dependent on human harvesting and planting, *O. nivara*, a wild annual species, and *O. rufipogon*, a wild perennial.

O. nivara grows in seasonally wet and dry savanna marshes, and *O. rufipogon* grows in more deeply and longer flooded swamps and river deltas. *O. sativa* crosses readily with both wild species when brought into contact, and by simple textbook definitions, the three would be considered a single species. They are in effect fairly discrete gene pools because hybrids are at a disadvantage in both natural habitats and rice fields. (The situation is somewhat comparable to that of the wolf and dog.)

Most of the multitude of landraces of *O. sativa* are evidently descended primarily from *O. nivara*. The most important and evidently the most ancient of these are the paddy or wet rice races. The so-called dry rices or hill rices have various characteristics suggesting derivation from paddy rice. They can be grown without irrigation in a sufficiently rainy climate. These are very widespread in tropical shifting cultivation; they are relatively low yielding. The so-called floating rices, grown where water levels fluctuate widely, have more *O. rufipogon* genes.

O. nivara and *O. rufipogon* have enormous, largely overlapping ranges extending from the Indus Valley eastward across all of Southeast Asia to southern China and southward to Ceylon, the East Indies, including the Philippines and New Guinea, and tropical Australia. This huge region was the homeland not only of rice but also of other ancient tropical crop plants and domestic animals, including water buffalo, intimately linked to rice cultivation for thousands of years. Unfortunately in this region, the available archaeological record of dated and identified crop remains is exceedingly poor. Impressions of rice glumes have been reported on pottery from Thailand dated at 3500 B.C. and from India dated before 2500 B.C., but there is no reason to believe this documents cultivation. There is evidence from New Guinea of swamp agriculture beginning at about 7000 B.C., but it is not known whether rice was among the crops planted. For all that is known, rice may have been taken into cultivation repeatedly at widely separated times and places in Southeast Asia and the East Indies.

The earliest coherent archaeological evidence on early rice cultivation is from outside the range of its wild ancestors. This was in eastern China north of the Yangtze River. Domesticated rice has been found in various Neolithic sites in this region dated from about 5000 B.C. until after 3000 B.C. Previously, Neolithic cultures had developed in semi-arid regions of central China, perhaps with *Setaria* and *Panicum* millets as their earliest grain crops. As these cultures expanded into warmer and wetter regions, they probably acquired rice as a crop from neighbors to the south.

After more than 2000 years of cultivation in temperate China, *O. sativa* was introduced to Japan about 300 B.C. The artificial aquatic ecosystem of the paddies helped the expansion out of the tropics, but even so, long evolution under human and natural selection was probably required.

At the other end of the range of the wild progenitor, in northwestern India and modern Pakistan, archaeological evidence suggests that rice was cultivated only after the arrival of a complex of other crops introduced from the west. These included wheat and barley, known to have been domesticated thousands of years earlier in the Near East, and grain sorghum and other crops native to Africa. The highly advanced Harappan civilization of the Indus Valley evidently had seafaring commerce with both the Near East and Africa. A couple of centuries after wheat and other western grains, rice appeared in the Indus archaeological record. By about 2000 B.C., it was very widespread across northern India. Conceivably it was domesticated there, but more likely was obtained from neighbors to the east where archaeological records are lacking.

Rice was known to the Greeks and Romans in classical times, but as a trade commodity from the Near East rather than a crop. By the 2nd century B.C., both Chinese and Greek sources reported rice cultivation in Persia and Mesopotamia. After 700 A.D., irrigated rice was widely grown in early Islamic agriculture, e.g., in Mesopotamia, in Egyptian oases and the Nile Valley, and in Anatolia. From the 10th century on, rice (*eruz* in Arabic) was repeatedly mentioned in Moorish Spain. Irrigated *O. sativa*, along with its old associate the water buffalo, has persisted in many parts of Mediterranean Europe up to the present. The earliest cultivation in the Pontine and other Italian marshes evidently dates only from the 15th century. Dry rice planting began only in the late 1700s in Valencia.

An amazing prehistoric introduction of rice was evidently accomplished by East Indian voyagers who crossed the whole Indian Ocean to colonize Madagascar. In the 1st millenium A.D., their descendants became the Malagasy people, racially and ethnically related to Sumatra rather than Africa. Their irrigated rice paddies remain a characteristic part of the landscape.

Post-Columbian introductions have established *O. sativa* as a significant crop in many places in the New World, as will be noted below. It is exceptional among major grain crops, however, in having its present production so heavily concentrated within its ancient geographic range. Asia still produces and consumes over 90% of the world's rice crop.

In addition to its palatability, digestibility, and versatility in cookery, rice is an excellent source of carbohydrates for calories and a good source, if not overly milled and polished, of protein, fat, and vitamins. It is also probably a very ancient source of alcoholic beverages. As in the rest of the world, the process depends on yeasts of the genus *Saccharomyces*, which ferment sugar anaerobically to produce ethyl alcohol and CO_2. Yeasts cannot use starch, however, and brewing with rice, as with all cereal grains, requires preliminary digestion of the starch to sugar. Like other cereals, rice can be malted, i.e., the living seeds can be germinated with the developing embryo digesting the starch in the endosperm. In Asia, however, instead of malting, traditional brewing uses rice flour or cooked rice, which is inoculated with cultures of fungi that digest the starch to sugar. In Japan, pure cultures of *Aspergillus oryzae* are used for modern saki brewing. In the rest of the region, from China through tropical Asia and the East Indies, various species of *Mucor* are mainly used, although in folk brewing various other fungi may be mixed in. A great variety of other ingredients are added in folk brewing, including drug and spice plants. As in the case of hops in western beer, these probably help suppress unwanted bacteria as well as giving flavor.

For thousands of years, the positive feedback between expanding rice production and expanding human population has been continuous and is accelerating. The population of Asia includes most of the world's people and has more than doubled in the last 40 years. The main reason the billions of people in Asia are not starving is the phenomenal productivity they have achieved in wet rice cultivation. Other crops, including introductions from the New World, have helped, of course, as noted in other sections, but rice remains basic. Asia was blessed with great mountains and heavy monsoon rains that supply the meters of water required to grow each crop of wet rice. But it was work and skill that created artificial marshes with controlled water flows and managed nutrient cycles. Organic manuring and nitrogen fixing bacteria are crucial parts of the artificial aquatic ecosystem. In some regions, water ferns with symbiotic cyanobacteria are maintained in the rice fields. Commonly ducks and fish, which helped control mosquitoes, were harvested along with the rice. Where the climate allowed, two or three crops are grown each year. Tight control of the artificial marshes and their water budgets may have been developed earlier in temperate East Asia than in rice's tropical homelands. In tropical Southeast Asia and the East Indies, until the late 19th century, human population and wet rice acreage were only a small fraction of what they are today. Modern engineering projects to control water flow in the great floodplains, deltas, and coastal swamplands have greatly expanded wet rice cultivation. During the present century, every generation has added millions of hectares of rice fields and tens of millions of humans.

A significant part of the increasing harvest of *O. sativa* in Asia during the present century is due to scientific breeding. A key element in such breeding has been the bringing together and hybridizing of formerly geographically separated races. The breeders have been fortunate in having a heritage of thousands of local varieties or landraces, the product of long evolution under selection by farmers and by environmental variables. These largely self-pollinated and inbred races provided the raw material for rapid yield increases by genetic recombination in new hybrids. Japan was the first country to embark on rice breeding by technical experts. Following the Meiji Restoration of 1868, the central government established agricultural research stations in different regions of Japan. These stations collected thousands of samples of farmer-bred landraces, mostly domestic but a few from abroad. Following comparative trials and classification, selection of pure lines and controlled hybridization were begun. By 1925, most of the paddy rice planted in Japan consisted of the new varieties, and the takeoff in yield was well underway. By the 1930s, Japanese rice yields per hectare were approximately double their late 19th century yields and triple the overall Asian average. This was accomplished without changing the traditional pattern of labor-intensive small farms. In fact, the average size of Japanese rice farms was declining to less than 1 ha. These farms did adopt one western-style innovation, the use of synthetic nitrogen fertilizer. Japanese rice breeders were aware early on that they needed to develop new varieties that did not grow too tall and lodge, i.e., fall over when heavily fertilized. This was the beginning of the so-called Green Revolution, which came later to the rest of Asia and the West.

On the subtropical island of Taiwan, scientific rice breeding began in the 1930s and involved hybridization between tall native varieties and semi-dwarf varieties from Japan and China. In tropical Asia, some early hybridization experiments were conducted in the Dutch East Indies but without notable success.

The major breakthrough in new hybrids for tropical Asia came after establishment of the International Rice Research Institute (IRRI) in the Philippines in 1962. Desirable traits from a semi-dwarf Taiwanese variety and a tall Indonesian variety were combined in a new high-yielding hybrid, the famous IR8. In 1983, about 40% of the rice acreage of tropical Asia, including the East Indies, was planted to IR8 and subsequently developed hybrids. Their high yields permitted the grain supply to more or less keep up for two decades with human population increase. However, similar yield increases are not predicted for the future. The new rice varieties have done well only on the better lands where water levels are tightly controlled and heavy nitrogen fertilization is profitable. The IRRI and collaborating agencies are, of course, attempting to breed varieties for more marginal lands, but results are expected to be slower and more modest than those in the 1960s and are not expected to keep up with projected increases in population. Thailand still had a surplus of rice for export during the 1980s, but tropical Asia as a whole is again a rice importer with the U.S. a leading source.

Rice from Spain regularly went along as provisions for the early explorers and colonists of the Americas, and some was promptly planted there, e.g., in Hispaniola under Columbus and Puerto Rico under Ponce de Leon. It is unlikely that these first introductions took hold. By the early 1600s, rice cultivation for local use was reported in Cuba, Jamaica, and the non-Spanish colony of Saint Christopher in the Lesser Antilles.

On the American mainland, rice was being grown under irrigation in the Cauca Valley of Colombia by about 1575. By about 1600, Panama was exporting rice to Peru every year. Rice was being planted soon after 1600 in Peru, Venezuela, and Brazil. Early tropical American planting was usually in coastal lowlands and worked by blacks. In the 1700s, Jesuit missions thrust rice cultivation on the native Indians in the Orinoco and Amazon basins. Meanwhile, *O. sativa* had been independently introduced in the late 1600s to the British colonies in southeastern North America, perhaps initially from West Africa. In 1699, a ship from Madagascar brought to Carolina a variety with larger grain. The earliest North American records refer to upland rice planting, but irrigated rice planting began on a small scale in the early 1700s, perhaps using skills of slaves from Madagascar and Africa.

Before the Revolution, new varieties of upland rice were introduced to the Carolinas from Cochin China and China. Jefferson had upland rice brought from Egypt, and there have been innumerable subsequent introductions from Asia and Africa.

In tropical America, *O. sativa* has also been enriched by innumerable introductions subsequent to the original ones from Spain. Presumably the Manila galleon brought some uneaten rice to Mexico. The Dutch introduced rice varieties from the Carolinas to the Guianas about 1700 as a slave plantation crop. With the end of slavery in the 19th century, Dutch, French, and English sugar planters in the Guianas, Trinidad, and elsewhere brought in indentured labor from the East Indies, Indochina, India, and other rice growing Asian countries. These new populations brought a huge increase in the demand for rice. They probably also brought additional rice varieties and improved techniques of cultivation, including irrigation. Since the mid-19th century, rice has become a common, cheap daily food of Latin Americans in general, not just those with an Old World heritage. Upland rice became very widely grown by shifting cultivation in slash-burn clearings. Irrigated rice has also expanded during the 20th century, especially in the Guianas where, since independence, small farmers have taken it over from the plantation system. Heavy capital investment has recently been attracted to several irrigated rice schemes in various Latin American countries, including Brazil, where Japanese colonists brought intensive modern methods to rice production in the Amazon basin. Brazil is not self-sufficient in rice and imports large quantities mainly from Uruguay and Argentina, which have extensive, well-watered alluvial plains.

Meanwhile, in the U.S. after the Revolution, commercial planting expanded greatly in coastal marshes of the Carolinas and Georgia. After emancipation, rice planting moved westward to the lower Mississippi Valley and Gulf coastal plains, which are still the main U.S. rice region. A secondary U.S. rice center developed in the Central Valley of California since 1912 with seed from Japan and China. Cultivation techniques are, of course, quite unlike those in Asia, relying heavily on expensive machinery and petroleum rather than human labor and water buffalo. In California, most of the operations, including seeding and application of agricultural chemicals, are done with airplanes, and the man hours required to produce a ton of rice are about 1% of those in Asia. However, yields per unit of area and time are roughly equal, which testifies to the robustness of this aquatic grass. U.S. rice growing also contrasts with Asia in that much of the crop is exported. Amounts fluctuate wildly with price and water supplies, but average over 5 million tons a year.

In the U.S. breeding of new hybrid rice varieties began in the 1920s and involved crosses between Japanese, Chinese, Indian, and Philippine varieties. Some of the U.S. bred varieties have become important in Australia and Latin America. Current breeding involves much use of IRRI germ plasm.

ZIZANIA — NORTHERN WILDRICE (Aiken et al. 1988; Hayes et al. 1989)

Zizania is a genus related to *Oryza* with one species *Z. latifolia* native to eastern Asia and two to eastern North America. One of the latter, *Z. texana*, is a perennial marsh grass endemic to Texas. The other, *Z. aquatica* (incl. Z.

palustris), is an annual with a wide native range; it is currently an incipient domesticate as a grain crop.

The native range of *Z. aquatica*, sensu lato, extends from Idaho and Manitoba to the maritime provinces of Canada and New England and southward to Texas and Florida. Its main habitats are shallow lakes and slow-moving streams with silty or muddy bottoms. It is commonly associated with cattails and water lilies. Along the seaboard, it sometimes grows in brackish marshes by river mouths. Suitable habitats are highly disjunct, and wildrice, like most aquatic plants, is dependent on bird dispersal for its wide range. Wildrice is a regular food of waterfowl and various other birds. It is also attractive to moose and muskrat.

As in all cereal grains, the wildrice in natural stands has shattering inflorescences. Seed is not buoyant, but sinks to the bottom and remains dormant over winter or over several winters. Upon germination, ribbon-like leaves float to the surface in spring, followed in summer by the emergent stalks, which usually reach about 1 m, sometimes as much as 3 m in height.

Indian peoples began harvesting *Zizania* prehistorically, especially in the northern glaciated terrain where it is most abundant. Traditionally, two people harvested the ripening grain from a canoe, one propelling the canoe while the other used sticks to bend the inflorescence over and gently dislodge the ripe grain. Most of the grain escaped harvest, even if a stand was reworked several times, and Indian exploitation probably had no more bad effects than waterfowl feeding; neither did it result in domestication. By 1900, wildrice bought from Indian gatherers had entered commerce as an expensive delicacy. Ingenious harvesting machines were invented, mounted on boats or pontoons. In both Canada and the U.S., exploitation was generally limited by government regulation, and much of it remained in Indian hands. There have, of course, been many attempts to expand the stands. During the 1930s, for example, the Indian branch of the U.S. Civilian Conservation Corps carried out extensive seeding attempts. As a rule, attempts to manage and expand natural stands had little effect, one way or the other. The species already occupied all suitable habitat, and artificial planting was a waste of seed.

In the 1950s in northeastern Minnesota, a different approach was begun, namely planting of wildrice in artificial paddies separate from the natural stands. The planting stock was still of the wild, shattering type, and the stands regenerated annually from the seed that escaped harvest.

Zizania grain at maturity has a remarkably high water content, usually over 40% by weight, and remains viable for only a few days if exposed to air drying. Grain harvested for consumption is usually promptly killed in the early stages of processing by drying and parching. Saving grain for artificial planting requires storage in cold water. This peculiar adaptation, the need for *Zizania* grain to be kept constantly in cold water to remain viable, has been a real obstacle to domestication, which has not yet been overcome. Even when the crop is grown in an artificial paddy, farmers have commonly continued to rely on natural reproduction from shattered seed stored over winter in the mud. A crucial process in the domestication of other grain crops was therefore blocked:

automatic selection of mutations for nonshattering was not begun by planting harvested grain.

Finally during the 1960s, deliberate selection of mutations that reduced shattering was begun in Minnesota, first by private entrepreneurs and then by a Univcrsity of Minnesota breeding program. These cultivars were rapidly adopted for planting artificial wildrice paddies in northeastern Minnesota. They allowed machines to harvest in one operation three times the grain obtained by repeated harvests of the shattering forms. However, the farmers usually continued to rely on reproduction by the grain that escaped harvest.

In the 1980s, *Zizania* cultivation on a much larger scale was begun far outside the native range in California's Sacramento Valley. This is a region where paddy cultivation of Oriental rice has long been important. Some new shatter-resistant cultivars were selected locally. More importantly, the California farmers generally practiced cold storage and sowing of harvested grain. The paddies, in a region of expensive irrigation water and high evaporation, were not maintained as permanent artificial marshes, but were drained and allowed to dry between wildrice crops. Sometimes the land was rotated to Oriental rice or other crops. As a result, intense automatic selection for less shattering and less dormancy began. It should be interesting to see how rapidly this may produce a truly domesticated cultivar, dependent on harvesting and sowing.

By the mid-1980s, California growers had more than 6000 ha of wildrice, with average harvests ten times the maximum obtained from native stands and accounting for about half of the world crop. This was too much wildrice for the existing market demand, and large stocks remained in storage with reduced planting as a result. The crop is costly to grow, largely because viable seed is expensive and much of it remains dormant after sowing. Also, much of the wildrice crop is harvested by flocks of redwing blackbirds. Whether *Zizania* will ever join *Oryza* as a staple rather than a luxury remains to be seen.

***HORDEUM* — BARLEY** (Bar-Yosef and Kislev 1989; Bender 1975; Harlan 1975, 1976; Murray 1970; Wiehe 1968; Zohary and Hopf 1988)

Hordeum includes about 25 species native to the north temperate regions, mainly in open, dry habitats. Some are tolerant of salinity and alkalinity. Most are weedy and have spread in artificially disturbed sites. A single wild species *H. spontaneum* is the progenitor of all the cultivated barleys.

H. spontaneum (=*H. vulgare* subsp. *spontaneum*) today ranges widely through the Levant and southwestern Asia as an agricultural weed. However, truly wild populations occupying natural habitats are concentrated in a coherent arc, the so-called Fertile Crescent, extending from the Jordan Valley northward to the Anatolia-Syria border region and then curving eastward along the Iran-Iraq border region. Wild barley and wheats are important members of the winter annual grassland and oak parkland of this Mediterranean climatic region. As the long summer drought sets in, they quickly mature a large crop of grain.

Fossil pollen evidence shows that the climate of this region was colder and drier during the last glacial period and the vegetation was predominantly a sagebrush and saltbush steppe. Presumably the Mediterranean type grasslands and oak parkland species were restricted to very local, coastal sites. About 9000 B.C., at the beginning of the Holocene period, the climate changed abruptly to approximately its present pattern, and wild barley and wheat probably became abundantly available.

The first indications in the archaeological record that humans had begun harvesting wild barley and wheat are indirect. They come from changes in the stone tools rather than from plant remains. *Homo sapiens* proper, as distinct from Neanderthals, first appeared in the Near East by 20,000 B.C. Like their Neanderthal precursors, these people were nomadic big-game hunters, leaving their chipped stone artifacts and bones of their prey in caves and campsites. Their artifacts changed little for 10,000 years until near the end of the Late Glacial stage. Then, here and there in the Near East, their tool kits suggest increased gathering and grinding of seeds. The first cohesive culture with definite evidence of cereal harvesting was the Natufian of the Jordan Valley region, known from dozens of sites dated between 9000 and 8000 B.C. The Natufians made wooden sickles with rows of sharp stones as blades. They left storage bins, although the ones excavated so far were empty. Wild wheat and barley would have been harvested as they ripened before the inflorescences shattered and scattered the grain.

As in other cereal crops, domestication of barley is documented archaeologically by the evolution of nonshattering forms.

Hordeum distichum (=*H. vulgare* subsp. *distichum*) was the first domesticated barley and the progenitor of all the rest. Human selection promptly developed not only the nonbrittle inflorescence but also nondormant seed, which germinated according to the farmers' or brewers' plans rather than on its own schedule. Barley is predominantly self-pollinated, so the cultigens rarely interbred with the wild species. Also, various pure lines developed within the crop, which could be grown together without pooling their genetic diversity. As a result, hundreds of slightly different races have been developed. Between 8500 and 6000 B.C., *H. distichum* was being cultivated at such classic sites as Hacilar in Anatolia, Jericho in Jordan, Jarmo in Iraq, and Ali Kosh in Iran. Like its wild progenitor, its inflorescence produced two rows of grain, and the grain was invested with husks or hulls that remained attached after threshing. *H. distichum* has persisted in cultivation in its ancient area and has been introduced elsewhere, but was soon overshadowed by its derivative *H. vulgare*.

H. vulgare sensu stricto (*H. vulgare* subsp. *vulgare*) originated by human selection of mutations that increased the number of rows of grain to six. This was soon followed by selection of other mutations that gave naked grain on threshing, free from the hulls or husks of the primitive crop. Although two-rowed and six-rowed barleys can be crossed, the hybrids are often morphologically peculiar and lower yielding than pure lines or intragroup crosses.

Archaeological remains of the six-rowed *H. vulgare* are reported from the famous Anatolian site of Çatal Hüyuk by 6000 B.C. Thereafter, *H. vulgare* increased in importance in the Near East in general, while *H. distichum* decreased. Also, *H. vulgare* generally left *H. distichum* behind as agriculture spread from the Near Eastern center into other regions. Some *H. distichum* is still grown in northern Europe, however.

The domestication of barley is inextricably intertwined with that of wheat, several legumes, sheep, goats, pigs, and cattle. Current evidence does not suggest a single place or time of domestication but rather a process involving regional cultural changes and pooling of cultigens among interacting peoples.

By 6000 B.C., Near Eastern Neolithic agriculture was moving down from the hill lands, where rainfall was adequate for dry farming, to the alluvial lowlands, where irrigation was needed. Barley was the staple grain of the great Potamic civilizations that developed there. The earliest cuneiform and hieroglyphic writings record barley as the most abundant and cheapest grain of the ancient city states of the Near East in general, including Babylonia, Sumeria, and dynastic Egypt. When irrigated lands in Mesopotamia suffered from salinity in the 3rd millenium B.C., barley became even more important relative to wheat.

Meanwhile, *Hordeum* cultivation had spread eastward to become a major crop in the Indus Valley in the 6th and 5th millenia B.C. at Mehrgarh, the oldest known agriculture in the Indian subcontinent. It continued to be grown by the Harappan civilization of the Indus Valley before the arrival of rice. Farther east, barley, mainly *H. vulgare*, became important toward the upper limits of agriculture in the Himalaya and toward the northern limits in China where it had arrived by 1000 B.C.

Barley also became a dominant crop in the highlands of Ethiopia, perhaps introduced about 3000 B.C. by Hamitic immigrants.

The westward spread of barley, almost always led by free-threshing, six-rowed *H. vulgare*, is well documented archaeologically. Along with wheat and other Near Eastern plant and animal domesticates, *H. vulgare* led the Neolithic Revolution across Europe. It spread rapidly through the Mediterranean, through Greece during the 6th millenium, and all the way to the Iberian peninsula by 4000 B.C. Here also, it was more important than the primitive wheats then available. The shift from barley to wheat as the staff of life for the common people of Greece and Rome came during the 1st millenium B.C., as recorded in written history. From the Mediterranean, *H. vulgare* spread northward through the Balkans and up the Danube into barbarian northern Europe. By 4000 B.C., it was being grown on the lower Rhine and by 3000 B.C. in the British Isles and Scandinavia. The first Neolithic cultures of northern Europe were evidently dependent primarily on their herds rather than their crops, but barley eventually became a staple food and beer source all across northern Europe.

Barley went overseas with the first European colonists, initially to unsuitable tropical climates, as when Columbus planted it on Hispaniola along with wheat. The crop was successfully introduced to the Andean region where it was being grown by both Spanish colonists and Indians by the mid-16th century. The Spaniards also introduced barley quickly to Argentina, the highlands of Mexico, and subtropical Baja California. Barley is still extensively dry-farmed in Baja California where other crops are only grown under irrigation. The barley crop often fails, but in winters with above normal rainfall, the yield makes up for it. The Spanish missions introduced barley and wheat to what is now the southwestern U.S. starting about 1700.

The California climate was ideal for winter barley, and after statehood, California's first major export crop was barley. It was largely shipped to Britain for malting. Although later surpassed by wheat and irrigated crops, barley has continued to be successful in California and has spread into the Pacific Northwest. The original Spanish introduction was the progenitor of a common modern variety, now called Coast. Its identity was discovered by agronomists who carried out a historical archaeological study of plant remains in mission adobe structures of known date. There has, of course, been modern breeding of barley varieties, but the improved malting varieties are descended directly from Coast. Recent introductions from northern Africa and elsewhere have been used in breeding feed barleys. Currently, California breeders are trying to emulate Israeli breeders in developing a barley that can be irrigated with seawater.

Introduction of barley to eastern North America was a different and more complex story, which will be only touched on here. French, English, and Dutch colonists introduced their own varieties to Quebec, New England, and New York in the 1600s. Planted in the spring, these were adapted to the severe regional climate and went west with pioneer settlers into the Midwest. A lot of barley has been grown for feeding livestock. Historical changes in the geographic distribution of the crop are complex, being partly related to location of major breweries but also affected by other factors, such as introduction of new varieties, particularly winter barleys. These presumably originated in southern Europe, Switzerland and the Balkans being sometimes suggested. In any case, a mixture of winter varieties was grown in a small way in mountain regions of the southeastern U.S. In the 1920s, locally bred new varieties were released in various southeastern states; during the 1930s, winter barley planting expanded westward and northwestward. Improved winter-hardiness has allowed these varieties to displace spring barley in more southern parts of the Midwest.

***TRITICUM* — WHEAT** (Feldman 1976; Harlan 1975; Renfrew 1973; Zohary and Hopf 1988)

Triticum, like several other important crop genera discussed above, includes both diploid and allopolyploid species. The latter are the result of hybridization between species with chromosome sets (genomes) too dissimilar to pair normally during meiosis. Such hybrids are sterile and lead nowhere unless a freak doubling of chromosome sets occurs. Then homologous chromosomes are available for meiotic pairing, and the hybrid becomes sexually fertile, capable of breeding with its own kind but not with either parent species. The freak doubling may occur in gametes, i.e., sperm and egg are diploid, or it may occur after formation of the hybrid zygote. It makes no difference in the result. This mode of speciation is widespread in many seed plant families, both in nature and in cultivation.

In *Triticum*, the basic haploid genome chromsome complement is 7, and the diploid species have 2n=14. There are allopolyploids that are tetraploids (4x=28) and others that are hexaploids (6x=42). There are five wild diploid species of *Triticum*, each with a different genome. Wheat geneticists internationally use the same notation for these genomes; the usage is more stable than the Latin names of the species. Thus a diploid may be AA, DD, and so on; the important tetraploids are AABB, and the hexaploids are all AABBDD.

Wheat domestication began in the Near East about 8500 B.C. Bar-Yosef and Kislev (1989) propose that barley was the first domesticate and that wild wheat grew as a weed in barley fields until mutations for nonshattering led to acceptance as a secondary crop. In any case, histories of the two have been intimately intertwined from the earliest, prepottery Neolithic.

Two wild wheat species were domesticated in the Fertile Crescent in the early Neolithic. A wild diploid *T. boeticum* (AA) was the progenitor of the only cultivated diploid, einkorn or *T. monococum*. A wild tetraploid *T. dicoccoides* (AABB) was the progenitor of the first cultivated tetraploid, emmer or *T. dicoccum* (=*T. turgidum* subsp. *dicoccum*). In both cases, as in cereal crops generally, human selection for nonshattering inflorescences that retained the ripe grain and for nondormant seed that germinated promptly resulted in drastic evolutionary divergence from wild *Triticum* species. As in *Hordeum*, this divergence was greatly accelerated by the predominance of self-pollination. Also as in barley, the primitive domesticated wheat species retained a wild characteristic, namely grain with hulls not removed by threshing.

Einkorn and emmer, along with barley, were generally the first grain crops wherever the Neolithic crop complex spread. Both were carried east to the Indus region and west through Greece and the Balkans during the 6th millenium. By the 5th millenium, both were in Spain and the Baltic region. Einkorn was commonly less abundant than emmer in Neolithic sites, and much of the spread of einkorn may have been as a tolerated volunteer or weed in the preferred cereal crops. It is hardy but relatively low-yielding. It occasionally was the main cereal, but only in hill country of the Near East and in northern Europe. Einkorn was absent from the low, hot regions of the Near East, including Eqypt. Emmer was the preferred food and beer grain of the great civilizations of the ancient Near East and Mediterranean.

Einkorn and emmer were both gradually replaced by newer kinds of wheat starting in the Bronze Age. Einkorn was still widely cultivated here and there in European peasant agriculture until early in the 20th century. Emmer is a minor crop in scattered places in Europe and western Asia; it is still important in Ethiopia. Einkorn was an evolutionary dead end, but emmer became the progenitor of all the other important cultivated wheats, both tetraploid (AABB) and hexaploid (AABBDD).

By far the most important modern tetraploid is *T. durum* (=*T. turgidum* subsp. *durum*), sometimes called macaroni wheat. *T. durum* has a large, hard, low gluten grain good for making firm pasta such as spaghetti. *T. durum* is also free-threshing, i.e., the grain is naked and free from husks after threshing, which makes a finer flour. Several other kinds of free-threshing AABB wheats were derived from emmer, which have been variously treated as species or as subspecies of *T. turgidum*, but they have never been of much importance.

Free-threshing *durum*-type wheats began appearing in the archaeological record early in the Neolithic, but they were evidently not generally preferred over the hulled einkorn and emmer. An exception was the western Mediterranean where *durum*-type wheat was already preferred by the 5th millenium B.C. During the Bronze Age, roughly 3000 to 1000 B.C., *T. durum* gradually replaced einkorn and emmer in the eastern Mediterranean and Levant. Mesopotamian records from about 2400 B.C. show that naked wheat was about twice as expensive as barley or emmer, but yielded only about half as much grain per unit area. Human selection, concentrating on the preferred grain, eventually greatly improved its yield. The replacement of hulled by naked wheat proceeded very unevenly. In Egypt, it did not take place until about 300 B.C.

Meanwhile, hexaploid wheat species (AABBDD) unknown in the wild had evolved. The D-genome was contributed by a wild diploid, *T. tauschii* (=*Aegilops squarrosa*), which never became a crop on its own. The native range of *T. tasuchii* was well north of the early Neolithic agricultural center in a more extreme, continental climate. Its natural range extends from the southern Caspian region eastward across northern Iran and northern Afghanistan and on into central Asia. Around the periphery of this range, the species has spread as a weed of cultivation, but it is not believed to have been in contact with tetraploid (AABB) wheat until cultivated emmer was introduced into the Caspian region. This region is much less well known archaeologically than the Fertile Crescent, and the exact date of introduction of emmer and origin of the hexaploid hybrid is uncertain.

On morphological and genetic grounds, it is generally believed that the first hexaploid wheat was *T. spelta* (=*T. aestivum* subsp. *spelta*). Until recently, there was no archaelogical record of *T. spelta* dated earlier than about 2000 B.C. Recently, however, *T. spelta* has been reported in sites dated at about 4500 B.C. west of the Caspian in Transcaucasia, right where the AABB crop may have first contacted the wild DD species. *T. spelta* has also recently been reported from a Neolithic site in Moldavia dated at about 4700 B.C. The early finds of *T. spelta* are usually as a minor admixture in mainly emmer and einkorn harvests, and the species was probably initially merely a tolerated weed.

The addition of the D-genome greatly broadened the climatic adaptibility of *Triticum* allowing it to withstand severe winters and wet summers, unlike the Mediterranean climate to which emmer was best suited. Archaeological finds indicate that *T. spelta* was increasingly abundant during the late Neolithic in the Balkans and central Europe and here and there was becoming a crop by itself. During the Bronze and Iron ages, *T. spelta* increased in abundance all over eastern, central, and northern Europe. *T. spelta* remains a minor crop in a few places in Europe and western Asia, but in most northern wheat areas, *T. spelta* has been replaced by another hexaploid (AABBDD) wheat *T. aestivum*.

T. aestivum (=*T. aestivum* subsp. *aestivum*, *T. vulgare*), common or bread wheat, has sticky gluten, which entraps CO_2 bubbles released by yeast or bacterial fermentation in rising dough, thus making for lighter baked goods. The simplest currently tenable hypothesis on the origin of bread wheat is simple selection from *T. spelta*. Two mutations give bread wheat free-threshing grain rather than the husked grain of *T. spelta*. Artificial selection would thus have developed this trait independently in both tetraploid and hexaploid wheat, as well as in barley.

In the archaeological record, it is difficult or impossible to discriminate between the tetraploid and hexaploid free-threshing wheats. There are thousands of varieties of the two, and the naked grain looks similar in most varieties, regardless of ploidy. It seems likely, however, that the early Neolithic naked wheats were all tetraploid *T. durum*, that by the 5th millenium B.C. some hexaploid *T. aestivum* was present, and that from the Bronze Age on, remains of naked wheat from finds outside the mild Mediterranean climate region were mostly hexaploid *T. aestivum*.

After Columbus, both *T. durum* and *T. aestivum* were rapidly and widely introduced to the New World. Strenuous Spanish efforts to grow wheat in Caribbean lowlands failed, but success at higher elevations and latitudes soon followed. By the mid-16th century, flour mills were operating in various Spanish New World colonies. Wheat was introduced to North America over and over again by European colonists of all nations. It arrived in Australia with the First Fleet in 1788. The spread of different varieties would, if traceable, be a stupefyingly complex story.

Evolution and dissemination of new varieties of wheat have escalated enormously during the 20th century as scientific breeding programs were developed. Breeding material has been pooled worldwide with innumerable new hybrids produced every year and progeny exchanged for trial and selection. The semi-dwarf wheats bred in Mexico can provide an example.

In 1941, a joint cereal breeding program was established by the Mexican Ministry of Agriculture and the Rockefeller Foundation. This eventually developed into the global program of CIMMYT (Centro Internacional de Mejoramiento de Maiz y Trigo). As the name indicates, maize improvement was put first and wheat second, but the wheat program proved more spectacular. Initially wheat breeding concentrated on improving disease resistance. This has traditionally been the chief task of wheat breeders worldwide because they are engaged in an endless evolutionary race with parasitic fungi. In 1954, a more radical wheat breeding program was begun in Mexico under direction of Norman Borlaug, who would later receive the Nobel Prize for his part in the Green Revolution. This program developed the famous semi-dwarf Mexican bread wheats, which broke the previous yield ceiling by standing erect and not lodging when heavily fertilized. The crucial dwarfing gene had a long international pedigree in different varieties of *T. aestivum* before reaching Mexico. It originated by mutation in a Japanese landrace. Starting in 1917, Japanese experiment stations crossed this variety with winter wheat varieties obtained from North America, including Turkey Red, which had been introduced to Kansas in the 1870s by Mennonite immigrants from the Crimea. After several cycles of selection of the hybrid progeny, the Japanese government registered a new variety, Norin 10, and released it to farmers in 1935. After World War II, an observant agricultural adviser to the American Occupation in Japan noticed Japanese farmers growing this short wheat that stood up under heavy fertilizer and introduced it to the U.S. in 1946. Norin 10 was not adapted to direct use under American conditions, but was the progenitor of a series of new hybrid varieties, including the one supplied to Mexico. The introduction of Norin 10 offspring into Mexican wheats led to a whole series of new varieties, which were released to Mexican farmers starting in 1962; great increases in yields followed. International diffusion of the new Mexican varieties began immediately, first on an experimental level. By 1964, India and Pakistan were buying Mexican seed wheat by the hundreds of tons, and by 1965, those two countries increased their orders to 60,000 tons. Most intercontinental crop introductions have been measured in ounces or pounds of seed.

Subsequent international movements of new *T. aestivum* varieties are far too complex to describe here. In the 1980s, CIMMYT alone has repeatedly sent new bread wheat varieties to over 100 countries, and CIMMYT is now just one of a great global system of wheat breeding programs. Hybridization and selection for regional and local needs are spread through hundreds of experiment stations. Moreover, the rate of diffusion of new kinds of wheat is multiplied by distribution of mixtures of genetically diverse cultivars, called nurseries, to be grown together. The present policy is to deliberately retain genetic diversity within the crop instead of breeding pure lines, as was formerly standard practice. In a way, this is reverting to more primitive wheat farming. Although pure lines may be superior if planted within the optimum environment for each, they are less adaptable and dependable under nonuniform and marginal conditions.

CIMMYT and wheat breeding networks in general have given *T. durum* less attention than *T. aestivum*. *T. durum* is grown on about 30 million ha as compared to over 500 million of *T. aestivum* in the world. However, *T. durum* is a major crop not only in its old Mediterranean-Mideastern region, but in the Andean countries and the wheat belts of temperate North and South America. CIMMYT began a *T. durum* breeding program in 1968, which has since been linked to international trials. New varieties of *T. durum* have been developed that equal or outyield the best *T. aestivum* varieties (≤10 metric tons/ha). By 1984, 50 of these varieties had been released after local selection in 15 countries.

SECALE — RYE (Evans, G. 1976; Renfrew 1973; Zohary and Hopf 1988)

Although *Secale* is native to some of the same regions as barley and wheat, the time, place, and mode of its domestication were completely different. There are three wild species in southern Europe and southwestern Asia and another in South Africa. All are diploids and so is *S. cereale*, a complex species that includes both the domesticated rye and closely related grain field weeds.

Two wild species *S. montanum* and *S. ancestrale* have been suggested as ancestors of *S. cereale*, and it is likely that both have contributed genes to it, although not in equal amount. Both have shattering inflorescences, unlike *S. cereale*. All three species are normally cross-pollinated and form hybrids when growing together. *S. ancestrale* and *S. cereale* hybrids are fully fertile; whereas, *S. montanum* and *S. cereale* hybrids are partly but not entirely sterile. *S. montanum* is a low-growing, turf-forming grass widely distributed in highlands of southern Europe and southwestern Asia. It is especially abundant in Anatolia and the Caucasus region. It differs from *S. cereale* not only in having a shattering inflorescence, but in being a perennial and producing relatively small quantities of grain, like perennial grasses in general.

S. ancestrale (=*S. cereale* subsp. *ancestrale*, *S. vavilovii*) is known in natural habitats, such as volcanic ash and riverine sand, only in eastern Anatolia and adjacent Armenia. These truly wild populations are closer to the crop than *S. montanum* not only in being annuals but in producing more and larger grain. *S. ancestrale* also occurs in Anatolia as an orchard and vinyard weed, but not in grain fields.

In contrast to barley and wheat, there is no firm archaeological evidence that Neolithic peoples exploited wild rye or planted rye as a crop. Instead, *S. cereale* apparently originated in barley and wheat fields as a weed, spread with them into Europe and southwest Asia as an unwanted contaminant of the crops, and gradually evolved to resemble the crops in which it grew. Any mutations of *Secale* that made it resemble wheat and barley would have made it a more successful grain field weed. Nonshattering forms of rye had improved chances of being harvested with the crop. Efforts by the farmer to clean rye grain out of the harvest by winnowing evidently favored mutants mimicking the crop in grain size and weight. Eventually rye became such a successful mimic of the crop that it was accepted as a useful grain in its own right, especially in years with poor harvests of the preferred grains and in regions marginal for them. At higher elevations and higher latitudes, would-be wheat farmers probably found themselves becoming rye farmers without any conscious intent. On the high Anatolian plateau, modern farmers tolerate some rye in their wheat; in years with extreme cold and drought, when the wheat does not survive, rye provides what is sometimes called the "wheat of Allah".

There is one isolated archaeological record from Anatolia that shows that nonshattering weedy *S. cereale* had already evolved by early Neolithic time, but no subsequent Neolithic records of rye, either as a weed or a crop, have been found anywhere in the Near East. Thousands of years later, an isolated Bronze Age cache of rye in northern Anatolia shows that large-grained *S. cereale* was being cultivated; perhaps it was a bad year for wheat.

Outside the Near East, a few grains of rye, presumably from weeds, have been reported from Neolithic sites in Poland and Austria. The first good evidence of rye cultivation is from several Bronze Age sites, 1800 to 1500 B.C. in Czechoslovakia. During the Iron Age, after about 1000 B.C., rye appeared in scattered sites in northern and eastern Europe, usually mixed with barley or wheat. *S. cereale* also appeared in settlements in Iran about 1000 B.C. and seems to have been a staple crop there during the Iron Age. Historically rye was part of the Romans' grain agriculture in their northern European provinces, including Germany and Britain. Rye was pushed farther north in Europe than any other winter crop, although not as far as spring barley. Among many Slavic and Germanic tribes, it became the staple grain for black bread and crisp bread. It provides better lysine content than wheat or barley and more flavor. Also, occasional finds of sprouted rye grain in Iron Age sites show that it was malted for brewing. (Rye whiskey is more recent in origin.)

Rye was introduced overseas early in the colonial period and has again been successful mainly in high latitudes. In North America, rye bread has decreased in importance since the colonial period. So-called rye bread is now heavily adulterated with wheat, except for the more traditional pumpernickel type. North American rye is largely used for whiskey or livestock feed. Some modern rye breeding aims at improving the crop as a pasture instead of a grain crop. About 95% of the world's rye crop is still grown in its ancient north European homeland.

Rye has been comparatively little affected by scientific breeding. This is partly due to its minor rank, but there is also a biological reason. Unlike the major cereals, rye is not normally self-pollinated. Its outbreeding character has maintained the gene pool as more of a continuum instead of diverging into a multitude of landraces. Thus, rye does not offer the chance to produce radically new varieties by hybridization of divergent races.

The most important and dramatic use of rye in modern breeding is in producing the following.

***TRITICOSECALE* — TRITICALES** (Carney 1990; Larter 1976; Smith, N. 1983a)

This new cereal, a hybrid between wheat and rye, is the creation of scientific plant breeders. Wheat and rye hybrids have long been known, but the genomes of the parents are so different that the raw hybrids are completely sterile. Methods of chemical treatment to cause chromosome doubling, discovered in the 1930s, allowed production of fertile allopolyploids. Triticales have since been synthesized repeatedly by crossing various kinds of wheat and rye. Tetraploid wheats, usually *Triticum durum* (genome AABB) and hexaploid bread wheat *T. aestivum* (AABBDD), combine with rye (genome RR) to give hexaploid (AABBRR) and octoploid (AABBDDRR) triticales, respectively. These new species are so new that they are not yet labeled with Latin binomials.

Triticales offered plant breeders the hope of combining the best characteristics of the two parental genera, including the generally preferred taste of wheat and the higher quality protein of rye. Serious breeding began in Europe in the 1930s, focusing on the octoploids, and yields comparable to bread wheat were reported in the 1950s. Since then, new triticales have been bred that often outyield the top wheat varieties.

By the mid-1980s, over $1^1/_2$ million ha a year were being planted to triticale, mainly in Europe, where it is grown commercially in 15 countries, Australia, South Africa, Canada, and the U.S. Nearly all the crop is grown for forage or feed for livestock. So far, little of the grain has been used for human food, in spite of its superior nutritional quality.

Meanwhile, in the 1960s in Mexico, the program that produced the famous semi-dwarf wheats began triticale breeding in the hope of adding a human food crop for smallholder farming in the tropics. In collaboration with Canadian universities, triticale cultivars were developed for marginal highland and semi-arid tropical regions. CIMMYT distributed triticale cultivars for experimental evaluation to over 100 countries. By the mid-1980s, they had been adopted by a few farmers in Latin America, Asia, Africa, and Madagascar, but the total area planted was only about 100,000 ha. Again, the crop has been used mostly for livestock feed. There are still hopes that the quality and productivity of triticale as a food grain will eventually lead to wide acceptance.

PENNISETUM — MILLET, KIKUYU AND ELEPHANT GRASS (Harlan 1975, 1989; Munson 1980; Parsons 1972; Purseglove 1976; Vishnu-Mittre 1977)

Pennisetum is small genus of annual and perennial grasses, primarily of tropical Africa.

Pennisetum glaucum (=*P. americanum, P. typhoides*) — Pearl, Spiked, Bulrush, Candle, or Cattail Millet)

Millet is a name applied by Europeans to a diverse array of ancient African and Asian cereal crops. *P. glaucum* is an ancient African domesticate. The wild progenitor *P. violaceum* is native to a narrow zone running clear across Africa from the Atlantic to the Nile. This includes parts of the Sahara and the Sahel, the more arid part of the savanna belt lying between the Sahara and the Sudan. The species is a tall, stout annual, capable of extremely fast growth during the short season that moisture is available. It has C_4-type photosynthesis, an unusual system that has evolved independently in various grasses and a few other families. C_4 photosynthesis is more efficient than ordinary C_3 photosynthesis in bright sunshine and with limited moisture.

Domesticated *P. glaucum* shares this and other attributes of the wild progenitor. The domesticate differs primarily in having spectacularly larger inflorescences that bear many times as many flowers with larger grain and that retain the ripe grain rather than dispersing it. The domesticate has many different local races in the Sahel and the adjacent Sudan and also in the desert margins of southern Africa. It is the staple cereal of the driest margins of African agriculture. It is planted both by sedentary black farmers and by nomadic Arab pastoralists. At the start of the rains, the nomads commonly plant pearl millet in special habitats, e.g., fossil sand dunes that accumulate rainfall over an impermeable claypan. The crop grows unattended while they move their herds north into the Sahara and is ripe when they return south. The crop is partly harvested by flocks of birds before it is reaped.

The earliest known archaeological remains of pearl millet in Africa are from Mauritania, dated about 900 B.C., but there is no doubt that domestication had taken place far earlier. The crop spread in antiquity through the Near East to India in regions where there are no wild progenitors. It reached northwestern India, specifically the Rajasthan region of modern Pakistan, before 1500 B.C. *P. glaucum* long ago became a staple grain of the hot, dry savanna regions of southern Asia. In modern times, *P. glaucum* has been introduced to many other regions. Although it is an excellent grain for human use, it is usually grown in the U.S. for forage and bird seed.

Two other species of *Pennisetum* native to Africa are incipient domesticates as fodder and pasture grasses (Parsons 1972). These are *P. purpureum* (elephant grass) and *P. clandestinum* (Kikuyu grass). Both species are vigorous perennials, spreading by horizontal stolons or rhizomes. They do not produce much grain and are useless as cereals, but as forage, they are highly productive, palatable, and nutritious for hoofed animals, wild and domesticated. Presumably, their adaptation to survive under grazing results from long coevolution with African ungulates. Propagation of both species, for fodder or pasture, is by cuttings rather than seed.

Elephant grass is a native pioneer of riverbanks and other open sites in the seasonally wet lowlands of tropical Africa. Like pearl millet, it has C_4-type photosynthesis and is remarkably drought resistant. However, elephant grass does much better with plenty of rain. It was taken into regular cultivation for livestock feed in the 19th century outside its native range, first in the Transvaal and Rhodesia. It was introduced to Costa Rica in the late 19th century and is now commonly planted there where rain forest and seasonal forests have been cleared for beef cattle ranching. Early in the 20th century, it was widely introduced elsewhere in tropical America through experiment stations.

P. clandestinum is native to the highlands of East Africa where it was traditionally propagated as a pasture grass by the pastoral peoples. Although it has the usual C_3-type photosynthesis, it is very productive and aggressive.

The most important introductions of Kikuyu grass were to the highlands of tropical America, beginning about 1920 in Guatemala and Peru. The species became the economic base of highland dairy farming in tropical America. Spreading on its own, it has blanketed many mountain slopes and stopped erosion. To farmers of potatoes and other highland crops, it is a noxious weed, almost uncontrollable except by expensive herbicides.

During the 20th century, Kikuyu grass has been introduced to many other warm regions around the world, both as a forage crop and a tough lawn grass. In Australia, South Africa, California, and other places distant from its homeland, it has become one of the most troublesome weeds.

***SORGHUM* — SORGHUM** (de Wet and Huckabay 1967; de Wet et al. 1976; Doggett 1976; Harlan 1977, 1989; Mann et al. 1983; Patiño 1969; Winberry 1980)

The genus *Sorghum* is pantropical with a complex of species: annual and perennial, diploid and polyploid, in different continents. However, the cultivated grain sorghums, although very diverse, are all derived from a single annual diploid species native to Africa.

***Sorghum bicolor* — Grain Sorghum, Kaffir Corn, Durra, Milo Maize**

In some ways, the historical geography of grain sorghum parallels that of pearl millet. Sorghum was also domesticated in antiquity in Africa south of the Sahara from a robust grass with C_4-type photosynthesis. The crop is thus also adapted to thrive under intense sunlight with limited moisture. Again, the archaeological record does not document the early history of the crop. The crop emerged from obscurity only after being introduced far outside its homeland. It was cultivated in Oman by c. 2500 B.C. and in Rajasthan in modern Pakistan by about 1750 B.C.

As in most cereal crops, the domesticated population is commonly given a different binomial than the wild progenitor. This is convenient because, although the two populations can hybridize, hybrids are unlikely to survive in nature. They are often aggressive weeds in the grain fields, but are not included in the harvest. African farmers commonly go through the ripe grain selecting individual plants for next year's seed grain before the general harvest. Human selection has made the domesticated *S. bicolor* diverge from the wild *S. arundinaceum* in many ways. The most basic difference is the nonshattering inflorescence of the crop, retaining the ripe seed for harvest. The domesticate also has much more massive inflorescences, producing more and larger grain than the wild.

The geography of the wild *S. arundinaceum* helps little in delimiting the region of domestication because the range virtually blankets the whole continent south of the Sahara. Wild sorghum is mostly a desert margin or savanna grass, but there is a race that grows as a riparian pioneer clear through the wet forest region of the Guinea Coast and Congo Basin. The geographic races of wild *S. arundinaceum* have differences similar to those between the basic cultivated races. An early, still tenable, hypothesis was that the cultivated races derive from multiple domestications of wild sorghums. Current authorities favor a single center of domestication, with resemblances between wild and crop races explainable by interbreeding during spread of the crop. However, there is no consensus on where domestication began. Doggett suggests the Ethiopian highlands, with sorghum being a secondary domesticate after cereal cultivation based on wheat and barley was introduced from the Near East by Hamitic peoples; the developing crop would then have been spread through the Sudan to West Africa and southward through East Africa. Harlan suggests a primary center west of the Nile in the Sudan from which the crop was carried both eastward and westward as well as south through East Africa. The most primitive cultigen race, typical *S. bicolor*, is very widespread in Africa and India. More advanced cultigens are the Guinea race, adapted to moister climates of West Africa and highlands of East Africa, the *kafir* race concentrated in southern Africa and the *durra* in Islamic northern Africa and India. The route from Africa to India is unknown, whether overland or by sea. The crop spread eastward through southeast Asia to China at an unknown, but early date. In southeast Asia, it may have picked up some genes from a local wild species *S. propinquum*. Such introgression may have contributed characteristics of the Chinese *kaoliang* race.

At any rate, *S. bicolor* became a staple food and beer grain for hundreds of millions of people in semi-arid Africa and Asia. India alone has about 15 million ha of grain sorghum, mainly in regions with a long dry season and a short wet season.

S. bicolor was not adapted to the summer dry Mediterranean nor to other European climates. Its post-Columbian spread was not via Europe but initially directly from Africa and perhaps from the Canary Islands where it was being grown by the 1500s. It was undoubtedly repeatedly introduced to the New World with the slave trade, commonly being the main food provided the prisoners on their trans-Atlantic voyage. It rapidly became established as a grain crop in drier parts of tropical America, being reported in the 1600s in Spanish, French, and Dutch colonies in both the Greater and Lesser Antilles. In some, it was the main food of the slaves. By the 18th century, it was widely grown in drier regions of the mainland of Central and South America by both Mestizo and Indian farmers.

Grain sorghum is today an important crop in the hot, dry lowland regions of Latin America. Commonly it is interplanted with maize as insurance against drought. Although tasty and nutritious, sorghum grain is a low prestige food in Latin America, perhaps because of its links with slavery. Farmers have stated that they grew it only for feed for their chickens, but it makes good tortillas. Sorghum is excellent when popped and has largely replaced the traditional popped corn and popped amaranth among the highland Maya of Guatemala; balls of popped sorghum grain bound with syrup are a festive food.

S. bicolor races were repeatedly introduced to the U.S. since about 1850 by private and government agencies. During the late 19th century, special races were widely planted as minor crops. One, known as broom corn, supplied the makings of the familiar old-fashioned brooms, made from the inflorescence after the grain was threshed off for chicken feed. More important were sorghum races with sweet stalks used to make molasses. There were two important introductions of these in the 1850s, both via Europe. The so-called Chinese sugar cane came from Shanghai via France and was much grown in the Midwest Corn Belt. The other came from Natal and was the progenitor of farmer-selected races most popular in the deep South. Both were planted in small patches on general farms, and the syrup was made by simple methods for home use or local sale. The Civil War led to great reliance on sorghum molasses in both the North and South because of disruption of cane sugar imports. Sorghum molasses has a strong flavor, but tastes better than sugar cane molasses. By the 1880s, it was being produced by farmers over much of the U.S. at the rate of tens of millions of gallons a year. Hopes for large-scale commercial production of granular sugar were never realized because of competition from cane and beet sugar. Cheap glucose syrup with maple flavoring has taken over the commercial syrup market, and sorghum molasses has declined to a rarity appreciated by those with old-fashioned tastes.

Grain sorghum cultivation, by contrast, has had explosive growth in the U.S. during the 20th century, primarily for livestock feed. Its success was aided by development of dwarf varieties that can be harvested by huge combines. More recently, use of male sterile lines has resulted in replacement of landraces by F_1 hybrid seed bought each year from commercial seed companies. Since the 1970s, the U.S. has produced over half of the world crop of grain sorghum, largely in regions too dry for maize farming. Grain sorghum has also had dramatic recent growth as a feed crop in Brazil and Australia.

ZEA — MAIZE, INDIAN CORN (Anderson 1946; Beadle 1980; Bird 1980; Clawson and Hoy 1979; Cutler and Meyer 1965; Doebley 1990; Doebley et al. 1988; Finan 1948; Galinat 1985, 1988; Galinat and Pasapuleti 1982; Goodman and Brown 1988; Miracle 1966; Pickersgill and Heiser 1977; Roe 1974; Roosevelt 1980; Sarkar et al. 1974; Simoons 1990; Spencer 1975; Struever and Vickery 1973; Upham et al. 1987; Wilkes 1967, 1977; Yarnell 1976)

The genus has four wild species in Mexico and northern Central America: the perennials *Z. perennis* and *Z. diploperennis* and the annuals *Z. luxurians* and *Z. mexicana*. After long, tedious controversy, it is now generally recognized that *Z. mexicana*, commonly called teosinte, is the progenitor of domesticated *Z. mays*, maize or, in the U.S., simply corn.

Like maize, teosinte is a quick-growing annual with C_4-type photosynthesis, which means it can function well in bright sunlight and with limited water. Teosinte grows in woodland and pine and oak parkland vegetation. It can also grow on limestone outcrops and as a weed along roadsides and field margins. Teosinte is eagerly browsed by livestock and was probably more abundant before Spaniards introduced domestic livestock. Teosinte closely resembles maize in general morphology and shares with maize the unique unisexual inflorescences, the tassel and the ear. Differences between teosinte and maize are concentrated in the grain and the ear that bears it and are precisely what might be expected to result from domestication as a cereal crop. Some of the changes parallel those in other domesticated cereals: increase in grain size, loss of dormancy, and retention of the ripe grain on the ear rather than shattering of the inflorescence. Other changes are unique to maize domestication, e.g., loss of the hard case surrounding teosinte grain, which protected it during passage through an animal's digestive tract; doubling and redoubling of the two rows of grain on the teosinte ear; enclosing the maize ear in husks with enormously elongated styles emerging at the tip of the ear for pollination. There are other morphological changes in the maize ear that are the subject of a formidable body of technical literature. However, although the morphological changes are spectacular, they are relatively simple genetically, and the crucial mutations are believed to have been selected very rapidly at the beginning.

There is some molecular evidence from both isozyme and DNA studies (Doebley 1990) suggesting that maize was domesticated primarily from a single one of the five main geographical races or subspecies of teosinte, namely the Balsas race (=subsp. *parviglumis*). This race is now confined to the central Balsas River basin in the Michoacan-Guerrero border region of western Mexico. As maize cultivation spread, some hybridization with other teosinte races probably occurred. Doebley (1990) suggests that some of the ancient highland maize races of the Central Plateau region of Mexico have a few genes from the Chalco race of teosinte.

However, the process of speciation of *Z. mays* to form an independent gene pool was evidently completed thousands of years ago, with subsequent diversification of maize into a wonderful array of races quite independent of *Z. mexicana*. Teosinte pollen cannot normally function on the styles, or silks, of the maize ear. Some maize can pollinate teosinte and produce hybrid seed, but the maize genes make the progeny misfits in a natural habitat. Cases are known in marginal Mexican maize fields where teosinte hybrids were tolerated for their drought resistance but in normal Mexican practice, maize contaminated with teosinte genes would be rejected as planting stock.

The earliest known archaeological remains identified as *Z. mays* rather than *Z. mexicana* are from the famous Tehuacan caves in Puebla, Mexico, dated about 5000 B.C. This maize belonged to an extremely primitive race, intermediate between teosinte and all living maize races. As in teosinte, the ears are very small and slender, and the grains are tiny and hard and were probably prepared by popping. Unlike teosinte, the cobs are nonshattering and have mostly eight rows of kernels with a few four-rowed types. The Tehuacan site is outside the habitat of teosinte; the caves lie in arid valleys with thorn-scrub vegetation. The same caves have yielded the earliest remains of some other crops that could not have been initially domesticated there. If they were actually grown locally, not packed in, they must have been planted around springs and perhaps irrigated.

The most primitive living maize races, such as Nal-Tel of Yucatán and Chapalote of western Mexico, are only moderately different from the early Tehuacan maize. They have rather short ears with very slender cobs and very small, hard pop-type grains, sometimes brownish like teosinte. Nal-Tel and its closely related races, however, generally have about 12 rows of kernels. They do not compare in yield with most modern maize races, but are retained as a tough, reliable quick maturing crop. Primitive popcorns closely related to Nal-Tel and Chapalote are widespread archaeologically in Mexico and Central America. Also, the oldest known archaeological maize ears from South America belong to the Nal-Tel complex. In South America, maize appeared before 3000 B.C. in the dry Valdivia site in Ecuador and in the Ayacucho Basin of central Peru. By 2700 B.C., maize was present at Pichasca in northern Chile (Rivera 1991). By 1800 B.C., maize was being grown in coastal oases in north central Peru. In the tropical forest region of the Orinoco Basin, archaeological sites in the Parmana region record the arrival of maize between 400 B.C. and 100 A.D.

The oldest archaeological maize ears from north of the Tropics are also primitive, slender-eared, tiny grain popcorn, those found in La Perra rockshelter in northeastern Tamaulipas, dated 2500 B.C., and in Tornillo rockshelter in southern New Mexico, dated about 1200 B.C.

Very similar maize is one of the earliest, or perhaps the earliest, to appear in the midwestern U.S. There the archaeological remains of crops are sparse and the dating commonly controversial. Traces of maize have been reported from scattered Middle Woodland sites, beginning about 300 B.C. The first solidly dated maize in eastern North America is from Tennessee after 200 A.D. (Fritz 1990). There is no evidence that acquisition of these primitive small-eared popcorns had a revolutionary impact on the Indian economy.

The archaeological record in various regions of Mexico and North America shows that people did not switch from hunting and gathering to food production as soon as they had maize. Rather, they planted some corn, probably as more of a snack than a staple, and remained primarily reliant on wild food sources for centuries. This pattern was continued to modern times by some Indian tribes, e.g., the Apache, who often left a patch of corn to shift for itself while they went off hunting and gathering.

In southwestern North America, Indians planted the primitive small-eared popcorn for well over a millenium before intensive agriculture was developed by the Hohokam and Anasazi cultures. When the shift finally occurred, it involved not only drastic economic and cultural changes but also rapid change in the races of maize. In southwestern Arizona, the Hohokam began building their great irrigation systems shortly before 500 A.D. In the Anasazi region to the north, expansion of dry farming and irrigation began in the late Basketmaker period, between 500 and 700 A.D., soon followed by building of the great Pueblo period cliff dwellings and towns. During the late Basketmaker period, an exotic and very different race of maize spread into this region from western Mexico. This was Harinoso de Ocho, an eight-rowed flour corn with long cylindrical ears and large, soft kernels. Its place of origin is unknown, but it was probably in tropical America and perhaps northern South America. The Indians of western Mexico and the southwestern U.S. have continued growing small quantities of Harinoso de Ocho, along with their old Chapalote for special purposes up to modern times. However, hybridization between these two races gave rise to a complex of new races that became the main crop of the Hohokam and Anasazi. It survives with very little change in maize grown by several Arizona and Sonora Indian groups, including Pima-Papago, Yuma, and Mojave. This complex includes both soft flour and hard flint popcorn and both 8-rowed and 10- to14-rowed races. Later on, the Hopi and the New Mexican pueblos added other corn races derived from eastern Mexico.

In eastern North America, there was likewise a pause of over a millenium between the arrival of maize and its becoming the staple food. Fragmentary evidence suggests that the shift came with the spread of eight-rowed flour and flint corns from the Southwest across the northern plains.

Considering the Americas as a whole at the time of Columbus, *Z. mays* had become diversified into a wonderful array of races. They had evolved partly by adaptation to diverse environments as the crop was taken into regions unlike its homeland. Mainly, however, the diverse races had been produced by careful human selection for both striking and subtle variations in plants, ear, and grain. Within a given climatic region, a single village or household commonly grew several races of maize, different in rates of maturity, yield, and quality of grain. One might be preferred for boiling or roasting, others for popping or for tortillas or brewing. One might be planted only after a preferred race began having a bad year. It may seem strange that in a predominantly outcrossing, wind-pollinated species discrete races were maintained in the same territory. The Indians did this partly by isolating the races in different milpas, often with forest between. They were helped by some unique features of the crop. One was the extraordinarily long styles, or silks, down which the pollen tube has to grow to reach the future seed; if the style receives a mixture of pollen of different maize races, pollen of its own race tends to outgrow the other. Also, if foreign pollen does succeed in producing a hybrid grain, its hybridity is often detectible through a peculiarity of maize known as xenia. This refers to visible differences between individual grains on an ear due to different pollen parents. The embryo and endosperm in each grain normally have the same pollen parent. If a few grains with white endosperm appear on a yellow ear or if plump, starchy grains appear on a sweet corn ear, the farmer can see that foreign pollen has blown in and reject the hybrid grain for planting. If a hybrid grain is planted, it may produce a calico (segregating) ear, favored today for Halloween decorations. Such maize would be considered a disgrace in traditional Indian agriculture.

Pre-Columbian America had perhaps a dozen extremely different, ancient, basic maize races, each differing from all others in many characteristics of plant, cob, and grain. Innumerable intermediate races were developed by reshuffling the basic characteristics in stabilized hybrids. Some of the basic races remained geographically localized, but others spread widely. For example, some maize races that evolved in South America are believed to also be ancient in Mexico; however, the lack of archaeological evidence makes both the source regions and timing quite speculative (Bird 1984).

The possibility of pre-Columbian presence of maize in various regions of the Old World was actively debated during the 1960s and 1970s. Historical evidence was drawn from early reports now generally interpreted as references to grain sorghum. Archaeological remains were reported only from 15th century India, but the dating is questionable. New evidence has been drawn from stone carvings in 12th and 13th century temples in southern India that depict objects resembling maize ears (Johannessen and Parker 1989). The resemblances are intriguing, but other possible models have been suggested, including *Pandanus* fruits. Moreover, the carvings may not be as old as the temples. If maize was really present, direct archaeological evidence should exist; the cobs are unmistakeable, radiocarbon datable, and nearly indestructible in a dry site.

Columbus brought maize grain from the Greater Antilles, along with its Arawak name, *maiz*, back to the Spanish court, and plants were grown in Spain in 1493. Soon after 1500, maize was repeatedly recorded as being grown in other parts of Europe; the great Renaissance herbals contain accurate descriptions and illustrations. Multiple introductions from the New World to Europe are indicated by 16th century records of presence of different maize races.

In much of Europe, maize remained a minor crop, but in various regions of southern Europe, it revolutionized the agricultural patterns. The western Pyrenees offer a remarkable example (Gomez-Ibañez 1975). Maize was introduced there from Peru by Basque companions of Pizarro. It was adopted by some Basque farmers in the late 16th century and spread gradually through the region, initially for human consumption in flat cakes. It had to penetrate a time-hallowed system of open fields with 3-year rotation regulated by the community: a crop of winter wheat or other fall-planted crop, followed by other, spring sown crops and barley or millet, followed by a year of fallow. Planted in the spring, maize was found to greatly outyield the former spring crops and to give a much more reliable harvest. It was adopted into a new 2-year rotation, with the fallow eliminated, leading to a huge increase in food supply and in the standard of living. Furthermore, maize was interplanted with legumes, and the maize stalks and legumes provided an increased source of winter fodder. Also the increased grain supply allowed feeding of maize to livestock. As a result, instead of the traditional transhumance from mountain to lowland pastures in winter, cattle were stabled and fed, with an increase in manure for fertilizing the crops. Furthermore, swine were raised in greater numbers than before in spite of the reduction in forests in which they had formerly been herded to forage on mast. All this change took place gradually during the 17th and 18th centuries, a time of greatly increased agricultural productivity and growing Basque population.

However, maize proved so productive and so cheap that in parts of southern Europe many poor people, especially tenant farmers, ate little else. Maize is deficient in some essentials for a balanced diet, particularly lysine and niacin. White maize lacks carotene, converted to vitamin A in the body. Niacin deficiency results in a serious and often fatal disease, pellagra. Pellagra was first recognized in the mid-18th century in northern Spain and soon afterwards appeared in Italy, the Danube basin, Turkey, and other parts of the Old World that had overconcentrated on white maize without beans. In the U.S., it became the plague of sharecroppers engaged in growing nothing but cotton or other commercial monoculture. Pellagra is not known to have been a problem among American Indians. Although maize was the staple, its nutritional value was enhanced by the general practice of soaking the grain with lime or alkaline ash before grinding it to make *masa* for tortillas. Also, the American Indians grew a rich variety of other crops, as discussed in previous sections.

Maize may have been taken from the Mediterranean by Arab caravans to Africa south of the Sahara, but it undoubtedly was also introduced there from the New World directly and repeatedly in the 16th and 17th centuries. Some records from that period refer to planting by Europeans on offshore islands, e.g., by the Portuguese on the Cape Verde Islands and on Zanzibar and Pemba, and by the French on Reunion. Also, 16th and 17th century reports show maize had already become an important mainland crop along the Gulf of Guinea from Dahomey to the Gold Coast and near the mouth of the Congo. Oral tradition dates the spread of maize through the Congo basin in the 17th century. In East Africa, the first clear reference to maize is from Mozambique in 1561. Evidently African farmers readily accepted maize, partly because it could be grown and used almost exactly like their traditional grain sorghum. African folk cuisine prepares both in a great variety of ways, involving parching, popping, boiling, roasting, and fermenting. Except in drier regions, maize largely displaced sorghum as the staple grain. Its spread into the interior commonly outran European exploration although in some regions, e.g., parts of Uganda, it continued to spread through the late 19th century.

Maize was almost certainly carried on eastward to India and the East Indies by the Portuguese. The earliest clear record from India is in 1590. Much of the spread of maize in India, southeast Asia in general, and the East Indies escaped historical documentation, presumably because it took place outside the advanced agricultural systems focused on rice and wheat. Maize became peculiarly important among more primitive shifting cultivators in mountainous, often remote areas, including the so-called hill tribes. This is one of the reasons why antiquity in the region was inferred.

The extensive Chinese commercial network probably brought quick contact with maize introduced by the Portuguese and the Spaniards to Southeast Asia and the East Indies, including the Philippines. The first recorded Spanish planting of Mexican maize in the Philippines was in 1543, and there were repeated later introductions. From whatever sources, maize was rapidly adopted in parts of China, becoming a significant crop by the late 16th century. It never replaced rice or wheat, but probably competed with millets and sorghum. Wherever there was a warm, wet growing season of at least a few months, maize allowed major expansion of agriculture on steep hills and became the main cereal of the poorest people.

In the Philippines, maize evidently spread initially as a minor garden vegetable, the immature ears being roasted or boiled. Unlike the Chinese, the Filipinos lacked technology for milling hard, ripe grain. During the 17th and 18th centuries, Chinese in the Philippines became millers of maize, producing coarsely ground meal that could be cooked like rice. Eventually, maize milling was taken over by Filipino villagers, and the crop began spreading rapidly in the early 19th century. Maize became the main upland crop of the Philippines, leading to greatly expanded agricultural settlement.

Mexican maize was even more successful in Guam where it was introduced by Jesuit missionaries in the early 18th century. It rapidly became the principal food staple of the whole island. The worldwide introductions of *Z. mays* cannot all be followed here. It is probably a significant crop in more countries than any other species.

By far the greatest historical expansion of maize planting was in the so-called Cornbelt of the eastern U.S. Various prehistoric Indian tribes grew maize in this region, usually on floodplains or in temporary clearings in upland hardwood forests. They had developed the so-called Northern Flint, descended from the narrow-eared Maiz de Ocho of western Mexico, which was inherited by the white and black immigrants to the region. The newcomers not only ate the corn themselves, like the Indians, but also fed it to pigs and other farm animals. They also malted the grain, fermented it, and distilled it for Bourbon whiskey. In the 19th century, with draft animals and plows, the immigrants pushed maize across the prairie grasslands that had been just hunting grounds for the Indians. The black prairie soil became the heart of the commercial Cornbelt. Also during that century, Cornbelt farmers developed new maize varieties descended from hybrids between the old Northern Flint and the so-called Southern Dent. Southern Dent has not been recovered from archaeological sites in North America and was probably introduced to the southeast from Mexico by Spaniards. It is clearly derived from an important race of the central Mexican plateau, which has a pyramidal, multi-rowed ear. It is extremely different genetically from Northern Flint, and hybrid progeny show extraordinary vigor. This was evidently discovered accidentally by 19th century Cornbelt farmers, who brought the two races together. By careful selection of hybrid progeny, various farmers developed superior named varieties. These varieties provided the basic germplasm for modern Cornbelt hybrids. These are produced by controlled pollination rather than the former normal, open pollination.

The first step in developing commercial hybrid corn was selection of various pure, inbred lines, obtained by generations of self-pollination. The inbreds themselves are weak and unproductive, but hybrids between certain lines turned out to be at the other extreme. By trial and error, the seed companies discovered combinations of inbreds that produced vigorous F_1 hybrids. The separation of male flowers into the tassel and female flowers into the ear made controlled hybridization easy. Two lines of inbreds could be interplanted, and the one used to produce F_1 seed detasseled before any pollen was shed. Grain produced on the unemasculated line could be fed to the hogs. The process was repeated the following year using two different F_1 lines to produce double-cross hybrid seed for sale to the farmers. Initially, detasseling provided good summer jobs for lots of high school students. Unfortunately, a way to save on this labor was discovered by using a male-sterile mutant that did not need detasseling.

Farmers began buying commercial hybrid seed corn in 1933. Within 10 years, it was used for half the Cornbelt crop, and in another 10 years, the old farmer-selected, open-pollinated varieties were totally replaced.

The line from which the male sterile gene mutation had been obtained eventually proved to be vulnerable to a serious fungus disease. In the 1970s, it was found that virtually the entire Cornbelt crop had inherited that vulnerability. The Cornbelt crop fell catastrophically, temporarily relieving the chronic U.S. grain surplus, until new resistant hybrids were bred. This was an embarassing episode because plant breeding doctrine has long emphasized avoidance of a genetically narrow base.

In the tropics, scientific maize breeding was begun soon after the rise of hybrid corn in the Cornbelt. In Mexico in 1943, a major, sophisticated program was launched by the organization that led the Green Revolution in wheat. By the 1980s, CIMMYT was part of a massive global network in over 90 countries, which pooled maize germplasm and exchanged promising new cultivars for local evaluation and selection. Many new cultivars have been released to farmers but with limited success. Population growth in Mexico and many other countries has far outrun increases in maize yields. A fundamental problem has been the need to breed different kinds of maize for narrowly limited environments and for different uses. There seems to be no hope of a general-purpose, broadly tolerant new maize comparable to the Mexican dwarf wheats. In addition to breeding for specified environments, an even tougher problem is the peasant farmers' need for a variety of races to plant at different times and for different uses. CIMMYT found that a high-yielding cultivar that it had bred specifically for a certain region in Puebla was poorly accepted there. The seed all required the same length of growing season and had to be planted at the same time. The farmers preferred to guard against a total crop failure in a bad year by planting various landraces at different times in different microenvironments. Also, when germination of an early sown race was poor, they needed to plant other lower yielding but faster maturing races.

Discussion so far has dealt almost entirely with maize grown as a field crop for its ripe grain. Only passing mention was made of American Indian cultivars grown for roasting or boiling as a fresh vegetable. These *maiz dulce* or sweet corn cultivars shared a recessive mutation (*su* on chromosome #4) that impedes conversion of sugars into starch. The same mutation probably arose independently in different landraces and spread to others by hybridization.

In eastern North America, white farmers adopted sweet corn as a common garden vegetable. As in the case of field corn, stabilized open-pollinated cultivars were selected, including Country Gentleman and Golden Bantam, both released about 1900. During the 1940s, seed companies began selling F_1 hybrid seed from selected inbreds. The new hybrids rapidly took over both in home gardens and in production for canning and freezing. In the 1970s, they were in turn replaced by radically new hybrids based on inbreds developed by experiment stations, particularly in Illinois, Iowa, and Florida. These used new recessive mutants, *sh2* on chromosome #3 and others, which result in kernels that are not only sweeter but also keep their sweetness and flavor much longer after harvest (Kaukis and Davis 1986). North American produce markets now supply corn on the cob that is comparable in taste to what was formerly only available to home gardeners and their friends. The *sh2* hybrids now dominate the sweet corn market in the Far East also.

SACCHARUM — SUGAR CANE (Brockway 1979; Burkill 1935; Galloway 1989; Kimber 1988; Merrill 1958; Ortiz 1947; Patiño 1969; Purseglove 1972; C.O. Sauer 1969; Simmonds 1976d; Simoons 1990; Watson 1983; Watts 1966)

There are two wild *Saccharum* species, both of which have contributed to the domesticated sugar canes. *S. robustum* is primarily a streambank pioneer ranging from Borneo to the New Hebrides with a center of diversity in New Guinea. *S. spontaneum* is a more generalized and aggressive pioneer of various open and disturbed habitats with a range from Nigeria across the Sudan throughout southern Asia and the East Indies to the Solomon Islands, Taiwan, and Japan; its center of diversity is in India. Both species are high polyploids believed to have originated in prehuman time as interspecific hybrids. Chromosome numbers vary within each species, especially *S. spontaneum*, different clones having numbers ranging from 40 to 128. Both are tall perennials, dispersed by tiny wind-borne seeds and capable of vegetative reproduction by rooting of stem fragments. Like *Sorghum* and *Zea*, both have C_4-type photosynthesis, which is highly productive under intense sunlight. Both are adapted to seasonally wet and dry climates. Both are tolerant of a wide range of soils, a trait inherited by domesticated sugar cane, which is grown successfully on soils derived from limestone, basalt, and granite as well as on alluvium. *Saccharum* can survive on leached soils with low organic content, evidently partly due to symbiotic nitrogen-fixing bacteria *Beijerinckia*, associated with the cane root system. Neither accumulates much sugar in the canes, which are hard enough at maturity to be used for arrows, fences, and house roofing. Shifting cultivators in their native ranges have sometimes deliberately created *S. spontaneum* savannas as pastureland by repeatedly burning abandoned clearings to prevent forest regrowth; after burning the *Saccharum*, clumps resprout from the roots.

Saccharum officinarum (incl. *S. barberi, S. sinense*) — Domesticated Sugar Cane

This is not a species in the textbook sense of a sexually interbreeding population; rather it is an array of vegetatively propagated clones, some very ancient, which differ genetically because of initial differences between the original seedlings and subsequent mutations. Cane plantations never use true, sexual seeds. Usually the crop is ratooned, i.e., the cane is allowed to resprout from the roots after cutting. When yields begin to decline after repeated ratooning, the fields are worked over to remove weeds and pests and replanted with cuttings from the last harvest. Flowering is unwanted because it diverts sugar from the cane. Flower initiation is sensitive to day-length, and some cultivars are grown in latitudes where they do not flower. Also, many cultivars are at least partly sterile sexually. It was generally believed until modern times that *S. officinarum* produced no viable seed.

Some cultivar clones, namely the so-called noble canes, are believed to be descended simply from *S. robustum*; these mostly have 80 chromosomes. Most cultivars are primarily derived from *S. robustum*, but have variable admixtures of *S. spontaneum* germ plasm with chromosome numbers varying from 82 to 124. These include both ancient hybrids derived from chance seedlings, the so-called thin canes, and also modern hybrids produced by controlled breeding. The sugar canes known to Europeans at the time Linnaeus bestowed the binomial *S. officinarum* were all thin canes, i.e., *S. robustum* and *S. spontaneum* hybrids. The names *S. sinense* and *S. barberi* pertain to other clones of thin cane.

Domestication of sugar cane is believed to have begun in very ancient times in New Guinea when Papuan gardeners began selecting clones of *S. robustum* with unusually soft sweet stalks, not for making granular sugar but for chewing raw. Eventually, such selection in ancient Papuan gardens produced cultivars with greatly increased juice and sucrose content, in other words, noble canes, which already possessed the essential characteristics that make sugar cane such a successful and productive commercial crop.

The noble canes spread prehistorically as an important garden crop eastward through the Pacific islands as far as Hawaii. Chewing canes also were taken northward prehistorically to Indochina and tropical China and westward through the East Indies to India. The stalks are still a standard component of sea stores on local trading vessels in Malaysia. The stalks remained viable planting material even on long voyages, and any remnants would have served to introduce cultivar clones rapidly. During prehistoric times, techniques were developed in Southeast Asia for crushing the cane and pressing out the juice for fermentation to make wine or for boiling down to a storable syrup.

Meanwhile, a major evolutionary change took place in cultivated sugar cane as noble cane spread through territory of weedy *S. spontaneum* and hybridization took place, probably repeatedly. Such hybrids would have had harder, less sweet, and less robust canes than the noble canes. They were, however, hardier and capable of being grown in subtropical regions of the Asiatic mainland, perhaps first in India. These so-called thin canes are also less susceptible to white ants, rats, and jackals that feed on the soft noble cane. In short, the thin canes were better suited to become a widespread subtropical field crop grown in large quantity, and it was on thin canes that the sugar making industry developed.

Written history of cane planting and sugar manufacture begins in northern India in Sanskrit literature. The earliest precise and secure date for manufacture of sugar from sugar cane is slightly before 300 B.C., although there are references to cane planting starting about 1000 B.C. In its simplest form, sugar making involves simply boiling down the cane juice until the sucrose crystallizes, a process believed to have been earlier used in India to make palm sugar. The resulting granular sludge or magma is allowed to sit for a few weeks while some of the uncrystallizable molasses drains off. The result is a sticky brown sugarloaf known in India as *gur* and later by Europeans as *muscovado*. By 300 B.C., in addition to *gur*, two other kinds of sugar were mentioned under the Sanskrit names *sarkara* and *khanda*, names that have come down in English as *sugar* and *candy*, indicating that processes for refining the raw sugar or *gur* had been developed, presumably by simply washing out the molasses with water. By about 400 A.D., commercial production of sugar was widespread in the Ganges valley.

In Indochina, either by cultural diffusion from India or by independent invention, production of syrup and wine from sugar cane juice was going on by the beginning of the Christian era according to Chinese trade and tribute records. By the 3rd century A.D., the Chinese were getting hard sugar cakes in the form of animals and men from Indochina, referred to as "stone money". By shortly after 500 A.D., the Chinese were manufacturing granular sugar in Kwangtung. In 647 A.D., the Chinese emperor sent a mission to India to study its evidently superior sugar technology. After 1000 A.D., sugar became an important peasant crop in southern China, particularly Fukien, Kwangtung, and Szechwan.

Sugar cane planting did not spread west of India until medieval time, although sugar was known as a rare and costly trade good in the Roman empire. Finally, about 600 A.D., cultivation of thin canes began in Iran, and the crop then was taken relatively rapidly westward with the expansion of Islam. It was introduced into Mesopotamia, the Levant in general, and Egypt before 700 A.D. and continued westward right behind the Arab conquest of North Africa, Sicily, other Mediterranean islands, and Spain. The westward spread of sugar cane, like various other tropical and subtropical Asian crops, depended on development of irrigation. Irrigation works expanded greatly under Islamic rule, particularly in the 10th and 11th centuries. It was a period of relative peace and stability, growth of cities and markets. Traditional land tenure patterns had been broken, and land could be bought and sold. Enterprising innovators could put together estates to produce cash crops. Islam was bringing westward the germ of a sugar plantation system, commonly with an imported labor force, that would grow monstrous during the European colonial expansion.

Europeans had various chances to learn sugar planting from the Arabs, not only in Moorish Spain but with the Norman conquest of Sicily in the 11th century and during the Crusades. After the fall of the Crusader kingdoms in the Levant, Venetian merchant-bankers invested heavily in sugar planting and manufacturing in Cyprus and Crete, which became important sugar exporters in the 14th century. A single Venetian sugar plantation in Cyprus during the 15th century had 400 workers. During the 15th century, Spanish and Portuguese colonists developed an export sugar industry in their Atlantic islands: the Canaries, Azores, Madeira, and Cape Verdes. Venetians and Genoese continued to be heavily involved.

Columbus was married to the daughter of a Madeira sugar planter. On his second voyage to the New World, Columbus introduced sugar cane from the Canary Islands to Hispaniola. Multiple other introductions to the New World followed from Iberia and the Atlantic islands. This marked the beginning of the end of commercial sugar production in the Mediterranean and Atlantic islands. Cane had been temporarily profitable there, but the climate was marginal, expensive terracing and irrigation had been required, and productivity was low. In the humid New World tropics, conditions were ideal, and abundant fertile valleys were free for the conquerors who quickly grasped the opportunity. By the 1540s, 24 plantations, each with its own mill, were producing sugar near Santo Domingo on Hispaniola. The first European commercial plantations had invaded the New World.

The technicians that started sugar manufacture in Hispaniola and other early Spanish Caribbean colonies came mainly from the Canaries. The technology was basically learned from the medieval Arabs and would change only slowly. The cane was crushed with millstones, usually driven by draft animals, in a few lucky locations by water power. The juice was pressed from the ground cane and boiled in a series of cauldrons over wood fires. After crystallization, part of the molasses was allowed to drain off, and the raw brown sugar was molded into hard loaves. For their time, the integrated plantation-factory industrial operations were big, costly, and if properly managed, amazingly profitable. From the outset, Caribbean sugar plantations depended on slave labor. The Greater Antillean Indians were rapidly exterminated by introduced diseases, slavery, and cultural collapse, and the planters promptly turned to African slaves. Some of the first plantations on Hispaniola had labor forces of over 100 Africans. Colonial era sugar plantations were typically large estates with sufficient forest to supply firewood for the cauldrons, provision grounds to feed the labor force, and land for the livestock, in some cases two or three thousand head kept for work and meat. Little needed to be bought except slaves and luxury goods for the great house.

By the mid-16th century, Spanish commercial sugar planting had begun in all the Greater Antilles, which remained the main Spanish source of export sugar throughout the colonial period. In the 16th century, sugar planting also began on the mainland, from Mexico to Peru, but mainly for local needs. In Peru, the Spaniards took over the great Indian irrigation works in the coastal oases for cane plantations. In Mexico, cane was first introduced to the humid Gulf Coast, but during the 17th century with the growth of Mexico City, irrigated cane plantations developed in the nearby warm valleys around Cuernavaca.

In Brazil, the Portuguese began sugar planting on the humid tropical coast between Natal and Rio de Janeiro in the 1520s, and by the 1580s, there were over a hundred estates with sugar mills. The pattern was generally similar to that in the Spanish colonies with large, private estates and slave labor. However, there were also some smaller estates without mills and some tenant farmers owning a few slaves. The Brazilian Indians died more slowly than the Island Arawak, and most of the slaves were still Indians until the 1570s. By soon after 1600, they were nearly all African.

In the early 1600s, a technical innovation appeared in Brazilian sugar factories, namely the three roller mill, which was to have a profound impact on the industry. It appeared about the same time in Peru. It has been suggested that the roller mill originated in China and was brought to South America by Jesuits, but the story is unclear. In any case, it greatly increased the efficiency of juice extraction, eliminating the need for two stages of grinding and pressing. It also produced a by-product, bagasse, which was easily combustible. Bagasse, the fibrous cane stalk minus the juice, soon replaced firewood in the boiling process and much later became the power source for the whole mill operation.

In the early 1600s, another major new factor entered the Brazilian sugar industry, namely participation by the Dutch and by Sephardic Jews, often as allies. In the early 1600s, the Dutch had gained independence from Spain and had developed the leading merchant marine of Europe. Dutch ships led in transporting sugar from Brazil to Lisbon. They also carried Brazilian and Caribbean sugar to Amsterdam, a center of sugar refining for Europe. Many Sephardim from Portugal had moved to Amsterdam and Brazil, among other places, and became involved in commercial sugar. Then the Dutch invaded Brazil and occupied the great sugar plantation region of Pernambuco from 1630 to 1660. After the Dutch and Sephardim were expelled from Brazil, they played a crucial role in developing British and French sugar colonies in the Caribbean, initially in the Lesser Antilles.

The Indians were already growing sugar cane in these islands before the European colonization. Various tropical American Indian tribes had begun planting sugar cane for chewing and syrup very soon after they met with it in Portuguese and Spanish plantations. As in Malaysia, the canes made good provisions for seafaring with leftovers viable for planting. Sugar cane was in Island Arawak and Island Carib gardens on the Leeward and Windward Islands and Barbados before the British and French began settling there in the 1620s. Cane is still planted in Carib gardens on Dominica, the only island where any of them survive. Generally sugar cane died out, along with the Indians, soon after European settlement. The first British and French settlers were not interested in sugar. They tried a variety of cash crops — tobacco, coffee, cacao, cotton — and subsistence crops, not very successfully.

The switch to sugar began as Dutch and Sephardic entrepreneurs moved out of Brazil and began organizing Caribbean sugar production and export. They began with Barbados in the 1640s where a dense population of English smallholders was struggling with a variety of crops. With Dutch and Sephardic expertise, capital, and shipping, Barbados was rapidly transformed into a long-enduring sugar monoculture. Small holdings were consolidated into estates, although these were minute compared to those of Brazil. The Barbados pattern was quickly followed by British and French colonists on other islands in the Lesser Antilles. On Jamaica, British sugar plantations developed on a grander scale after that island was taken from Spain in the 1650s. On Hispaniola, the French took full possession of the western end by 1700 and developed, with the help of irrigation, the most productive sugar colony in the Caribbean. The sugar industry was, of course, a major element in the chronic political struggle and warfare between the European colonial powers. The prevailing mercantilist policies restricted commerce between empires, but illicit trade usually thrived. Throughout the 18th century, West Indian plantations dominated sugar exports to Europe and North America, and the wealth of the planters became legendary. Meanwhile, the nonelite European colonials, the smallholders and indentured labor force, declined in numbers by emigration, often to become buccaneers or pirates. They were replaced by African slaves, whose descendants would one day, with poetic justice, inherit the islands.

So far, ever since sugar cane spread westward from India, the whole history of the crop has been based on thin canes. These had been clonally propagated by cuttings with no known sexual reproduction by seed during all the westward spread through the Mediterranean to the New World. These thin canes came to be known in the American tropics as creole canes. The plot finally thickened in the 1780s when a different kind of *Saccharum*, the ancient Papuan noble canes, came into European hands. Noble cane was collected in Tahiti in the 1760s by Bougainville on his voyage around the world. He introduced it to the French Indian Ocean islands of Reunion, then called Bourbon, and Mauritius, then called Ile de France. From there, it was transmitted to the French Caribbean colonies of Martinique and Cayenne under the names of Bourbon or Otaheite cane. About 1780, the Dutch independently introduced noble cane from Java to the Lesser Antilles and Guianas. In 1791, Captain Bligh introduced noble cane from Timor and Tahiti to Jamaica and St. Vincent, a more valuable gift than his breadfruit. In the humid tropics, noble cane grows more luxuriantly and produces more sugar than the thin creole canes and rapidly displaced them. By 1800, Alexander von Humboldt noted that Otaheite cane had increased sugar plantation yields in Mexico by $^1/_3$ and in Cuba by $^1/_4$. He also noted that it yielded better bagasse fuel for boiling the juice.

Around 1800, several things were happening that were to finally change the pattern of the European sugar colonies, which had remained remarkably stable for 300 years. The most dramatic was the great slave uprising on Hispaniola, beginning in 1791 on the western end, which had been a rich French sugar colony for a century. Afraid of the possible precedent for their own slave plantations, Britain and Spain sent troops to aid the French, but in the turmoil of the French Revolution and the Napoleonic wars, Haiti emerged as an independent republic, the second in the New World. The government of the new nation tried to rebuild the totally wrecked sugar industry, but the former plantation laborers preferred independent subsistence farming, and Haiti was permanently out of the sugar business (although Haiti still exports a little rum, perhaps the world's best). Sugar production was temporarily wrecked on the rest of Hispaniola, the present Dominican Republic, during over 50 years of chaos due to invasion by Haitian armies and revolts against Spanish rule.

The collapse of the sugar production on Hispaniola led to expansion of the previously quiescent plantations on Cuba. Partly this was triggered by arrival of experienced French planters, refugees from Haiti. Also, the Spanish government opened the Cuban ports to slavers of all nations and abolished import duties on sugar mill machinery of foreign manufacture. The large landowners took full advantage of the opportunity to expand plantation acreage and modernize their mills, and Cuba was soon on its way to becoming the world's greatest sugar producer.

Brazil became the refuge of the king of Portugal in 1808, after Napoleon invaded Portugal. The king opened Brazil to foreign commerce, and sugar production promptly soared.

The Napoleonic period saw development of new sugar industries in such far-flung places as Louisiana, Trinidad, Guiana, and Mauritius.

On the Gulf coast of North America, a little sugar cane had been planted since the 17th century by Spanish and French settlers, but the climate made its productivity too low to compete with West Indian plantations. The situation changed with the purchase of Louisiana from France in 1803 and the arrival of refugee French planters from Hispaniola about the same time. Thanks to inclusion behind the U.S. tariff wall, sugar planting became highly profitable, even in the marginal climate. By 1817, newly introduced clones from Java proved more cold resistant than the old creole and Otaheite clones. By 1850, there were over 1500 slave sugar plantations in Louisiana, mainly on the natural levees of the Mississippi Delta floodplain. Only a few were large enough to have steam-powered mills.

In 1802, Trinidad was ceded to England, and in 1803, the British recaptured what would become British Guiana from the Dutch for the last time. A great expansion of sugar planting soon followed.

Mauritius and the neighboring island of Reunion in the western Indian Ocean were the last European colonies to establish sugar monoculture under the old pattern of private estates with resident planters, slave labor, and individual crude mills. Both islands had been uninhabited until colonized by the French East India Company in the early 18th century, not for sugar planting but as strategic bases on the main route between the Cape of Good Hope and India. During the mid-18th century, both islands had famous botanical gardens with many plant introductions, including sugar cane from India and later Bougainville's Otaheite cane. The French colonists with their Malagasy and African slaves were engaged mainly in growing provisions for passing ships and experimenting with spices, dyewoods, and other tree crops. Export sugar production was often precluded by British blockades. In 1810, Mauritius was taken by the British as spoils of war, Reunion remaining a French colony. The French planters on Mauritius were left in possession of their estates, gradually being joined by a few British. During the next 20 years, Mauritius sugar production increased nearly 100-fold.

By mid-19th century, the cane sugar plantations all over the world were running into severe problems from four new elements: epidemic plant disease, competition from the new beet sugar industry, technological innovations, and loss of the slave labor force.

In its long migration westward from India to the New World, the thin cane had left far behind related grasses with their reservoir of fungus and virus diseases. Moreover, there had evidently been only the one original sugar cane introduction to the west before Bougainville's Otaheite cane. Mauritius and Reunion, however, were in frequent direct contact with the East and their botanical gardens, and planters were constantly importing new clones. It was inevitable that plant diseases would go along. In the 1840s on Mauritius and Reunion, Otaheite cane plantations encountered disastrous disease problems. They were able to replace the failing clone with another noble cane Cheribon, obtained from Java. Mauritius continued to import cultivars from southern Asia, the East Indies, and the Pacific islands as a hedge against new disease outbreaks, and by the 1880s had what was probably the most complete *Saccharum* collection in the world. This was, of course, a two-edged sword because until modern strict quarantine and testing procedures were developed, undetected diseases rode along. This dilemma became worldwide during the late 19th century with expanding exchanges of sugar cane cultivars. For example, the red rot fungus was introduced to the West Indies on cane sent from Mauritius and soon attacked the Otaheite cane plantations, while mosaic disease was introduced to South America with cane sent from Java.

Beet sugar manufacture began on the continent of Europe at the end of the 18th century and expanded explosively during the Napoleonic wars and British blockades. With the help of heavy subsidies, beet sugar eventually became a major export from the continent, especially to Britain. In the 1880s, German beet sugar even displaced Peruvian cane sugar in the Chilean market. On a worldwide basis, beet sugar exports surpassed cane sugar exports during part of the late 19th century. However, the world population explosion and per capita consumption increase were so great that cane sugar production in most regions increased through much of the 19th century.

During the 19th century, the cane sugar industry began to adopt steam power and other technological innovations of the industrial revolution, but the change came slowly and painfully. Steam-powered mills for crushing the cane began to be manufactured in England soon after 1800. They had much heavier rollers than the old roller mills driven by animal, water, or wind power. They were, of course, faster, extracted more juice, and produced drier bagasse, which was superior as fuel for both crushing the cane and boiling the juice. They were also much more expensive to buy and maintain. They were not suitable for the average sugar estate, and their general adoption involved consolidation of estates or cooperation between estates to supply fewer, larger mills. Two other technological innovations introduced early in the 19th century required even more capital. Vacuum pans allowed boiling down the juice at lower temperature, thus greatly economizing on fuel. Centrifuges allowed quicker, more complete removal of molasses than the old gravity drainage. Centrifugation produced a dry, tan, granular sugar. (Refining to white sugar continued to be concentrated in the major consuming regions.) Both vacuum pans and centrifuges were adopted early by the beet sugar industry. French manufacturers of machinery for the beet sugar industry tried to expand their market to cane. In the 1840s, they established some modern factories in Guadeloupe and other French colonies. However, the cost placed such factories completely out of reach of traditional family-owned sugar estates.

The shift from small estate mills to large central factories began in the mid-19th century in a few regions where there were large areas of undeveloped arable land, notably central and eastern Cuba, Trinidad, and British Guiana. In the older sugar regions, the shift did not generally take place until late in the 19th and early in the 20th century.

Meanwhile, the long era of slave labor was finally drawing to a close. In the British empire, the slave trade was outlawed in 1807, emancipation was proclaimed in 1834, and the slaves were actually freed two years later. Emancipation in the French empire followed in 1846, in the U.S. in the 1860s, and finally in the Spanish colonies and Brazil in the 1880s. The consequences were quite diverse. In some of the old British sugar islands in the Lesser Antilles, abolition made curiously little difference. Barbados was an extreme example.

This low, limestone island had long since been almost totally partitioned among relatively small sugar estates, each with its own wind-powered mill and resident labor force. Short of emigration, the freed slaves had little choice but to keep on working for whatever wages the planters would pay. Gradually a few estates were subdivided for subsistence farming by free villages, but the great majority of the estates survived with very little change into the 20th century. The story was much the same on Antigua, St. Kitts, and Nevis. Virtually all the arable land was in sugar cane, and there was no escape from continuing to work on the estates for low wages except emigration. Some people did leave for Trinidad and British Guiana where expanding sugar plantations offered higher wages.

By contrast, in Jamaica, there was much unoccupied land suitable for small farms, and only the more prosperous planters were able to retain a labor force by paying better wages. By the end of the century, more than $^3/_4$ of the estates were abandoned or subdivided for peasant farming. By then, there were over 100,000 smallholders on the island, growing both subsistence crops and cash crops, such as coffee and bananas. Here and there one of the old great houses was restored as a historic monument and tourist attraction.

In Mauritius, as in Jamaica, there was much unoccupied land to which the freed slaves escaped to become smallholders. Here, however, the decline in sugar planting was temporary. The difference was due to exploitation of a new source of labor, namely India. In place of the 70,000 black slaves freed in 1836, Mauritian sugar planters brought in 450,000 indentured workers from India during the succeeding 70 years. Most of them became permanent residents when their term of labor expired. They brought wives from India. Before Mauritius became an independent nation in 1968, some descendants of these people had become owners of sugar estates and mills. Indians now govern the new nation.

Indentured labor has had profound effects on the demography of various other sugar colonies. Reunion, Trinidad, and British Guiana together imported over 500,000 Indian laborers, most of whom became permanent residents, Imported indentured labor was also basic to establishment of plantations in Africa. In Africa, the Portuguese had begun planting some sugar in their early colonies on both the Atlantic and Indian Ocean coasts, but the British started a much greater industry in Natal after they took that colony away from the Boer peasant farmers in 1843. Would-be British sugar planters were unable to draft many African peasant farmers as plantation labor, so they brought in over 150,000 indentured laborers from India.

Over 200,000 Chinese contract laborers, many of them recruited in Portuguese Macao, were brought to Cuba and Peru during the mid-19th century, mainly for sugar plantations. Many other sugar colonies imported sufficient numbers of indentured workers to make a permanent imprint on the ethnic character of their populations.

Indentured sugar cane labor played an anomalous role in Australia. Starting in the 1860s, plantations were established along the humid tropical and subtropical northeast coast, employing indentured Melanesians, locally called kanakas. Eventually, the government required all of them to be repatriated, and Australian sugar plantations now depend entirely on white labor. Much of the cane is grown on relatively small private landholdings, but the landscape is that of a great monoculture with the separate properties commonly tied to a few great central factories by narrow-gauge tramways. The mills are partly corporate, partly government enterprises. Enough sugar is produced to supply domestic needs and have a large surplus for export.

In Brazil, the transition from slave to free labor was under way by 1850, well before Abolition in 1888. In the late 18th and early 19th century, sugar planting had expanded from the old core on the humid tropical coast, westward with the frontier and southward into subtropical Sao Paulo. The plantations commonly relied on part-time labor of free men, with their own small farms and on cane produced on contracts with tenants, a pattern easily expanded after Abolition.

In Argentina, a remarkably late and geographically unusual cane sugar industry developed in the 1870s, long after independence from Spain. Some cane had long been grown in subtropical Tucuman for local use. The region was isolated from the Pacific coast by the Andes and over 1000 km from the metropolitan market of Buenos Aires. With the end of the long period of civil wars and construction of the railroad system, an explosive development of commercial plantations and modern factories began. The entrepreneurs were Argentinian, although they received some technical help from French machinery manufacturers. Cane harvest and peak labor demand came in the dry season when most other farming operations had a slack season. Labor was recruited from the Argentine lowlands and from the Andean Indian villages. Before 1900, Argentina had a sugar surplus, but was unable to compete profitably on the export market.

Up to this point, we have been following constant expansion of sugar plantations into new regions. There remains a final chapter, the circling back of commercial sugar planting into the ancient homeland of the crop in the East Indies and Pacific islands. With the end of slavery in the British West Indies in the 1830s, some failing estate owners had the idea of starting plantations in India using the relatively advanced West Indian technology and exploiting the abundant cheap labor. British planters were successful with other commercial crops in 19th-century India, but those were crops new to India that could be grown on lands that the Indians considered submarginal. Cane had to compete for land suitable for intensive native agriculture. The British planters were never able to produce sugar as cheaply as the Indian smallholders did.

In contrast, the Dutch were phenomenally successful in developing commercial sugar planting in Java. A minor export when the Dutch government took direct control of the island, sugar became the main export during the 19th century. This was not done by importing the West Indian estate pattern but by a unique system. The Javanese villages had been growing patches of sugar cane since antiquity. When European trading vessels began arriving, they could buy small quantities of brown sugar. Chinese entrepreneurs expanded the manufacture, using their own technology, and by 1800, they were operating over a hundred small mills in Java. Under Dutch rule, the Javanese villages retained their lands and kept growing rice, but they were required to plant about 20% of their land to export crops, mostly coffee and sugar, which the Dutch took as tribute in lieu of taxes. The rules were modified somewhat with time, and sugar estates were developed on lands leased from the villages with villagers working part-time on the estates. Dutch corporations built large modern factories. Extensive irrigation works expanded the area in cultivation for both sugar and rice. The remarkable mixture of export sugar with traditional subsistence agriculture supported an exploding Javanese population throughout the Dutch colonial period. The Dutch also developed huge monoculture sugar plantations in the East Indies, but these were on the lightly populated outer islands, particularly Sumatra, not on Java.

In the Philippines also, sugar was an ancient garden crop, which was initially developed by Chinese entrepreneurs largely for export to China. In the 1880s, there were about 5000 animal powered sugar mills, a few dozen water-powered mills, and a few hundred steam mills. After 1898, with U.S. occupation and duty-free admission of sugar to the U.S., plantations were greatly expanded and mills modernized. Most of the capital was American, but some Filipino families became large sugar planters.

In Formosa, now Taiwan, small-scale sugar cane planting was well established before 1624 when a Dutch trading enclave was established on the southwestern coast. Exports, mainly to Japan, increased during the 60 years of Dutch activity and continued intermittently under Manchu rule after expulsion of the Dutch. Japan grew a little cane of its own in the Ryukyus, Kyushu, and Shikoku, but imports from Taiwan expanded greatly with the opening of Japan to foreign commerce in 1867. By 1880, Taiwan had about 1500 small sugar mills owned by landlords and moneylenders with cane bought from peasants and tenant farmers. Taiwan was occupied by Japan in 1895 and remained virtually a Japanese colony until being returned to China after World War II. Meanwhile, sugar production expanded enormously as it became a staple rather than a luxury with modernization of Japan. Cane planting remained mainly in the hands of Taiwanese farmers, but manufacture and trade were taken over by Japanese entrepreneurs, including members of the imperial household. By World War I, production was concentrated in 35 modern central factories owned by 13 Japanese companies.

Geographic expansion of commercial sugar plantations was completed in the late 19th century with two tropical Pacific island groups joining the major exporters. Both had indigenous populations who had grown chewing canes in their gardens since prehistoric times, Melanesians in Fiji, Polynesians in Hawaii. In both archipelagos, the native peoples had suffered catastrophic mortality with exposure to diseases introduced by Europeans. European colonists had access to huge expanses of volcanic soils ideal for cane growing, but were unable to recruit a local labor force to exploit the opportunity.

In Fiji, the British governor suggested in 1875 that the planters bring in indentured labor from India, a pattern he had seen work in Trinidad and Mauritius. Over 60,000 Indians were brought in during the next 40 years. Starting in 1880, the Colonial Sugar Refining Company, which already had successful operations in Australia, began buying large tracts of land in Fiji and building modern factories. By the 1920s, it had bought out all its competitors. Fiji now has two landscapes with almost completely separate economies and cultures: traditional Melanesian villages with subsistence crops and gardens vs. sugar monoculture owned by a single corporation with its own modern factories and narrow-gauge railways employing ethnic Indians. The ethnic antagonism and political conflict remain very active.

In Hawaii, commercial sugar planting began slowly in the mid 19th century while the islands were still technically an independent Polynesian kingdom. The labor force was initially mainly native Hawaiians, but that population had dwindled to such low numbers that expansion depended on importation of indentured labor, in this case mainly from China and Japan, with a smaller but significant number from Portugal. Starting in 1875, Hawaiian sugar was admitted duty-free to the U.S., and from then until annexation of the islands in 1898, planting expanded greatly. Exports were controlled by a few wealthy merchant families, some founded by early missionaries from New England. A few families were German, including that of Claus Spreckels, a wealthy entrepreneur with experience in beet sugar refining in Germany and California. Spreckels pioneered vertically integrated operations from a large, irrigated plantation on Maui through a central factory, a shipping line, and a refinery in California. By the time Hawaii was annexed to the U.S. in 1898, its sugar exports were controlled by half a dozen companies. By then annual exports amounted to hundreds of thousands of tons, second only to Cuba.

Deliberate breeding of sugar cane was not begun until the late 19th century. Plantations had always depended on vegetatively propagated clones, not seedlings, and it was universally believed that the cultivars did not produce viable seed. This was in fact true of the traditional cultivars; the thin canes and noble canes are usually male sterile with nonviable pollen and hence set no seed. However, in 1888, fertile clones producing seedlings were discovered simultaneously in Barbados and Java. By 1900, scientific breeding had begun in various countries. Significant results came in the 1920s in Java where Dutch breeders produced a new clone POJ (from Proefstation Oost-Java) 2878, the first great "nobilized" cane. The nobilized canes, which became standard for plantations, were obtained by essentially reenacting the accidental hybridizations that occurred in ancient Asia when the cultivated noble canes met wild *S. spontaneum* to produce the hybrid thin canes. The nobilized canes are high polyploids with most of their genomes derived from *S. robustum* and a variable but small percentage from *S. spontaneum*. They combine the sugar yield of the former with the hardiness, disease resistance, and ratooning capacity of the latter. Modern breeding has been invaluable to the planters, by control of major diseases without chemicals. However, the genetic base of the crop remains very narrow with a few old clones recurring in the pedigrees of nearly all cultivars.

Compared to the 19th century's profound changes, changes in the sugar cane industry during the 20th century have been rather mild. There have, of course, been plenty of quantitative changes, abrupt ones due to wars, revolutions, and dissolutions of empires, gradual ones due to economic and demographic processes. Enumeration of these would be inappropriate, but a general pervasive process during this century deserves mention. This was the ongoing reduction in number of small mills and concentration of milling in huge central factories called simply *centrales* in Spanish. For example, by 1950, a single Cuban *central* could take in over 10,000 tons of cane a day, requiring daily harvests of several hundred hectares. This centralization was not universal. There are places in Central America, for instance, where oxcart loads of cane are still processed to brown sugar loaves in mills run by a few part-time workers. But sugar exports during the 20th century have been taken over more and more by regions with large central factories. Some former exporters that were unable to make this shift have had to abandon sugar planting. In the Lesser Antilles, sugar was Antigua's main export for three centuries, but the rather low and unreliable rainfall made productivity too uncertain to justify investment in central factories, and the old wind mills were simply abandoned. Nevis was simply too small to support a modern mill. Saint Vincent had excellent climate and productive cane lands, but they were scattered in isolated valleys, and the ruins of the old water mills are now surrounded by banana plantations.

On the whole, however, *S. officinarum* has been a resilient survivor in the face of cultural and economic changes. The plant is such an amazingly efficient photosynthetic machine that, where the climate suits it, other crops are usually unable to compete. This advantage may be highly immune to political changes. In Cuba, sugar monoculture thrives as well under Fidel Castro as it did when plantations belonged to Spanish landed families or to North American corporations.

Cane is also a tough crop to displace because it requires little labor for the value of the product. Unless irrigated, it usually needs no care between planting and harvest. The crop is a model of regenerative agriculture. Several successive crops can be cut before replanting. The canes grow so fast and so densely that the crop suppresses most weeds. Its canopy and litter prevent erosion. Cane has been grown continuously without rotation for over a century on many plantations. Minerals are recycled by returning bagasse ash and sometimes surplus molasses from the mills to the cane fields. The root bacteria provide nitrogen compounds. All the elements in the exported sucrose and alcohol come from air and water.

The bagasse fuel provides all the energy needed to run the mill and often more. Some central sugar factories sell electric power by transmission lines. In some regions, sugar cane planters do rely on imported agricultural chemicals and petroleum for agricultural machinery and hauling. Hawaii is a notorious example. Subsidized by artificially supported sugar prices and a protected market, Hawaiian plantations could afford to have extravagant costs for producing cane. Rather than being cut close to the ground with macheté, Hawaiian cane is commonly bulldozed and loaded on huge trucks, dirt and all, by draglines. Copious water is used at the mill to wash the soil and rocks off the cane and out to sea.

It is possible for cane plantations to be net producers rather than consumers of motor fuel. This has been demonstrated on a grand scale in Brazil. Since 1931, Brazil has been producing ethanol for fuel by fermenting and distilling the whole cane juice, not just the molasses by-product. In the 1970s, the government provided strong financial incentives for the switch from gasoline to reduce costly petroleum imports. By the late 1980s, nearly 4 million cars were running on pure alcohol fuel while Brazil's other 5 million cars were running on a gasoline-alcohol blend. The alcohol was produced by hundreds of distilleries and provided about 500,000 jobs. Also, the switch alleviated a severe smog problem. The program would have been a shining example of relying on a renewable energy source except that periods of overpumping of petroleum and collapse of crude oil prices have made alcohol more expensive than gasoline. Also, new offshore oil fields have glutted Brazil with petroleum. There are so many beneficiaries of the cane alcohol program, however, that the Brazilian government has had to continue it in spite of the cost. Unfortunately, it was a good idea ahead of its time.

Gymnosperms — Conifers and Allies

Gymnosperms, the first plants to reproduce by seed rather than unicellular spores, appear in the fossil record in the Paleozoic period about 350 million years before *Homo sapiens.* Like the reptiles, the gymnosperms peaked in diversity and dominance during the Mesozoic period. Many families are known only as fossils. Members of various surviving families — ginkgo, cycad, yew, cypress, araucaria — are widely planted as ornamentals and for landscaping and yield some useful products, especially wood and edible seeds. However, two other families — pine and redwood — stand out economically, both as forest and plantation trees.

CONIFERS AND ALLIES

Pinaceae — Pine Family

The pine family originated and diversified in the Mesozoic supercontinent of Laurasia. Various of its genera have remained widespread in the present northern hemisphere vegetation: *Abies*, *Picea*, *Larix*, *Tsuga*, as well as the two genera discussed individually, *Pinus* and *Pseudotsuga*. Many species grow rapidly into tall trees. The wood is comparatively homogeneous, soft, strong for its weight, and has been indispensable to many ancient and modern tribes and nations.

PINUS — PINES

The pine genus became widespread all around the northern hemisphere in the mid-Cretaceous period about 100 million years ago. By the Tertiary, many fossil pines closely resembled living species. There are about 100 living species; their overlapping ranges include most northern hemisphere forests and woodlands from arctic alpine timberline to desert mountains and tropical savannas. A single species reaches the equator (in Sumatra). Pines are generally ecological pioneers that colonize naturally open and disturbed sites. Although most species are not adapted to reproduce under a forest canopy, they can survive long periods between disturbances by towering over shade-tolerant hardwoods. When growing in dense stands, shading out of the lower branches results in pine trunks with straight-grained, knot-free wood.

Several pine species are of human interest for their production of pitch or resins. These are secreted by specialized ducts or canals within the vascular tissues. The pitch exuded by wounded pines presumably serves as a defense against invasion by boring beetles and against other injury. In ancient times, different peoples independently discovered how to tap pines for pitch. Traditional uses include waterproofing Apache basketry and caulking Greek wine casks and the hulls of wooden ships. The archaic term "naval stores" is retained today for turpentine and rosin distilled from pine pitch.

Several pine species (European stone pines, North American pinyons) are also of human interest for their large, sweet seeds, rich in fat and protein. The so-called pine nuts attract seed-eating birds, such as jays and nuthatches, which cache some of the nuts in open ground. Although such dispersal is costly to the plant species, it was evidently worth the cost by increasing the disperal radius beyond that of the usual wind-dispersed conifer seeds.

Human use of pine wood, pitch, and nuts has depended mainly on wild trees. Five species, however, are at least incipient domesticates and will be discussed individually.

***Pinus pinea* — Italian Stone Pine** (Lanner 1981; Mirov 1967; Rikli 1943–1948)

This species grows along seacoasts and in other naturally open habitats of the Mediterranean region from Portugal and Spain to Turkey and Lebanon. Nobody knows how much of this range is due to natural dispersal because the species has been commonly planted for centuries. According to one legend, it was introduced to Italy from the east during Etruscan times. In various languages, its common name means domesticated or planted pine.

Like several other Eurasian stone pine species, *P. pinea* is valued mainly for its edible nuts, the kernel containing over 80% protein and fat. The shells are typically harder to crack than North American pinyons, but a soft, thin-shelled variety is known, probably due to long human selection.

P. pinea is also planted for shade and ornament. Its low, much branched trunk and spreading umbrella-shaped crown make the wood of little value as lumber. Both spontaneous and planted stands are tapped in southern Europe for naval stores.

P. pinea nuts are imported to the U.S., usually shelled, under the name of pignolia and marketed in competition with nuts gathered from the native wild pinyons.

***Pinus sylvestris* — Scotch Pine** (Bertsch 1953; Förskell 1963; F. Hesse 1938; Spain 1966)

Despite the common English name, Scotland holds only a minor outpost of this species, the world's widest ranging pine. It is abundant over huge areas from Germany and Scandinavia clear across Eurasia to eastern Siberia. It has scattered populations in the mountains of southern Europe from Spain to the Caucasus. In various languages, it is called simply the wild pine or the pine of the forests.

P. sylvestris was even more widespread and abundant at the end of the last Glacial period when it moved northward into former tundra and glaciated regions, outrunning other forest trees. These gradually overtook and replaced the pine on mesic sites, but the pine had a resurgence with Neolithic forest disturbance by shifting cultivators.

The trees can grow fairly tall, commonly to 30 m, and have been exploited since prehistoric time for general purpose timber. The pitch provided northern Europe with an early source of naval stores.

In Germany, sporadic cases of planting of *P. sylvestris* were recorded from about 1400 to 1800 A.D., but during that period, most forestry was so-called soft silviculture, i.e., logging was done in ways that promoted spontaneous restocking with the desired species.

Near the end of this period, German soldiers returning from the American Revolution brought back news and seed of pines much larger than *P. sylvestris*. For a while, German foresters developed a virtual mania for planting these exotics, particularly *P. strobus*, the magnificent white pine of the Great Lakes and New England. However, because of climate, soil, and disease problems, the exotics generally failed to displace *P. sylvestris*.

The Napoleonic wars left German forests in bad condition. During the mid-19th century, German foresters established sustained yield silviculture in state, communal, and private forests. In the case of *P. sylvestris*, this usually meant clear cutting and artificial replanting. In 19th century Germany, the substitution of coal for firewood and the demand for more conifer lumber and pulpwood led to wholesale replacement of the natural mixed hardwood forests with monocultures of *P. sylvestris* and the native spruce *Picea abies*.

Long before acid rain became a threat, leading German foresters warned that these simple stands were vulnerable to disease, parasites, and soil deterioration. However, even the once romantic Schwarzwald became converted to even-aged rows of pure conifer plantations. By World War II, about 45% of the German forest area was under *Pinus sylvestris*, about 20% under *Picea abies*. By a national law, commercial *Pinus sylvestris* seed had to be certified as coming from high quality stands within the same climatic region as the plantation.

Planting of *P. sylvestris* and, on a smaller scale, *Picea abies* developed similarly in other northern European countries. For example, in Sweden some planting began in the 18th century. For a long time, unselected seed was used, generally whatever was cheapest. Little attention was paid to geographic origin, and tons of *Pinus sylvestris* seed were imported. Pines planted with exotic seed often grew poorly, especially at higher latitudes. In 1905, the Swedish government began compelling reforestation after logging and became involved in supplying seeds and seedlings. However, genetic seed selection was then far in the future, and for a long time, unselected seed was routinely collected during logging. In the 1920s and 1930s, Swedish foresters began tackling the formidable job of studying hereditary variation in *P. sylvestris*. As in any organisms, genetic study requires experimenting with large progenies of known ancestry. In the case of trees, such experiments take lots of time and space. By the 1940s, Swedish foresters had learned enough about the genetics of *P. sylvestris* to begin practical applications to pine plantations in royal, county, and private forests. Pine seed orchards, isolated from wild pine stands to escape uncontrolled pollination, were planned for different regions of Sweden. Elite individual trees had been selected after fairly complete survey of natural stands throughout the country. Some of the most promising were propagated clonally by grafting for establishing seed orchards. By the 1960s, *P. sylvestris* had become an incipient domesticate in Sweden with planting of unknown parentage gradually decreasing.

Planting of *P. sylvestris* has developed independently in other parts of its native range. For example, in mountainous regions of northern Spain, 175,000 ha of the species were planted between 1945 and 1965, and in Siberia, extensive plantings have recently been reported, but details are not available.

***Pinus radiata* — Monterey Pine** (Axelrod 1980; Carron 1980; Eldridge 1979; Libby 1990; McIntyre 1959; C.W. Scott 1960; Spain 1966; Streets 1962; White 1987)

The earliest fossils assignable to *P. radiata* grew in southern California about 15 million years ago. The species ranged widely in California during the Pleistocene period, but in Post-glacial time retreated to tiny relic populations on the central California coast and on islands off the west coast of Baja, California. *P. radiata* was evidently driven out of the rest of its former range by the rise of modern plant communities in which it is a poor competitor. It barely escaped extinction by hanging on to a few coastal habitats where there was little competition.

Since Spanish times, the small native stands have been used locally around Monterey for cheap lumber and firewood. Monterey pine is occasionally planted in California for shade, landscaping, or shelter, but has never been considered worth planting for timber. In its marginal refuges, the species is usually stunted and misshapen and does not show how vigorously it can grow when planted away from the sea and freed from competition. *P. radiata* retains much broader climatic and edaphic tolerances than its natural distribution would suggest.

The species is an important plantation tree in modern Spain, but there is no record that it was taken there while California belonged to Spain and Monterey was its capital. Its introduction to Europe is attributed to David Douglas, a famous Scottish plant explorer. He visited Monterey in 1831 when it belonged to Mexico and collected *P. radiata* seed for the British Royal Horticultural Society. Although Monterey pine never became silviculturally significant in Britain, trees planted there in parks and gardens may have provided the seed for Spain. A famous old individual of *P. radiata* was planted near San Sebastian on Spain's Bay of Biscay coast about 1860. The species spread slowly until after 1900 when it was recognized as capable of producing commercial wood crops faster than the native conifers. It soon was widely planted in the Basque Provinces, generally for pulp or low cost timber on a 16- to 30-year rotation. Between 1940 and 1956, about 50,000 ha of *P. radiata* were planted on private lands. So much land was being taken out of food crops for Monterey pine that the Spanish government imposed limitations on the area planted.

In the southern hemisphere, *P. radiata* has become the main timber crop of Australia, New Zealand, and Chile. All three countries have magnificent native conifers, particularly *Araucaria* and *Podocarpus* species, but these have generally been destructively exploited rather than regularly planted. In experimental plantings, they have not compared well with Monterey pine in wood production.

P. radiata was introduced to Australia and New Zealand in the mid-19th century, perhaps repeatedly and both via England and directly from California. There were various possible carriers, including American whalers, sealers, and miners leaving the California gold fields for Australian gold after 1856.

P. radiata was growing in the botanical garden at Melbourne, Victoria, by 1857, and commercial planting was recorded near there at Mount Macedon in 1880. In both Australia and New Zealand, planting was very tentative well into the 20th century, *P. radiata* being just one of various native and exotic conifers that were under trial. A long period of trial and error was needed to learn the species' broad but specific environmental tolerances and how phenomenally productive it could be when planted in the right regions.

In New Zealand, massive planting began in the 1920s and 1930s on both public and private lands. A spectacular early success with Monterey pine plantations was afforestation of a great barren area on the North Island known as Pumiceland. This was an area of geologically recent volcanic activity that supported only worthless scrub and sparse grass. Here and elsewhere in New Zealand, *P. radiata* plantations proved more productive than any other conifers, native or exotic.

Planting accelerated in the 1940s and from 1975 to 1985 averaged 55,000 ha/year, about 45,000 new and 10,000 replanted. New Zealand now has about a million hectares planted to Monterey pine about half being on public lands, mostly on the relatively warm and dry North Island. Thinnings and short rotation harvests are used for pulpwood. Grown to larger size on a 30 or so year rotation, Monterey pine provides about 85% of the country's saw timber. Production is expected to double by the year 2000, and about 75% of the wood is expected to be exported by then.

Australia has not planted Monterey pine quite so fast. The area increased from about 10,000 ha in 1957 to about 50,000 in 1983. Monterey pine is the main conifer planted in all the states except Queensland where another North American species *P. elliottii* predominates. The plantations are a disaster for the native plant communities they replace. In the Snowy Mountains of New South Wales, for example, whole mountainsides have been cleared of magnificent, complex native forests to make way for even-aged stands of this single species.

P. radiata had been introduced to Chile by 1885, but was not heavily planted until after 1935. It then rapidly expanded in the Concepcion region, which has a climate much like Monterey. For decades, 10,000 ha or so were planted in Chile each year, mainly on private lands. Plantation pine has taken over the burden the natural *Araucaria* forests once bore of supplying Chile with abundant, cheap soft wood for domestic use and export. Current highway expansion, which is pushing the frontier of settlement far southward, is accompanied by explosive expansion of the pine plantations.

Genetic selection and breeding of Monterey pine is currently being given careful attention in both Australia and New Zealand. A major goal is developing trees with straighter trunks for better saw logs. Considering that the pine seed originally introduced to the southern hemisphere probably had a narrow geographic and genetic base, pine plantings there have displayed surprisingly great diversity for breeding purposes. However, crop breeders always want the maximum possible gene pool, if for no other reason than to reduce vulnerability to plant diseases. Thus, they have sought fresh introductions from native stands. The most ambitious of these was organized in 1978 by forestry agencies of Australia, New Zealand, and the U.S. Four tons of cones were collected by sampling all the populations in California and on the Baja, California islands. The seed was extracted at the Institute of Forest Genetics, Placerville, California, which has been operated by the U.S. Forest Service since 1935. Since survival of the native populations is precarious, a subsample will be kept in cold storage and renewed when viability declines by planting at Placerville. Bulk trials have begun at many sites in Australia. In all three countries, genetic variation of Monterey pine is now being studied by isozyme analysis.

***Pinus taeda* — Loblolly Pine, and *P. elliottii* — Slash Pine** (Carron 1980; Healy 1985; McDonald and Krugman 1986; Pinkett 1980; Powers 1986; Schreiner 1969; Walker 1980; Wells and Lambeth 1983; Zobel 1964, 1971)

The southeastern U.S. was endowed with eight closely related species of yellow pines, all of which have similar histories of exploitation as volunteer stands. At present, *P. taeda* and *P. elliottii* are leading the other species in being domesticated and planted on a large scale. The other species will not be treated individually.

P. taeda has a huge range in coastal plains and foothills from Oklahoma and Texas east to Delaware and central Florida. *P. elliottii* has a more limited range mainly in the coastal plains from Mississippi and South Carolina down through Florida. Both species are adapted to grow in poorly drained podzols with high water tables, but can also grow on well-drained slopes given enough rainfall. They are strong ecological pioneers in both primary succession, as in freshwater marshes, and in secondary succession, as in burns and clearings. The pines colonizing such openings commonly form an even-aged stand with an understory of hardwoods. Although the pines do not reproduce in an undisturbed forest, they grow taller than the hardwoods, so some trees are able to survive for a long time.

Both loblolly and slash pines can grow to fairly large sizes, over 30-m tall, with trunks over a meter in diameter at the base and tapering gradually to yield a long, straight saw log. When growing in a dense stand, the trees are self-pruning, and the wood is consequently free of knots. The wood is of mediocre quality, being brittle, resinous, and coarse-grained.

European colonists encountered the so-called First Forest of southern yellow pines, partly natural stands, partly regrowth on former Indian maize fields. The Indians' shifting agriculture involved long forest fallow with hardwood regeneration and soil development before the next brief clearing. The Europeans unintentionally created the Second Forest of yellow pines. Their attempts to maintain permanent clearings for cotton, tobacco, and other clean-cultivated crops led in many cases, particularly in the Piedmont, to catastrophic soil erosion and permanent abandonment. The pines that took over, the spontaneous Second Forest, became the major resource of much of the southeastern U.S. for a long period before the development of the deliberately planted Third Forest.

Until after the Civil War, the Second Forest provided the South with a superabundance of pine. Exploitation of the trees, both by tapping for naval stores and by logging, was generally a small scale, low paying occupation, often by part-time farmers. The naval stores were good quality and in wide demand. The lumber, however, could not compare with that of the magnificent white pine *P. strobus*, native to northeastern North America.

Instead of the great boom and bust logging industry of the northeast, 19th century logging in the southeast was typically a low-capital, diffuse industry with thousands of small sawmills; the intake of pine logs by such mills was often less than the rate of growth in their localities.

After the Civil War, as the white pine began to run out, some of the northern lumber barons began acquiring old growth southern pine stands. After 1880, lumbermen bought millions of acres of federal land in the South for a dollar or so an acre to be logged with the same cut and get out philosophy that had been so profitable in the northeast. During the first quarter of the 20th century, southern yellow pines were the country's main source of lumber for domestic use and export to Europe. Before the southern pines were depleted, the Panama Canal was opened. The great virgin Douglas fir forests of the Pacific Northwest became the main source of lumber, and the less valuable southern pines continued to be logged at a reduced rate for more local markets.

However, it was southern yellow pines, not Douglas fir that led to the first large scale silviculture developed in the U.S., but this did not come until the latter half of the 20th century. The reasons for the slow development were clearly economic rather than conceptual or technical.

Since the middle of the 19th century, foresters trained in Europe had repeatedly exposed the South to their ideas on both soft and hard silviculture. The most famous case was the Vanderbilt estate in North Carolina where Gifford Pinchot and his successors attempted to introduce European silvicultural methods. Pine plantations were readily established, but evidently not perceived as profitable.

In 1906, the Great Southern Lumber Company made another premature attempt to establish pine plantations. It built a large, modern sawmill at Bogalusa, Louisiana, planted pines on lands of its own, and encouraged planting by private landholders, hoping for a steady supply of saw logs. However, at the time there was generally little incentive for pine planting because of an overabundance of volunteer stands on old fields and logged areas. Young pine stands were so little valued, even if vigorous, that they were commonly abandoned to avoid paying taxes. During the long depression in the region during the 1920s and 1930s, many millions of hectares of prime pine lands were auctioned off by local governments as tax delinquent. Much of this land was bought by the federal government to patch together new National Forests, quite unlike the old-growth National Forests that had been carved out of the public domain in the western states. Depression era federal programs, such as the Tennessee Valley Authority and the Civilian Conservation Corps, engaged in extensive planting of southern yellow pines, but the intent was to prevent erosion, protect watersheds, and create jobs, not to develop an economic timber crop.

In the 1930s, more or less academic research on pine genetics was begun in various southern universities and in national and state forests.

However, the revolution in pine forestry in the 1950s that started the Third Forest was not led by the academic or government agencies. It was led by big paper companies that found loblolly and slash pines to be the best source of cheap pulpwood. Commercial paper making had begun in the South in an inconspicuous way when the big Bogalusa lumber mill, mentioned above, began manufacturing paper from sawmill waste by the Kraft process. By 1915, this mill was producing 50 tons of brown Kraft paper a day as a lumber by-product. The Kraft process had been developed in Europe for making pulp from Scotch pine to supplement the traditional spruce. The chemical liquor in which the pine wood is cooked dissolves the oleoresins, which are unacceptable in paper. Thus, naval stores can be recovered as a by-product of paper manufacture instead of tapping live pines. Kraft paper is very strong, whence its German name, but it was suitable only for brown bags and wrapping paper. By 1930, most of the U.S. Kraft paper production was from southern pines. At that time, the U.S. was importing 70% of its pulpwood, mainly Canadian spruce.

Research in Georgia in the 1930s led to economical production of white paper from yellow pines, and soon U.S. paper manufacturing underwent a massive geographic shift until the bulk of its production came from the South. A modern paper mill is one of the most expensive and technologically advanced of industrial installations. Unlike the old-time sawmills, it is intended to be permanent. It has to be fed by an enormous, constant supply of water, energy, and pulpwood. Various of the big paper companies, some originally from New England and Canada, bought hundreds of thousands of hectares of land in the South and set out to establish sustained yield from short rotation pine crops. Large blocks of forest, generally mixtures of pines and hardwoods, were clear cut and hardwood regeneration prevented by mechanical cultivation, herbicides, and, after the pines were old enough, controlled burning and grazing. Rather than relying on spontaneous pine seedlings, the industrial forests generally moved rapidly into planting of seedling pines started in nurseries with a rotation time of 20 to 30 years.

Much applied research was done on planting density, thinning, pruning, fertilizing, and all the other practices used to bring a wild species under cultivation. The most difficult research dealt with selection and breeding aimed at genetic control of the plantations. Since the 1950s, this work has involved an increasingly complex network of foresters working in universities, industry, and state government. The programs were initially influenced by Swedish work with Scotch pine. Spontaneous stands were searched to find elite trees, careful attention being paid to geographic proveniences; their progenies were grown and evaluated; selected individuals were further propagated by seed and as grafted clones. As the researchers learned more about heredity in loblolly, slash, and other native pines, they moved into breeding for specific goals. For example, depending on the kind of paper to be produced — Kraft, newsprint, tissue, high quality white, cardboard — different seed orchards for plantation stock were established. As another example, to broaden the genetic base for future breeding, loblolly pines from widely separated geographic proveniences have been hybridized. Some of the breeding programs rapidly produced very appreciable increases in quality and quantity of wood produced by the pine plantations. By 1971, a single breeding cooperative in North Carolina had about 1000 ha of seed orchards and was supplying its members with about 100 million genetically improved seedlings a year.

The success of southern pine silviculture for pulpwood led to renewed interest in growing pines for lumber and even for plywood, which requires larger diameter logs than saw logs. Plywood was formerly a monopoly of the Pacific Northwest, but production began in the South in 1962 and expanded rapidly.

As the cream was skimmed off the fine old ancient conifer forests of the Pacific Northwest, the lumber industry has become much more interested in the plebeian southern pines. In the region's long, hot, wet summers, they commonly grow to saw timber size in 40 to 60 years and produce a peeler log for plywood in 65 years. Industrial forests in the South are no longer restricted to former Second Forest pine lands, but have been expanding on a grand scale with clearing of natural hardwood forest for new pine plantations. By the mid-1980s, combined replanting and new plantings totaled above a million hectares a year in 13 southern states; half of the pines planted were genetically improved stock.

Pines are now the South's most valuable crop, worth more than traditional crops or livestock. Large stretches of country have become monotonous landscapes of a single pine species. How permanent this land use will be is uncertain. Some foresters are doubtful that short-rotation, constant monoculture of pines can be sustained indefinitely without loss of soil structure and other problems. Also, the economics of a crop that takes decades between planting and harvest is a tricky problem. Profitability may depend more on tax structure, interest rates, and land values than on skillful forestry. The big paper and lumber companies evidently believe southern pine plantations are worth the risk. However, many private landholders do not agree. In 1981, less than half the private landowners bothered to replant their logged areas, even though the states offer pine seedlings, subsidies, and technical advice.

Although the largest slash and loblolly pine plantations are located in their native region, both are being widely planted abroad in regions with a climate similar to the southeastern U.S. Slash pine was introduced to China from the U.S. in 1946, and loblolly pine arrived later. They out performed the Asiatic *P. massoniana* in growth rate and trunk form. By the mid-1980s, the two species were successfully established on over 1 million ha in 14 provinces of southeastern China, and additional plantations were being made at the rate of 40,000 ha/year (Richardson 1990). Chinese seed orchards cannot keep up with the demand for planting.

Slash and loblolly pines are also the most widely planted conifers in subtropical regions of the southern hemisphere where they grow faster than the higher quality native *Araucaria* species. They were introduced to subtropical Queensland in the 1920s and, after it was learned that phosphate fertilizer was required, became extensively planted on both government and private land. In 1953, Queensland began a long-range breeding program, which has advanced faster than in the pines' homeland because of freedom from the uncontrolled pollination that occurs where wild stands are present. Seed orchards for planting stock were established in the 1960s.

Large scale slash and loblolly planting is also underway in South Africa, southern Brazil, and Argentina.

PSEUDOTSUGA — DOUGLAS FIR (Hagenstein 1952; Hermann 1987; D.R.W. Scott 1980; Weyerhaeuser 1981)

Pseudotsuga is a genus of six species restricted to eastern Asia and western North America. The genus appeared in the fossil record in both regions in the Tertiary. In their growth form, the species resemble firs, spruces, and hemlocks more than they do pines.

P. menziesii, the Douglas fir proper, is by far the most important species. It is one of the dominants in the great conifer forests of the maritime Pacific Northwest from British Columbia to northern California. Farther inland, it has a highly disjunct range in mountain forests from western Canada down to central Mexico. Douglas firs are capable of extremely rapid growth and a very long life, culminating in occasional trees over 100-m tall and over 4 m in trunk diameter. The Pacific Northwest had extensive stands of old-growth Douglas firs in nearly even-aged stands that dated from catastrophic crown fires during rare drought years. The species rapidly colonizes burns with its light, wind-borne seed. If there is no other fire for a few centuries, Douglas fir is gradually replaced by more shade-tolerant conifers, such as hemlock.

Douglas fir provides a good, general purpose lumber, used much like pine. Logging for local use began in the mid-19th century. The federal government gave the Northern Pacific and other railroads huge tracts of Douglas fir to subsidize construction. Foreseeing the end of the Great Lakes pine logging, Weyerhaeuser and other eastern lumbermen acquired Douglas fir stands from the railroads and homesteads on a grand scale. With access to eastern markets after construction of the railroads and the Panama Canal, Douglas fir became the country's main source of ordinary lumber and supplied nearly all the large dimension timber and plywood with pulp as a by-product. For a long time, the old cut and get out pattern was continued. This was by no means solely the fault of the lumbermen. As in the east, public opinion saw clearing the forest as a desirable frontier episode, preparing the land for fields and pastures. Until well into the 20th century, federal and state taxes were prohibitive for permanent private land ownership for sustained lumber production.

However, much of the cutover land proved submarginal for nonforest uses. As the old stands were depleted, attitudes began to change. In the late 1940s, Weyerhaeuser and other lumber companies began planning tree farms. At first, this meant simply soft silviculture with spontaneous regeneration of Douglas fir encouraged. Clear-cuts were checkerboarded, leaving uncut blocks temporarily as seed sources until the first clearings had been restocked. The results were usually slow and uneven for various reasons. Cone crops are often poor for years on end, and seedling mortality from drought, insects, and browsing animals was severe. Red alder, *Alnus oregana*, which was mainly confined to streambanks in old forests, often colonized clear-cuts in masses and competed with the Douglas fir seedlings. Committed to tree farming, the lumber companies predictably experimented with increasingly intensive forestry: site preparation, herbicides to control alder and other competition, pesticides, nitrogen fertilizer, planting of Douglas fir seed, and thinning of young stands. Direct seeding was tried first and is still used on very steep sites, seed being dropped from helicopters. Direct seeding is still used also when the supply of nursery grown seedlings is insufficient for planting logged areas.

Use of chemicals is under increasing restriction. Also, attempts are being made to manage mixtures of Douglas fir and alder. Nitrogen fixing bacteria on the alder roots contribute to soil fertility as in a natural burn. Since 1970, planting of nursery grown seedlings has rapidly expanded. Until the late 1960s, most Douglas fir seed for nursery planting was obtained by robbing caches of cones harvested by red squirrels. The squirrels selected cones for edible seed quality. Foresters subsequently decided that it pays to harvest cones from trees selected for timber quality. Attempts to propagate elite trees clonally have been frustrated by unsuccessful grafts, so seed orchards are generally simply progenies of elite trees. By the late 1970s, a great many breeding programs had been initiated, involving networks of cooperation between industrial and government foresters. Massive research is under way because of the great diversity of Douglas fir races and ecotypes.

In short, domestication is in an incipient stage, far behind the process of domestication of southeastern pines. In the long run, it seems that farming of Douglas fir may be important. On good sites, a rotation of about 50 years may produce saw logs, although they will not be comparable to old-growth trunks.

In northwestern European forest plantations, Douglas fir has become the most important non-native conifer. Seed collected by David Douglas near the Columbia River mouth was sent to Britain in the 1820s. Along with various other western North American conifers, Douglas fir was widely planted in

European parks and arboreta during the 19th century. Douglas fir's impressive growth when compared to the European native conifers attracted foresters' attention. The earliest plantations in the British Isles were established in the 1850s in Perthshire, Scotland, probably from the second generation of seed sent by David Douglas. The species has performed well, particularly in Scotland, when planted on sites with some shelter from the wind. It is grown on about 50,000 ha in Great Britain.

German foresters began experimenting seriously with Douglas fir in the 1880s and since 1910 have done continued research on performance of seed collected in different parts of the native range. Planting only began in earnest after World War II.

France has more extensive Douglas fir plantations than any other European country, about 220,000 ha, compared to 80,000 in Germany. Although the species was introduced to France before 1850, forest plantations developed slowly until about 1950. At that time, France had only about 10,000 ha of Douglas fir plantations. About that much is currently being planted each year, and large future expansion is planned. Just as in the Pacific Northwest, large timber harvests from plantation Douglas fir in France are decades in the future.

Taxodiaceae — Redwood Family

The family ranged over most of the world during the Mesozoic period, but is now widely disjunct in relic areas of the northern and southern hemispheres. There are 10 living genera, including the redwoods (*Sequoia*, *Sequoiadendron*, and *Metasequoia*).

The Taxodiaceae, like Pinaceae and many other conifers, are heliophiles, intolerant of shade. They outstrip competition by growing fast, straight, and tall. Like other conifers, they produce relatively homogeneous, light wood. Also like other conifers, they are of economic interest to mankind primarily as sources of large, straight logs, easily cut into light lumber that is very strong for its weight. The Taxodiaceae include some of the most massive and long-lived of all trees. Their lumber is exceptionally decay-resistant. If it were not, they could not continue to stand for centuries with repeated injury from wind and lightning. Other members of the family, particularly the California coast redwood, are heavily exploited for their premium lumber, but only one has been domesticated.

CRYPTOMERIA — **SUGI** (Brouard 1963; Nuttonson 1951; Totman 1985; Tsukada 1982)

This genus contains only one living species, *C. japonica* — sugi. A graceful evergreen, to 50-m tall, sugi is the tallest member of the complex mixed temperate conifer and hardwood forests of Japan.

Until about 5 million years ago, this species had a wide range in Asia, but thereafter it disappeared from the fossil record except in Japan. During the Pleistocene, it repeatedly retreated to small refuge areas and spread out again over Japan during interglacials. There is no evidence that sugi survived the last glacial anywhere except in southern Honshu and Shikoku. Fossil evidence suggests that sugi was introduced to other Japanese islands less than 2000 years ago. It is unknown whether stands in Korea, Taiwan, and China are truly wild or derived from human introduction from Japan.

The post-glacial fossil record suggests that sugi stands were over-exploited after the introduction of rice farming about 2300 years ago. Sugi boards and posts were decay resistant and much used to dike rice fields, being easily split with primitive tools. Sugi wood was used in innumerable artifacts. It also played a key role in the development of Japanese architecture, peasant houses, castles, and temples. Shrines and teahouses are traditionally fashioned of sugi lumber and roofed with slabs of sugi bark. According to Japanese lore, planting of sugi and some other conifer species has been going on for over a millenium. Tsukada's studies of fossil *Cryptomeria* pollen support this. He found that in parts of central and southern Japan there had been a long decline in abundance of sugi pollen, presumably due to over exploitation. This was reversed about 1000 years ago, suggesting conservation and actual planting. This indicates silviculture on a grand scale, not just local preservation of revered ancient trees.

The whole country did not quickly switch to silviculture, however. Totman's history of forestry in Akita Prefecture shows how halting and drawn out the switch could be. Akita is in northwestern Honshu, remote from the main population centers and timber markets. In the 16th century, its virgin forests held abundant old-growth *Cryptomeria* trees mixed with other conifers and hardwoods. Intensive commercial logging began in the 1590s for shipment south. Selective logging of old-growth sugi continued through the 17th century, generally leaving the hardwoods to take over because of poor sugi reproduction. Depletion of sugi led to a series of government regulations in the late 17th century and through the 18th century. It became illegal for commoners to use sugi lumber for house construction, in rice paddies, or to cut sugi for fuel, but they were allowed access to controlled forests to cut hardwoods. In the 18th century, removal of hardwood competition gradually led to a resurgence of spontaneous sugi, but before it was old enough to cut, Akita had to import sugi lumber from other prefectures that had started sugi plantations early.

Although Akita had an official policy of promoting sugi planting ever since 1712 and had tried various incentives, neither the feudal lords nor the villages engaged in any regular planting until the 1790s. Planting expanded rapidly in the early 19th century, both small peasant plantings and large scale samurai planting. With the Meiji era, starting in 1868, the newly centralized Japanese government converted some Akita forests to national forests supervised by professionals influenced by German forestry.

During World War II, forests throughout Japan were subjected to extreme overcutting, but subsequently Japan has taken advantage of a cheap supply of American lumber to allow rebuilding its forests. Under domestication, sugi has produced high yields of premium lumber over a rather broad range of ecological conditions.

By 1950, about 1/3 of Japan's forest area was in plantations and about half of this area was planted to sugi. Several hundred thousands of hectares are logged per year, mainly on millions of private holdings. *Cryptomeria*, like *Sequoia*, resprouts from logged stumps, but to produce a better stand, the land is usually cleared and replanted with nursery grown seedlings or vegetatively propagated sugi cuttings. It is common practice to grow food crops on the logged land for a few years before replanting sugi. The species is extremely productive, has wide tolerances, and the lumber is premium quality. The species now grows in pure stands instead of the original mixed stands, higher up the mountains as its native lowlands have mainly been cleared for permanent agriculture.

Cryptomeria has been widely planted in eastern and southern China. It has also been introduced from Japan to many distant regions since the late 19th century and in a few cases has been successfully planted as a timber tree. For example, in 1880 *Cryptomeria* planting was begun by the British colonial government of Mauritius; this tropical Indian Ocean island had no native conifers. The plantings were on crown lands in the rugged volcanic mountains. They did well in spite of hurricanes. By the 1930s, *Cryptomeria* was making a significant contribution to the colony's timber supply. Since independence in 1980, the Mauritius government has maintained these forest plantations, about 500 ha being under *Cryptomeria*. It is logged for timber on a 40- to 50-year rotation. *Cryptomeria* seedlings are interplanted with pines for shade; the pines are removed after about 20 years, leaving the *Cryptomeria* to grow large.

Discussions and Conclusions

So far, no attempt has been made to fit the natural and cultural histories of individual species into a bigger picture or relate them to any theoretical models.

The gross outlines of a bigger picture, seen simply as an historical narrative, can be sketched fairly easily. We start with a dynamic planet upon which seed plants and their animal dependents had been evolving and migrating for 350 million years. During the last Interglacial period, about 100,000 years ago, we introduce into Africa a new species *Homo sapiens* in the strict sense, i.e., people physically resembling living humans and distinct from precursors, such as Neanderthals, known only as fossils. In the middle of the last Glacial period, these people spread (out of Africa?) through Eurasia, gradually eliminating the Neanderthals. Before the end of the Glacial period, they had moved into continents where no hominids had been before, Australia and the Americas. They hunted and gathered wild plants, but domesticated nothing other than the dog until after the Glacial period ended about 10,000 years ago. Then here and there in five continents, people rather abruptly began planting and harvesting a few species, not always staple foods. A positive feedback developed between spreading agriculture, growing human populations, and increasing cultural complexity. New crops were domesticated in more places. Exchanges began prehistorically between Europe, Asia, Africa, and within the Americas. Exchanges abruptly became global after Columbus, and additional domesticates have been acquired at an accelerating rate.

This simplistic outline conceals many puzzling questions. For example, the world has five major regions with a subtropical climate of winter rain and summer drought. Of these, the Near Eastern-Mediterranean region has contributed a wonderful array of domesticates over a long time span, prehistoric and historic. The Chilean region had prehistoric agriculture, but it was based on crops domesticated elsewhere; Chilean Indians did domesticate a strawberry. The other regions — in California, Australia, and South Africa — all had exceptionally rich native floras, which were skillfully exploited by hunter-gatherer groups, but none of the species were domesticated, and agriculture began only with European colonization. Other sets of climatically and vegetationally similar regions also had quite unlike histories, e.g., eastern North America and eastern Asia.

An environmental deterministic explanation of these geographic patterns is obviously untenable. A more seductive model postulates the other extreme, namely that plant domestication began with a rare flash of inspiration by exceptional individuals. In a much-quoted essay, Charles Darwin (1896) visualized plant domestication as an invention or discovery: it began when some "wise old savage" noticed an especially fine plant, perhaps on the village refuse heap, and transplanted it or sowed its seeds. Recent studies of surviving hunting-gathering cultures have led anthropologists to downplay the role of inspiration in agricultural origins. No old genius seems to be needed; almost any member of the group is familiar enough with the reproductive biology of the plants they harvest to know how to propagate them if it were considered worthwhile. Actually hunters and gatherers commonly manipulate the habitat, particularly by digging or burning, to increase the abundance of useful species. Within a species' native range, spontaneous propagation is likely to be so effective that human planting is redundant. However, casual experimental planting may be as ancient as *Homo sapiens*. It could have occurred innumerable times without leading to domestication. The Australian Aborigines' practice, noted above, of planting drift coconuts beyond the reach of waves is an example. Such desultory planting by nonfarming peoples would be unlikely to be detected in the archaeological record. An exception is the case of planting bottle-gourds by preagricultural, preceramic cultures, also noted above; it was detected because of a fortuitous conjunction of nonperishable remains far outside the native range of the species.

Currently, general theories on plant domestication focus on motivation rather than inspiration. Why did certain foragers choose to remain in comfortable equilbrium with their wild plant resources with stable populations, good nutrition, and leisure? Why did others abandon this way of life and choose the treadmill of constantly increasing population, constantly increasing labor input, and the dynamic cultural and economic changes leading to modern civilization? Some of the deductive models have invoked necessity — hunger due to climatic deterioration or population pressure — but much recent theory places population growth, malnutrition, and poverty after rather than before the switch to farming.

Meanwhile, botanists and geographers have developed spatial models of plant and animal domestication. Some of the classic hypotheses suggest a few major centers of ancient domestication and generalized pathways of diffusion. Harlan (1975) critically reviews these deductive models. I cannot hope to match his extraordinarily insightful and readable treatment and will simply quote some of his conclusions:

> Every model proposed so far for agricultural origins or plant domestication has generated evidence against it. . . Some crops arose in the Vavilovian centers and others did not ... Some people were sedentary long before agriculture; others maintained a nomadic way of life long after plants were domesticated ... There is no model with universal or very wide application yet most of them contribute, in some degree, to an understanding of the problem ... A search for a single, overriding cause of human behavior is likely to be frustrating and fruitless.

I believe this conclusion still stands, although vigorous discussion of general deductive concepts of plant domestication has continued (e.g., Blumler and Byrne 1991; Cohen 1977; D. Harris 1989; Hutchinson 1977; Rindos 1984).

It is a formidable problem for modern people to try to think about crop plants the way people did in the Stone Age. We tend to think of rational choices between different inputs and outputs, always with the benefits of hindsight. Stone Age people could not have foreseen how productive farming would become with advances in technology and genetic changes in the crops. They might have selected the best grain for planting, sending a message to the earth gods that they wanted more like that, but they could have had no concept of crop improvement by selection. During early stages of domestication, yields must have often been very poor. In some regions, the archaeological evidence suggests a long period of desultory crop planting before a threshold was crossed and subsistence farming became a viable way of life.

Crossing this threshold was not guaranteed, and much experimentation with planting probably led nowhere. For better or worse, those human groups that made the transition started the positive feedback between farming, population growth, and technological advances, and they were the ones who took over most of the earth. However, the patterns of this so-called Neolithic Revolution were heterogeneous in different regions.

The only region where the story of agricultural beginnings is at all well known is the Near East. An incomparable amount of scholarship has been devoted to the archaeological record of the spread of agriculture from the Near East to Europe. The story seems remarkably coherent: the basic staple crops — barley, wheat, lentils, peas — were domesticated fairly close together in space and time. They spread together as part of an integrated complex, including domestic animals, ceramics, and village organization, into the European wilderness. A similarly simple, coherent pattern may have occurred in a few other regions where uninhabited or lightly inhabited regions were invaded by agricultural peoples, e.g., the Polynesians in the Pacific or the Arawak in the Antilles. The same pattern cannot be extrapolated to most other regions, however. In China, for example, the basic crops were partly local domesticates, partly exotic introductions, and the advanced cultures had different geographic sources. In the New World, the maize-beans-squash complex was assembled from different geographic sources rather than migrating as a bloc.

Thus, no single model applies even to the relatively small roster of staple Neolithic food crops. As time goes on, the number of crops expands, and as cultural, economic, and social institutions become more complex and diverse, the quest for universal generalizations about crop-human interactions becomes hopeless. Universal generalizations tend to be banalities, such as increased labor input and yields for farming compared to exploiting wild plants.

The fact that crop histories are highly heterogeneous does not imply that interesting generalizations cannot be made about subsets. Before noting examples of such limited generalizations, some ways in which case histories have differed will be outlined. Different types of abstract patterns, both geographic and genetic, can be categorized, both at the time of initial domestication and during later migrations and evolution. (A few crops are cited parenthetically to recall examples of each situation; these are not meant to enumerate all crops in each category.)

HETEROGENEITY OF GEOGRAPHIC PATTERNS OF INCIPIENT DOMESTICATES

Domestication can begin within, adjacent to, or distant from the native range of the wild progenitors. The archaeological record is not detailed enough to distinguish between domestications beginning *in situ* and those starting with introduction to adjacent regions. Commonly the record begins in arid regions where the crops had to be cultivated under irrigation and where they were first planted is unknown (grapes in Near East, maize in Mexico, peanut in Peru). Some ancient crops were taken into cultivation far from their native ranges. Seed may have arrived by long-range ocean drift (bottle gourd outside Africa, perhaps sweet potato in Polynesia), by migratory birds (perhaps tomato in Mexico, *arabica* coffee in Arabia), or by human trade in wild products (cotton from Africa to India, *pepo* gourds into eastern North America). Historic domestications clearly exhibit this hetereogeneity: *in situ* starts (cranberries in New England bogs, Douglas fir in the Pacific Northwest), in adjacent areas (pecans in the southeastern U.S., vanilla in Mexico), or remote (Pará rubber in Asia, Monterey pine outside California).

Among ancient crops, there are cases of both direct domestication from the wild (emmer wheat, cacao) and from weedy volunteers (rye, apples). A few weeds may have been domesticated even before food crop agriculture developed, perhaps from dump heaps and camp margins (bottle gourd, cotton). Modern domestications also include examples of both truly wild ancestry (North American wildrice, cranberry) and weedy ancestry (slash pine, brown mustard).

Individual crop species may have been domesticated only once or at multiple places and times. Wide ranging progenitors, wild or weedy, have often been repeatedly domesticated (kidney beans, Douglas fir); in some cases for multiple purposes (beets, mustards). On the other hand, some fairly wide ranging wild species may have been domesticated only once (teosinte to maize, Pará rubber).

Spread of crops following domestication has varied strikingly in rate, even over time for a given crop, and in ultimate extent. Some ancient domesticates never spread far from home: these generally escape mention in this book except as congeners of important crops (most Andean potatos, Polynesian bananas). Before Columbus, some important crops had been widely exchanged between Europe and Asia (wheat, apricots), between Africa and Asia (sorghum, bananas), or between North and South America (maize, sweet potato), but most widespread New World crops were multiple domesticates (beans, avocado, manioc). Other species had spread widely within their native continents (soybean in Asia, olive in Europe, some chili peppers and squash species in North and South America).

Since Columbus, the global migrations of crops have followed enormously varied patterns. In the colonial period, some intercontinental introductions were unique or few (coffee, sugar cane), and others were repeated ad infinitum (beans, barley). Some crops spread mainly in commercial plantations (cotton, coffee), but others rapidly diffused from European colonies to indigenous agriculture (sugar cane, peaches). Newly introduced crops often outran European exploration (watermelons in America, maize in Africa, peanuts in Asia). Other introductions were accepted slowly and did not complete their spread until the 20th century (soybeans in the Americas, manioc in Africa).

Modern domesticates have shown equal individualism in spread, some remaining close to home (sugi in Japan, cranberries in North America) and others migrating fast and far (macadamia nuts, Monterey pine). At present, international networks of gene banks and governmental cooperation in plant exchanges have drastically affected the geography of certain crops (rice, wheat), but have had remarkably little effect on most.

HETEROGENEITY OF PATTERNS OF CROP DIVERSIFICATION

Genetic diversity that incipient domesticates inherited from their progenitors was quite uneven, and subsequent evolution under human selection followed different patterns.

Wild ancestors were in some cases primarily self-pollinated and hence tended to be homozygous (barley, peas), and others were obligately cross-pollinated and thus formed genetically diverse gene pools (Robusta coffee, rye). Diversity of incipient domesticates also varied, of course, with the size of the wild population sampled. In combination, narrow domestication of an inbred wild progenitor (lentil, *arabica* coffee) started with drastically less diversity than repeated domestication of an outbreeder (apples, Robusta coffee). Further evolutionary heterogeneity came when some domesticates were converted by human selection from outbreeders to inbreds (soybean).

Some crops have been genetically isolated from wild relatives, either by spatial segregation (tomato, Pará rubber) or by sterility barriers too complex to discuss here. Others have hybridized with wild and weedy relatives, sometimes locally (maize, kidney beans), sometimes over huge areas (sorghum, apples). The hybrid progeny have sometimes become accepted as new cultivars, some diploid (grain amaranths, mangoes), some as allopolyploids (bread wheat).

So far, only sexually reproducing populations have been noted. Some crops have been propagated since antiquity vegetatively (taro, sugar cane) or by apomictic seed (oranges, lemons). This allows proliferation as clones of sexually sterile mutants and hybrids (bananas, breadfruit, sisal).

These are merely a few examples of the enormously complex ways in which human propagation and selection of crops have interacted with their initially diverse genetic systems. The stories of some crops have recently become even more complex by scientific manipulation of variation, including radical hybridization (triticale, strawberries). This has not yet, however, had dramatic effects on more than a few crops, fewer than popular press on genetic engineering implies.

Valid, interesting generalizations have been made for subsets of crop plants. It has long been noted that a disproportionate number of crops are annuals, which are rather rare in natural floras except in tropical and subtropical deserts and adjacent regions with long, warm dry seasons. The dominance of annuals is strongest among crops grown for storable grain — cereal grasses, amaranths and chenopods, legumes, sunflowers, mustards and other oilseeds. For such crops, the annual habit offered farmers a syndrome of advantages — quick results after planting, heavy seed production in a sudden burst for a single harvest, escape from unfavorable cold or dry seasons by storage of seed indoors allowing spread into diverse climatic regions, and periods of fallow or rotation preventing buildup of pests and parasites. Generally annual habit is inherited from the wild, but in some cases, it developed in the cultigen. The most dramatic case is cotton — all the wild progenitors are shrubs, the annual habit evolving independently in four different cultivated species. Cotton is also a storable seed crop, although grown primarily for the seed hairs rather than the seed itself. A common evolutionary trend among annual seed crops has been farmer selection of mutations that break down the wild-type dispersal mechanisms, resulting in retention of the seed in nonshattering inflorescences and indehiscent pods. Also farmer selection has repeatedly developed cultigens in this subset that lack the wild-type seed dormancy, which would prevent uniform, rapid germination after sowing. These domesticates and their wild progenitors are typically self-pollinated and thus tend to form inbred, pure lines. However, these are occasionally shaken up by cross-pollination among cultivars or with weedy relatives, so there has been enough genetic variation for great diversification of landraces.

The group of trees, shrubs, and woody vines grown for their tender fleshy fruits has quite different general characteristics. They generally have indigestible, sometimes toxic seeds and depend on bird dispersal to colonize scattered, patchy habitats. Compared to the annual seed crops, this group is recently domesticated, and some members are extremely recent (e.g., cranberries, brambles). They retain seed dormancy and other wild-type characteristics. Attractiveness of fruits to birds has necessarily been retained, and volunteer seedlings are common around orchards, vineyards, and berry farms. The crops generally have flowers attractive to insects and retain wild-type cross-pollination. The domesticates have also continued to interbreed with wild relatives. Thus the cultigens have tended to be highly heterozygous and genetically diverse. Farmers have coped with this by propagation of elite individuals by cuttings or grafting, resulting in multiplication of clones. Selection has been focused primarily on fruit quality, often as a seasonal delicacy or for wine making. The fruit of some members is dried (raisins, dates) or preserved (olives) and provides important year-round sources of vitamins and calories.

A small subset of crops (pineapple, sisal, henequen) includes succulent xerophytes. They have a syndrome of morphological and physiological adaptations, including CAM photosynthesis discussed with the crops, that allows them to take in ephemeral water supplies and retain them tenaciously in spite of extreme drought and heat. This syndrome is shared with a vast array of wild species, not only in the Bromeliad and Agave families but in the Cactus and several other, unrelated families. Its disproportionately low representation among crops can be attributed to the slow growth rate and low productivity entailed by the xerophytic habit. The crops are grown for peculiar products unavailable from more ordinary species.

I will not multiply examples of subsets. Those noted above involve convergence between unrelated families. Many other subsets are closely related, e.g., the conifers planted for softwood timber, cucurbits grown for melons and squashes. The natural and cultural histories of some crops are so individualistic that they defy grouping above the species level. Coconut, vanilla, and alfalfa are examples. Of course, if looked at closely enough, every species is unique. Otherwise it could not exist.

Such inductive/empirical generalizations are the kind that are anticipated but not predicted in detail by Darwinian theory. The theory was originally developed by extrapolating what was known about evolution in domestic animals and plants to the undocumented realm of natural history. It assumes that evolution is channeled by selection, natural or artificial, but is not predetermined. The driving force, genetic mutation and recombination, is seen as largely random. Environmental selection, while not random, is seen as too complex for deterministic modeling, particularly when human choices are part of the plant environment. A corollary of the theory that environmental selection channels genetic patterning is that it also channels geographic patterning. The driving force, dispersal of seed and other propagules, again has a large random component with much trial and error, and the process of selection is again too complex for precise prediction or explanation.

It was stated in the Introduction that the historical geography of crops had become greatly clarified since mid-century. This is true in that we know much more about what happened — where and when they were domesticated and how they spread and diversified. This does not mean we can explain why processes happened exactly as they did rather than in various other ways. The hope of a grand predictive theory seems to have faded. I do not feel we should be unhappy about this situation. It makes learning actual histories more worthwhile.

As for the future history of world crops, some rather sweeping predictions are currently being made. For example, sources of new crops are being extinguished by destruction of natural ecosystems; diverse peasant crop complexes are being displaced by commercial monoculture; the wealth of landraces produced by the long process of folk selection is being threatened by high-yielding, scientifically bred cultivars. On the other hand, equally well-informed opinion predicts great advances by assemblage of international gene banks, radical interspecies hybridization, including embryo rescue to circumvent sterility barriers, and even more radical genetic engineering. In a way, these predictions are complementary rather than mutually exclusive. However, the safest prediction, now as at any time in the past, may be that the future historical geography of crops will develop in ways that are quite unplanned and unforeseen.

Glossary

Allopolyploid — see **Polyploid**.

Apomictic seed — a seed containing an embryo (or embryos) produced asexually and genetically identical to the mother plant.

Autopolyploid — see **Polyploid**.

Clone — the set of individual plants produced by asexual reproduction. If mutants are recognized, they may be considered a new clone.

Cultigen — a crop species or other taxon evolved under domestication to become distinct from its wild progenitor or progenitors.

Cultivar — a particular variety of a cultigen, usually named informally.

Diploid — cells with two chromosomes of each kind, the normal state in the seed plant body and reproductive organs except for some contents of the incipient seed and the pollen.

Domestication — evolutionary divergence from the wild progenitor under human selection.

Endosperm — food storage tissue within the seed drawn upon by the embryo during germination. In gymnosperms, the endosperm is haploid and genetically identical to the egg before fertilization. The angiosperm endosperm is triploid with the same genes as the associated diploid embryo plus an extra haploid set of the maternal genes.

Genome — the complete chromosome set of a given diploid species, usually designated by a capital letter. The term is mainly used in characterizing allopolyploids in which genomes of two or more parental species are combined.

Haploid — having a single chromosome of each kind, the normal state in the sexual life cycle of plants between reduction division (meiosis) and union of the sperm and egg, i.e., within the incipient seed and in the pollen.

Isozyme analysis — comparison of the spectra of enzymes produced by individual plants and, by extension, of the populations from which the plants were drawn. The number of shared enzymes (isozymes) and unshared enzymes (allozymes) are used as evidence of evolutionary relationships.

Landrace — a sexually reproducing crop variety developed under local natural and folk selection.

Polyploid — having more than two chromosome sets in the cells of the plant body. Polyploids originate by freak nuclear divisions, either somatic doubling or production of unreduced gametes. Polyploids are often exceptionally vigorous. Whether they are sexually fertile depends on chromosome complements. In **autopolyploids** (genomes AAA, AAAA, etc.), the multiple sets are structurally so similar that they form multivalents and univalents during meiosis and produce inviable gametes. In **allopolyploids**, the genomes are structurally so differentiated that they tend to form pairs instead of multivalents. However, allopolyploids with odd numbers of chromosome sets (AAB, etc.) also have to be propagated vegetatively. Allopolyploids with two sets of each genome (AABB, AABBSS, etc.) function sexually like diploids and can be grown from seed. Such allopolyploids can breed with their own kind but not with either parental species, resulting in instant speciation.

Seed — here broadly defined to include the true seed of botany textbooks, i.e., with the outer coat developed from a particular layer of maternal tissue (the integument) and dry, indehiscent one-seeded fruits, such as cereal grains and sunflower achenes, in which additional layers of maternal tissue surround the integument.

Self-incompatible — the pollen being impotent on its parental genotype.

References

Adams, R. E. W. 1990. Salvaging the past at Rio Azul, Guatemala. *Terra* 29(1):17–26.

Aiken, S. G., P. F. Lee, D. Punter, J. F. Stewart. 1988. Wild rice in Canada. *Agric. Can. Publ.* 1830:1–130.

Anderson, E. 1946. Maize in Mexico: A preliminary survey. *Mo. Bot. Garden Ann.* 33:147–247.

Anderson, E. 1950. Variation in avocadoes at the Rodiles grove. *Ceiba* 1:50–55.

Andres, T. C. 1990. Biosystematics, theories on the origin, and breeding potential of *Cucurbita ficifolia*. In *Biology and Utilization of the Cucurbitaceae,* D. M. Bates, R. W. Robinson, and C. Jeffrey, Eds. Cornell University Press, Ithaca, NY, pp. 101–133.

Andrews, J. 1984. *Peppers: The Domesticated Capsicums,* University of Texas Press, Austin, 170 pp.

Aschmann, H. 1957. The introduction of date palms into Baja California. *Econ. Bot.* 11:174–177.

Austin, D. F. 1977. Hybrid polyploids in *Ipomoea* sect. *batatas*. *J. Hered.* 68:259–260.

Austin, D. F. 1978. The *Ipomoea batatas* complex. I. Taxonomy. *Torrey Bot. Club Bull.* 106:114–129.

Austin, D. F. 1988. The taxonomy, evolution, and genetic diversity of sweet potatoes and related wild species. In *Report of the First Sweet Potato Planning Conference,* International Potato Center, Lima, Peru, pp. 27–59.

Axelrod, D. I. 1980. History of the maritime closed-cone pines, Alta and Baja California. *Univ. Calif. Publ. Geol. Sci.* 120:1–143.

Bahre, C. J., D. E. Bradbury. 1980. Manufacturing of mescal in Sonora, Mexico. *Econ. Bot.* 34:394–400.

Bailey, C. H., L. F. Hough. 1975. Apricots. In *Advances in Fruit Breeding,* J. Janick and J. N. Moore, Eds. Purdue University Press, West Lafayette, IN, pp. 367–383.

Bailey, L. H. 1911. *The Evolution of Our Native Fruits,* 3rd ed. Macmillan, New York, 472 pp.

Barlow, R. H. 1949. The extent of the empire of the Culhua Mexica. *Ibero-Americana* 28:1–141.

Bar-Yosef, O., M. Kislev. 1989. Early farming communities in the Jordan Valley. In *Foraging and Farming: The Evolution of Plant Exploitation,* D. R. Harris and G. C. Hillman, Eds. Unwin Hyman, London, pp. 621–631.

Bates, D. M., R. W. Robinson, C. Jeffrey, Eds. 1990. *Biology and Utilization of the Cucurbitaceae,* Cornell University Press, Ithaca, NY, 485 pp.

Beadle, G. W. 1980. The ancestor of corn. *Sci. Am.* 242(1):96–103.

Beard, B. H. 1981. The sunflower crop. *Sci. Am.* 244(5):150–159.

Bellwood, P. S. 1980. The peopling of the Pacific. *Sci. Am.* 243(5):138–147.

Bender, B. 1975. *Farming in Prehistory,* St. Martin's Press, New York, 268 pp.

Bergh, B. O. 1975. Avocadoes. In *Advances in Fruit Breeding,* J. Janick and J. N. Moore, Eds. Purdue University Press, West Lafayette, IN, pp. 541–567.

Bergh, B. O. 1976. Avocadoes. In *Evolution of Crop Plants,* N. W. Simmonds, Ed. Longman, London, pp. 148–151.

Bertsch, K. 1953. *Geschichte des deutschen Waldes,* 4th ed. Gustav Fischer, Jena, 124 pp.

Beutel, J. A., F. H. Winter, S. C. Manners, M. W. Miller. 1976. A new crop for California: Kiwifruit. *Calif. Agric.* 30(10):5–7.

Bird, R. M. 1980. Maize evolution from 500 B.C. to the present. *Biotropica* 12(1):30–41.

Bird, R. M. 1984. South American maize in Central America? In *Pre-Columbian Plant Migration,* Papers of the Peabody Museum of Archaeology and Ethnology, Harvard University, Vol. 76.

Blake, L. W. 1981. Early acceptance of watermelon by Indians of the United States. *J. Ethnobiol.* 1:193–199.

Blumler, M. A., A. R. Byrne. 1991. The ecological genetics of domestication and the origins of agriculture. *Curr. Anthropol.* 32:23–54.

Boardman, J. 1977. The olive in the Mediterranean: Its culture and use. In *The Early History of Agriculture,* J. Hutchinson, G. Clark, E. M. Jope, and R. Riley, Eds. Oxford University Press, New York, pp. 187–196.

Bohrer, V. L. 1970. Ethnobotanical aspects of Snaketown, a Hohokam village in southern Arizona. *Am. Antiq.* 35:413–430.

Bolton, J. L. 1962. *Alfalfa: Botany, Cultivation, Utilization*, Leonard Hill, London, 473 pp.

Bonavia, D. 1984. La importancia de los restos de papas y camotes do época precerámica hallados en el valle de Casma. *Journal de la Société des Améericanistes* 70:7–20.

Brand, D. D. 1971. The sweet potato: An exercise in methodology. In *Man Across the Sea: Problems of Pre-Columbian Contacts,* C. L. Riley, J. C. Kelley, C. W. Pennington, and R. L. Rands, Eds. University of Texas Press, Austin, pp. 343–365.

Brockway, L. J. 1979. *Science and Colonial Expansion: The Role of the British Royal Botanic Garden,* Academic Press, New York, 215 pp.

Brouard, N. R. 1963. *A History of Woods and Forests in Mauritius*, Mauritius Government Printer, Port Louis, 84 pp.

Brown, A. G. 1975. Apples. In *Advances in Fruit Breeding,* J. Janick and J. N. Moore, Eds. Purdue University Press, West Lafayette, IN, pp. 3–37.

Bruman, H. J. 1944. Some observations on the early history of the coconut in the New World. *Acta Americana* 2:220–243.

Bruman, H. J. 1948. The culture history of Mexican vanilla. *Hispanic Am. Hist. Rev.* 28:360–376.

Brush, S. B., H. J. Carney, Z. Huaman. 1981. Dynamics of Andean potato agriculture. *Econ. Bot.* 35:70–88.

Brücher, Heinz. 1985. Domestikation und Migration von *Solanum tuberosum. L. Kulturpflanze* 23:11–74.

Brücher, H. 1988. The wild ancestor of *Phaseolus vulgaris* in South America. In *Genetic Resources of Phaseolus Beans,* P. Gepts, Ed. Kluwer Academic Publishers, Dordrecht, pp. 185–214.

Buckley, R., H. Harries. 1984. Self-sown wild-type coconuts from Australia. *Biotropica* 16:148–151.

Burkill, I. H. 1935. *A Dictionary of the Economic Products of the Malay Peninsula,* Crown Agents for the Colonies, London, 402 pp.

Bye, R. A., Jr. 1972. Ethnobotany of the Southern Paiute Indians in the 1870's. In *Great Basin Cultural Ecology, a Symposium,* D. R. Fowler, Ed. Desert Research Institute, Reno, NV, pp. 87–104.

Bye, R. A., Jr. 1979. Incipient domestication of mustards in northwest Mexico. *Kiva* 44:237–256.

Cameron, J. W., R. K. Soost. 1976. Citrus. In *Evolution of Crop Plants,* N. W. Simmonds, Ed. Longman, London.

Campbell, G. K. G. 1976. Sugar beet. In *Evolution of Crop Plants,* N. W. Simmonds, Ed. Longman, London.

Carney, J. 1990. Triticale production in the central Mexican highlands: Smallholders' experiences and lessons for research. *CIMMYT Econ. Pap.* 2:1–48.

Carron, L. T. 1980. A history of forestry and forest products research in Australia. *Hist. Rec. Aust. Sci.* 5:7–57.

Carter, G. F. 1951. An early American description probably referring to *Phaseolus lunatus. Chron. Bot.* 12(4/6):155–161.

Castetter, E. F., W. H. Bell, A. R. Groves. 1938. The early utilization and the distribution of the *Agave* in the American Southwest. *Univ. N.M. Bull., Biol. Ser.* 5(4):1–92.

Chang, T. T. 1976. Rice: *Oryza sativa* and *O. glaberrima*. In *Evolution of Crop Plants,* N. W. Simmonds, Ed. Longman, London, pp. 98–104.

Chang, T. T. 1989. Domestication and spread of the cultivated rices. In *Foraging and Farming: The Evolution of Plant Exploitation,* D. R. Harris and G. C. Hillman, Eds. Unwin Hyman, London, pp. 408–417.

Clawson, D. L., D. R. Hoy. 1979. Nealtica, Mexico: A peasant community that rejected the Green Revolution. *Am. J. Econ. Sociol.* 38:371–387.

Clement, C. R. 1988. Domestication of the pejibaye palm (*Bactris gasipaes*): past and present. *Adv. Econ. Bot.* 6:155–174.

Cock, J. H. 1985. *Cassava; New Potential for a Neglected Crop,* Westview Press, Boulder, CO, 191 pp.

Cohen, M. N. 1977. *The Food Crisis in Prehistory; Overpopulation and the Origins of Agriculture,* Yale University Press, New Haven, CT.

Coons, M. P. 1982. Relationships of *Amaranthus caudatus*. *Econ. Bot.* 36:129–146.

Cope, F. W. 1976. Cacao. In *Evolution of Crop Plants,* N. W. Simmonds, Ed. Longman, London, pp. 284–289.

Corbet, S. A. 1987. Pollination of crops imported to new countries. *Brit. Ecol. Soc. Bull.* 18(1):22–23.

Cuatrecasas, J. 1964. Cacao and its allies: A taxonomic revision of the genus *Theobroma*. *Contrib. U.S. Natl. Herbarium* 25:375–605.

Cutler, H., W. Meyer. 1965. Corn and cucurbits from Wetherill Mesa. *Am. Antiq.* 31:136–152.

Cutler, H. C., T. W. Whitaker. 1961. History and distribution of the cultivated cucurbits in America. *Am. Antiq.* 26:468–485.

Cutler, H., T. W. Whitaker. 1967. Cucurbits from the Tehuacan caves. In *The Prehistory of the Tehuacan Valley,* Vol. 1, D. S. Byers, Ed. University of Texas Press, Austin, pp. 212–219.

Dalziel, J. M. 1948. *The Useful Plants of West Tropical Africa,* Crown Agents for the Colonies, London, 612 pp.

Darrow, G. M. 1975. Minor temperate fruits. In *Advances in Fruit Breeding,* J. Janick and J. N. Moore, Eds. Purdue University Press, West Lafayette, IN, pp. 269–284.

Darwin, C. 1896. *The Variation of Animals and Plants Under Domestication,* 2nd ed. D. Appleton & Co., New York.

Decker, D. S. 1988. Origin(s), evolution, and systematics of *Cucurbita pepo*. *Econ. Bot.* 42:4–15.

Decker-Walters, D. S. 1990. Evidence of multiple domestication of *Cucurbita pepo*. In *Biology and Utilization of the Cucurbitaceae,* D. M. Bates, R. W. Robinson, and C. Jeffrey, Eds. Cornell University Press, Ithaca, NY, pp. 96–101.

Dequaire, J. 1966. La culture et la preparation de la vanille. *J. Seychelles Soc.* 5:14–28.

de Wet, J. M. J., J. R. Harlan, E. G. Price. 1976. Variability in *Sorghum bicolor*. In *Origins of African Plant Domestication,* J. M. J. de Wet, J. R. Harlan, and E. G. Price, Eds. Mouton, The Hague, pp. 453–478.

de Wet, J. M. J., J. P. Huckabay. 1967. The origin of *Sorghum bicolor*: Distribution and domestication. *Evolution* 21:787–802.

Dickson, M. H., D. H. Wallace. 1986. Cabbage breeding. In *Breeding Vegetable Crops,* M. J. Bassett, Ed. AVI Publishing Company, Westport, CT, pp. 396–435.

Doebley, J. 1990. Molecular evidence and the evolution of maize. *Econ. Bot.* 44(3 — Supplement):6–27.

Doebley, J., J. D. Wendel, S. C. Smith, C. W. Stuber, Major M. Goodman. 1988. The origin of cornbelt maize: The isozyme story. *Econ. Bot.* 42:129–131.

Doggett, H. 1976. Sorghum: *Sorghum bicolor*. In *Evolution of Crop Plants,* N. W. Simmonds, Ed. Longman, London, pp. 112–117.

Dovring, F. 1974. Soybeans. *Sci. Am.* 230(2):14–20.

Einset, J., C. Pratt. 1975. Grapes. In *Advances in Fruit Breeding,* J. Janick and J. N. Moore, Eds. Purdue University Press, West Lafayette, IN, pp. 130–153.

Eldridge, K. G. 1979. Seed collection in California in 1978. In *Annual Report 1977–1978,* Australia CSIRO Division of Forest Research, pp. 8–17.

Evans, A. M. 1976. Beans: *Phaseolus* spp. In *Evolution of Crop Plants,* N. W. Simmonds, Ed. Longman, London, pp. 168–172.

Feldman, M. 1976. Wheats. In *Evolution of Crop Plants,* N. W. Simmonds, Ed. Longman, London, pp. 120–128.

Ferdon, E. N. 1988. A case for taro preceding *kumara* as the dominant domesticate in ancient New Zealand. *J. Ethnobiol.* 8:1–5.

Ferwerda, F. P. 1976. Coffees. In *Evolution of Crop Plants,* N. W. Simmonds, Ed. Longman, London, pp. 257–260.

Finan, J. J. 1948. Maize in the great herbals. *Mo. Bot. Garden Ann.* 35:148–191.

Fogle, H. N. 1975. Cherries. In *Advances in Fruit Breeding,* J. Janick and J. N. Moore, Eds. Purdue University Press, West Lafayette, IN, pp. 340–366.

Forde, H. I. 1975. Walnuts. In *Advances in Fruit Breeding,* J. Janick and J. N. Moore, Eds. Purdue University Press, West Lafayette, IN, pp. 439–455.

Förskell, W. P. 1963. Genetics in forest practice in Sweden. *Unasylva* 28:119–127.

Fritz, G. J. 1984. Identification of cultivated amaranth and chenopod from rock shelter sites in northwest Arkansas. *Am. Antiq.* 49:558–572.

Fritz, G. J. 1990. Multiple pathways to farming in pre-contact eastern North America. *J. World Prehistory* 4:387–433.

Fry, W. E., L. J. Speilman. 1991. Population biology *Phytophthora infestans. Adv. Plant Pathol.* 7:191–192.

Fryxell, P. A. 1979. *The Natural History of the Cotton Tribe.* Texas A&M University Press, College Station, 245 pp.

Fukunaga, E. T. 1967. The development of the Macadamia nut industry in Hawaii — a sequel to plant introduction. In *Proceedings of the International Symposium on Plant Introduction,* Escuela Agricola Panamericana, Tegucigalpa, pp. 129–135.

Gade, D. 1970. Ethnobotany of cañihua. *Econ. Bot.* 24:55–61.

Gade, D. 1972. Setting the stage for domestication: *Brassica* weeds in Andean peasant ecology. *Assoc. Am. Geogr. Proc.* 4:38–40.

Galinat, W. C. 1985. Domestication and diffusion of maize. In *Prehistoric Food Production in North America. Anthropological Papers 75*, R. I. Ford, Ed. University of Michigan, Museum of Anthropology, pp. 275–282.

Galinat, W. C. 1988. The origin of corn. In *Agronomy Monograph 18. Corn and Corn Improvement,* 3rd ed. G. F. Sprague, Ed. American Society of Agronomy, Crop Science Society of America, Madison, WI, pp. 1–31.

Galinat, W. C., G. V. Pasapuleti. 1982. *Zea diploperennis*: II. A review of its significance and potential value for maize improvement. *Maydica* 27:213–220.

Galletta, G. J. 1975. Blueberries and Cranberries. In *Advances in Fruit Breeding,* J. Janick and J. N. Moore, Eds. Purdue University Press, West Lafayette, IN, pp. 154–196.

Galloway, J. H. 1989. *The Sugar Cane Industry: An Historical Geography from its Origins to 1914,* Cambridge University Press, New York, 266 pp.

Gentry, H. S. 1969. Origin of the common bean, *Phaseolus vulgaris. Econ. Bot.* 23:55–69.

Gepts, P. 1988. Phaseolin as an evolutionary marker. In *Genetic Resources of Phaseolus Beans,* P. Gepts, Ed. Kluwer Academic Publishers, Dordrecht, pp. 215–241.

Gepts, P., T. C. Osborn, K. Rashka, F. A. Bliss. 1986. Phaseolin-protein variability in wild forms and landraces of the common bean (*Phaseolus vulgaris*): Evidence for multiple centers of domestication. *Econ. Bot.* 40:451–568.

Glendinning, D. R. 1983. Potato introduction and breeding up to the early 20th century. *New Phytol.* 94:479–505.

Gomez-Ibañez, D. 1975. *The Western Pyrenees: Differential Evolution of the Spanish and French Borderlands,* Clarendon Press, Oxford, 162 pp.

Goncalves de Lima, O. 1956. *El Maguey y el Pulque en los Codices Mexicanos,* Fondo de Cultura Economica, Mexico, 278 pp.

Goodman, Major M., W. L. Brown. 1988. Races of corn. In *Agronomy Monograph 18. Corn and Corn Improvement,* 3rd ed. G. F. Sprague, Ed. American Society of Agronomy, Crop Science Society of America, Madison, WI, pp. 33–79.

Gregory, W. C., M. P. Gregory. 1976. Groundnut. In *Evolution of Crop Plants,* N. W. Simmonds, Ed. Longman, London, pp. 151–154.

Griggs, W. H., B. L. Iwakiri. 1977. Asian pears in California. *Calif. Agric.* 31(1):8–12.

Gritton, E. T. 1986. Pea breeding. In *Breeding Vegetable Crops,* M. J. Bassett, Ed. AVI Publishing Company, Westport, CT, pp. 283–321.

Grun, P. 1990. The evolution of cultivated potatoes. *Econ. Bot.* 44(3 — Supplement):39–55.

Haarer, A. E. 1958. *Modern Coffee Production,* Leonard Hill, London, 467 pp.

Hagenstein, W. D. 1952. *Tree Farms — An American Approach to Forestry,* University of British Columbia, Vancouver.

Hanson, C. R., Ed. 1972. *Alfalfa Science and Technology,* American Society of Agronomy, Madison, WI, 812 pp.

Hardon, J. J. 1976. Oil palm. In *Evolution of Crop Plants,* N. W. Simmonds, Ed. Longman, London, pp. 225–229.

Hariot, T. 1590. *A Brief and True Report of the New Found Land of Virginia,* Frankfort.

Harlan, J. R. 1975. *Crops and Man,* American Society of Agronomy, Madison, WI, 295 pp. (Second edition, 1992)

Harlan, J. R. 1976. Barley: *Hordeum vulgare.* In *Evolution of Crop Plants,* N. W. Simmonds, Ed. Longman, London, pp. 93–98.

Harlan, J. R. 1977. The origins of cereal agriculture in the Old World. In *Origins of Agriculture,* C. A. Reed, Ed, Mouton, The Hague, pp. 358–383.

Harlan, J. R. 1982. Human interference with grass systematics. In *Grasses and Grasslands: Systematics and Ecology,* J. R. Estes, R. J. Tyrl and J. N. Brunken, Eds. University of Oklahoma Press, Norman, pp. 37–50.

Harlan, J. R. 1989. The tropical African cereals. In *Foraging and Farming: The Evolution of Plant Exploitation,* D. R. Harris and G. C. Hillman, Eds. Unwin Hyman, London, pp. 335–343.

Harries, H. 1978. The evolution, differentiation, and classification of *Cocos nucifera. Bot. Rev.* 44:265–320.

Harris, D. R. 1989. An evolutionary continuum of people-plant interactions. In *Foraging and Farming: The Evolution of Plant Exploitation,* D. R. Harris and G. C. Hillman, Eds. Unwin Hyman, London, pp. 11–26.

Hartley, C. W. S. 1977. *The Oil Palm,* 2nd ed., Longman, London, 806 pp.

Hawkes, J. G. 1990. *The Potato: Evolution, Biodiversity, and Genetic Resources,* Smithsonian Institution Press, Washington D.C., 259 pp.

Hawkes, J. G., J. Francisco-Ortega. 1992. The potato in Spain in the late 16th century. *Econ. Bot.* 46:86–97.

Hayes, P.M., R. E. Stucker, G. C. Wandrey. 1989. The domestication of American Wildrice. *Econ. Bot.* 43:203–214.

Healy, R. G. 1985. *Competition for Land in the American South,* The Conservation Foundation, Washington D.C.

Hedrick, U. P. 1917. The peaches of New York. *N.Y., Dep. Agric. Annu. Rep.* 24(2,II):1–39.

Hedrick, U. P., Ed. 1919. Sturtevant's notes on cultivated plants. *N.Y., Dep. Agric. Annu. Rep.* 27(2,II):1–686.

Heiser, C. B., Jr. 1965. Sunflowers, weeds, and cultivated plants. In *The Genetics of Colonizing Species,* H. G. Baker and G. Ledyard Stebbins, Eds. Academic Press, New York, pp. 391–401.

Heiser, C. B., Jr. 1969. *Nightshades, the Paradoxical Plants,* W. H. Freeman, San Francisco, CA, 200 pp.

Heiser, C. B., Jr. 1973. Variation in the bottle gourd. In *Tropical Forest Ecosystems in Africa and South America,* B. J. Meggers, E. S. Ayensu, and D. Duckworth, Eds. Smithsonian Institution Press, Washington D.C., pp. 121–128.

Heiser, C. B., Jr. 1976. Sunflowers. In *Evolution of Crop Plants,* N. W. Simmonds, Ed. Longman, London,

Heiser, C. B., Jr. 1978. Taxonomy of *Helianthus* and origin of domesticated sunflower. In *Sunflower Science and Technology,* J. F. Carter, Ed. American Society of Agronomy, Madison, WI, pp. 31–53.

Heiser, C. B., Jr. 1989. Domestication of Cucurbitaceae: *Cucurbita* and *Lagenaria.* In *Foraging and Farming: The Evolution of Plant Exploitation,* D. R. Harris and G. C. Hillman, Eds. Unwin Hyman, London, pp. 471–480.

Hemingway, J. S. 1976. Mustards. In *Evolution of Crop Plants,* N. W. Simmonds, Ed. Longman, London, pp. 56–59.

Hermann, R. K. 1987. North American tree species in Europe. *J. For.* 85(12):27–32.

Hesse, C. O. 1975. Peaches. In *Advances in Fruit Breeding,* J. Janick and J. N. Moore, Eds. Purdue University Press, West Lafayette, IN, pp. 285–335.

Hesse, F. 1938. *German Forestry,* Yale University Press, New Haven, CT, 342 pp.

Hobhouse, H. 1986. *Seeds of Change: Five Plants that Transformed Mankind,* Harper & Row, New York, 252 pp.

Hodgson, W., G. Nabhan, L. Ecker. 1989. Prehistoric fields in central Arizona: Conserving rediscovered *Agave* cultivars. *Agave: Q. Mag. Desert Bot. Garden, Phoenix* 3(3):9–11.

Hougas, R. W. 1956. Foreign potatoes, their introduction and importance. *Am. Potato J.* 33:190–198.

Hunter, J. R. 1990. The status of cacao (*Theobroma cacao*, Sterculiaceae) in the western hemisphere. *Econ. Bot.* 44:425–439.

Hunter, J. R. 1991. Cacao: Will it disappear from our diets? *Terra* 21(2):14–21.

Hunziker, A. T. 1943. Las especies alimenticias de *Amarantus* y *Chenopodium* cultivadas por los Indios de América. *Revista Argentina de Agronomía* 10:297–354.

Hutchinson, J. 1977. India: Local and introduced crops. In *The Early History of Agriculture,* J. Hutchinson, G. Clark, E. M. Jope, and R. Riley, Eds. Oxford University Press, New York, pp. 129–141.

Hutchison, C. R. 1946. *California Agriculture*, University of California Press, Berkeley, 443 pp.

Hymowitz, T. 1976. Soybeans. In *Evolution of Crop Plants,* N. W. Simmonds, Ed. Longman, London, pp. 159–162.

Hymowitz, T., J. R. Harlan. 1983. Introduction of soybean to North America by Samuel Bowen in 1765. *Econ. Bot.* 37:371–179.

Hymowitz, T., C. A. Newell. 1981. Taxonomy of the genus *Glycine*, domestication and uses of soybeans. *Econ. Bot.* 35:272–288.

Jenkins, J. A. 1948. The origin of the cultivated tomato. *Econ. Bot.* 2:379–392.

Jennings, D. L. 1976. Raspberries and blackberries. In *Evolution of Crop Plants*, N. W. Simmonds, Ed. Longman, London, pp. 251–254.

Johannessen, C. L. 1966. *Pejibaye* in commercial production. *Turrialba* 16:181–187.

Johannessen, C. L., A. Z. Parker. 1989. Maize ears sculptured in 12th and 13th Century A.D. India as indications of pre-Columbian diffusion. *Econ. Bot.* 43:164–180.

Johnson, D. V. 1972. *The Cashew of Northeast Brazil: A Geographical Study of a Tropical Tree Crop*, Ph.D. dissertation in Geography, University of California, Los Angeles. University Microfilms, Ann Arbor, 168 pp.

Jones, J. D. 1976. Strawberry. In *Evolution of Crop Plants,* N. W. Simmonds, Ed. Longman, London, pp. 237–242.

Jones, W. O. 1959. *Manioc in Africa,* Stanford University Press, 315 pp.

Kaplan, L. 1956. The cultivated beans of the prehistoric Southwest. *Mo. Bot. Garden Ann.* 43:189–251.

Kaplan, L. 1965. Archaeology and domestication in American *Phaseolus* (beans). *Econ. Bot.* 19:358–368.

Kaplan, L. 1981. What is the origin of the common bean? *Econ. Bot.* 35:240–254.

Kaplan, L., L. Kaplan. 1988. *Phaseolus* in archaeology. In *Genetic Resources of Phaseolus Beans*, P. Gepts, Ed. Kluwell Academic Publishers, Dordrecht, pp. 125–142.

Kauffman, C. S. 1992. Realizing the potential of grain amaranth. *Food Rev. Int.* 8:5–21.

Kaukis, K., D. A. W. Davis. 1986. Sweet corn breeding. In *Breeding Vegetable Crops,* M. J. Bassett, Ed. AVI Publishing Company, Westport, CT, pp. 475–519.

Kester, D. E., R. Asay. 1975. Almonds. In *Advances in Fruit Breeding,* J. Janick and J. N. Moore, Eds. Purdue University Press, West Lafayette, IN, pp. 387–419.

Killip, E. P., A. C. Smith. 1931. The use of fish poisons in South America. *Smithsonian Institution Annu. Rep. 1930,* pp. 401–408.

Kimber, C. 1988. *Martinique Revisited: The Changing Plant Geography of a West Indian Island,* Texas A&M University Press, College Station, 458 pp.

Kirch, P. V. 1991. Polynesian agricultural systems. In *Islands, Plants, and Polynesians: An Introduction to Polynesian Ethnobotany,* P. A. Cox and S. A. Banack, Eds. Dioscorides Press, Portland, OR, pp 113–133.

Kitamura, S. 1950. The cultivated *Brassicae* of China and Japan. *Univ. Kyoto, Col. Sci., Mem., Ser. B* 19(2–16):1–6.

Krapovickas, A. 1969. The origin, variability, and spread of the groundnut (*Arachis hypogaea*). In *The Domestication and Exploitation of Plants and Animals,* P. J. Ucko and J. W. Dimbleby, Eds. Aldine, Chicago, IL, pp. 427–441.

Krapovickas, A. 1973. Evolution of the genus *Arachis*. In *Agricultural Genetics: Selected Topics,* R. Moav, Ed. Israel, National Council for Research and Development, Jerusalem, pp. 135–151.

Kulakow, P. A., H. Hauptli, S. K. Jain. 1985. Genetics of grain amaranths I. Mendelian analyses of six color characteristics. *J. Hered.* 76:27–30.

Lanner, R. M. 1981. *The Piñon Pine: A Natural and Cultural History,* University of Nevada Press, Reno, 208 pp.

Larter, E. N. 1976. Triticale, *Triticosecale* spp. In *Evolution of Crop Plants,* N. W. Simmonds, Ed. Longman, London, pp. 118–120.

Laufer, B. 1938. American plant migration. Part 1. The potato. *Field Mus. Nat. Hist., Anthropol. Ser.* 28:1–132.

Layne, R. E. C., H. A. Quamme. 1975. Pears. In *Advances in Fruit Breeding,* J. Janick and J. N. Moore, Eds. Purdue University Press, West Lafayette, IN, pp. 38–70.

Lee, D. R. 1963. Date cultivation in the Coachella Valley, California. *Ohio J. Sci.* 63:82–87.

Lesins, K. A. 1976. Alfalfa, lucerne. In *Evolution of Crop Plants,* N. W. Simmonds, Ed. Longman, London, pp. 165–168.

Libby, W. J. 1990. Genetic conservation of Monterey pine and coast redwoods. *Fremontia* 18(2):15–21.

Lock, W. G. 1969. *Sisal: Thirty Years Sisal Research in Tanzania,* Longmans, Green and Company, London.

Luckwill, L. C. 1943. The genus *Lycopersicon*: An historical, biological, and taxonomic survey of the wild and cultivated tomatoes. *Aberdeen Univ. Stud.* 120:1–44.

Lynch, T. F., Ed. 1980. *Guitarrero Cave: Early Man in the Andes,* Academic Press, New York.

Mackie, W. W. 1943. Origin, dispersal, and variability of the Lima bean, *Phaseolus lunatus. Hilgardia* 15(1):1–24.

Madden, G. D., H. L. Malstrom. 1975. Pecans and hickories. In *Advances in Fruit Breeding,* J. Janick and J. N. Moore, Eds. Purdue University Press, West Lafayette, IN, pp. 425–438.

Mann, J. A., C. T. Kimber, F. R. Miller. 1983. The origin and early cultivation of sorghum in Africa. *Tex. A&M Exp. Stn. Bull.* 1454:1–21.

Martin, G. B., M. W. Adams. 1987. Landraces of *Phaseolus vulgaris* in northern Malawi. *Econ. Bot.* 41:190–215.

Mason, S. C. 1915. Dates of Egypt and the Sudan. *U.S. Dep. Agric. Bull.* 271:1–40.

Mason, S. C. 1923. The Saidy date of Egypt. *U.S. Dep. Agric. Bull.* 1125:1–35.

Massal, E., J. Barrau. 1956. Food plants of the South Pacific islands. *South Pac. Comm. Tech. Pap.* 94:1–51.

McCue, G. A. 1952. The history of the use of the tomato: An annotated bibliography. *Mo. Bot. Garden Ann.* 39:289–348.

McDonald, S. E., S. L. Krugman. 1986. Worldwide planting of southern pines. *J. For.* 84 (6):21–24.

McIntyre, M. P. 1959. The Monterey pine (*Pinus radiata*) in New Zealand. *Assoc. Pac. Coast Geogr. Yearb.* 21:59–66.

McNaughton, I. H. 1976. Turnips and relatives; Swedes and rapes. In *Evolution of Crop Plants,* N. W. Simmonds, Ed. Longman, London, pp. 45–48; 53–56.

Merlin, M. 1982. The origins and dispersals of true taro. In *Native Planters: Ho'okupu,* Vol. 1, J. Kennedy, Ed. University of Hawaii, Honolulu, pp. 6–16.

Merrick, L. C. 1990. Systematics and evolution of a domesticated squash, *Cucurbita argyrosperma*, and its wild and weedy relatives. In *Biology and Utilization of the Cucurbitaceae,* D. M. Bates, R. W. Robinson, and C. Jeffrey, Eds. Cornell University Press, Ithaca, NY, pp. 77–95.

Merrill, G. C. 1958. The historical geography of St. Kitts and Nevis, West Indies. *Instituto Panamericano de Geografía é Historia, México, Publ.* 23:1–145.

Miller, A. G., M. Morris. 1988. *Plants of Chofar, the Southern Region of Oman: Traditional, Economic, and Medicinal Uses,* Sultanate of Oman, Royal Court, 361 pp.

Miracle, M. 1966. *Maize in Tropical Africa,* University of Wisconsin Press, Madison, 327 pp.

Miracle, M. 1967. *Agriculture in the Congo Basin: Tradition and Change in African Rural Economies,* University of Wisconsin Press, Madison, 355 pp.

Mirov, N. T. 1967. *The Genus Pinus,* Ronald Press, New York, 602 pp.

Morinaga, T. 1960. Rice improvement in Japan. *Ninth Pac. Sci. Congr. Proc.* (Bangkok) 8:11–13.

Munson, P. J. 1980. Archaeological data on the origins of cultivation in the southwestern Sahara and their implications for West Africa. In *West African Cultural Dynamics,* B. K. Swartz, Jr. and R. E. Dummett, Eds. Mouton, The Hague, pp. 101–121.

Murray, J. 1970. *The First European Agriculture: A Study of the Osteological and Botanical Evidence until 2,000 B.C.,* The University Press, Edinburgh, 380 pp.

Nabhan, G. P. 1978. *Chilitepenes*: Wild spice of the American Southwest. *El Palacio* 84(2):30–34.

Nabhan, G. P. 1980. Amaranth cultivation in the U. S. Southwest and northwest Mexico. In *Proceedings of the Second Amaranth Conference,* Rodale Press, Emmaus, PA, pp. 129–133.

Narain, A. 1974. Rape and mustard. In *Evolutionary Studies in World Crops: Diversity and Change in the Indian Subcontinent,* J. Hutchinson, Ed. Cambridge University Press, New York, pp. 67–70.

Neal, M. C. 1965. *In Gardens of Hawaii,* Bernice P. Bishop Museum Press, Honolulu, 924 pp.

Nee, M. 1990. The domestication of *Cucurbita*. *Econ. Bot.* 44(3 Supplement):56–68.

Needham, J. 1986. *Science and Civilization in China, Volume 6. Biology and Biological Technology, Part 1. Botany,* Cambridge University Press, New York, 716 pp.

Nixon, R. S., J. B. Carpenter. 1978. Growing dates in the United States. *U.S. Dep. Agric., Agric. Infor. Bull.* 20:1–63.

Nuttonson, M. Y. 1951. *Ecological Crop Geography and Field Practices in Japan, Japan's Natural Vegetation, and Agro-climatic Analogues in North America,* American Institute of Crop Ecology, Washington D.C., 213 pp.

Nye, M. M. 1991. The mis-measure of manioc. *Econ. Bot.* 45:47–57.

O'Brien, P. J. 1972. The sweetpotato: Its origin and dispersal. *Am. Anthropol.* 74:342–365.

Olmo, H. P. 1976. Grapes. In *Evolution of Crop Plants,* N. W. Simmonds, Ed. Longman, London, pp. 294–298.

Ortiz, F. 1947. *Cuban Counterpoint: Tobacco and Sugar,* Translated by Harriet de Onis. Knopf, New York, 312 pp.

Oudejans, J. H. M. 1976. Date palm. In *Evolution of Crop Plants,* N. W. Simmonds, Ed. Longman, London, pp. 229–231.

Ourecky, D. K. 1975. Brambles. In *Advances in Fruit Breeding,* J. Janick and J. N. Moore, Eds. Purdue University Press, West Lafayette, IN, pp. 98–129.

Paris, H. S. 1989. Historical records, origins, and development of the edible cultivar groups of *Curcurbita pepo*. *Econ. Bot.* 43:423–443.

Parsons, J. J. 1962a. The acorn-hog economy of the oak woodlands of southwestern Spain. *Geogr. Rev.* 52:211–235.

Parsons, J. J. 1962b. The cork oak forests and the evolution of the cork industry in southern Spain. *Econ. Geogr.* 38:195–213.

Parsons, J. J. 1972. Spread of African pasture grasses in the American tropics. *J. Range Manage.* 25:12–17.

Patiño, V. M. 1963. *Plantas cultivadas y Animales Domésticos en América Equinoccial. Tomo 1. Frutales,* Imprenta Departmental, Cali, Colombia, 547 pp.

Patiño, V. M. 1964. *Plantas cultivadas y Animales Domésticos en América Equinoccial. Tomo II. Plantas Alimentícias,* Imprenta Departmental, Cali, Colombia, 364 pp.

Patiño, V. M. 1967. *Plantas cultivadas y Animales Domésticos en América Equinoccial. Tomo III. Fibras, Medicinales, Misceláneas,* Imprenta Departmental, Cali, Colombia, 569 pp.

Patiño, V. M. 1969. *Plantas cultivadas y Animales Domésticos en América Equinoccial. Tomo IV. Plantas Introducidas,* Imprenta Departmental, Cali, Colombia, 571 pp.

Peattie, D. C. 1950. *A Natural History of Trees of Eastern and Central North America,* Houghton Mifflin, Boston, MA, 606 pp.

Phillips, L. L. 1976. Cotton. In *Evolution of Crop Plants,* N. W. Simmonds, Ed. Longman, London, pp. 190–200.

Pickersgill, B. 1976. Pineapple. In *Evolution of Crop Plants,* N. W. Simmonds, Ed. Longman, London, pp. 14–18.

Pickersgill, B. 1988. The Genus *Capsicum*: a multi-disciplinary approach to the taxonomy of cultivated and wild plants. *Biologisches Zentralblatt* 107:281–387.

Pickersgill, B., C. B. Heiser, Jr. 1977. Origins and distributions of plants domesticated in the New World tropics. In *Origins of Agriculture,* C. A. Reed, Ed. Mouton, The Hague, pp. 803–835.

Pickersgill, B., C. B. Heiser, Jr., J. McNeill. 1979. Numerical taxonomic studies on variation and domestication in some species of *Capsicum.* In *The Biology and Taxonomy of the Solanaceae. Linnaean Society Symposium Series No. 3,* J. G. Hawkes, R. N. Lester, and A. D. Skelding, Eds. Academic Press, London, pp. 679–700.

Pinkett, H. P. 1980. Forestry comes to America. *Agric. Hist.* 54:4–10.

Pitte, J. R. 1983. Vignobles et vins du Japon. *Ann. Géogr.* 92(509):172–199.

Plaisted, R. I., R. W. Hoopes. 1989. The past record and future prospects for the use of exotic potato germplasm. *Am. Potato J.* 66:603–627.

Plucknett, D. L. 1976. Edible aroids. In *Evolution of Crop Plants,* N. W. Simmonds, Ed. Longman, London, pp. 10–12.

Popenoe, W. 1950. Central American fruit culture. *Ceiba* 1:269–367.

Powell, J. 1976. Ethnobotany. In *New Guinea Vegetation,* K. Peijmans, Ed. Australian National University Press, Canberra, pp. 100–184.

Powell, J. 1982. History of plant use and man's impact on the vegetation. In *Biogeography and Ecology in New Guinea,* J. L. Gressitt, Ed. Dr. W. Junk, The Hague, pp. 207–228.

Powers, H. R., Jr. 1986. Performance of Livingstone Parish loblolly pine in the Georgia Piedmont. *South. J. Appl. For.* 10:84–87.

Prance, G. T. 1984. The pejibaye or peach palm, *Guilielma gasipaes* (HBK.) Bailey. In *Pre-Columbian Plant Migration, Vol. 76*, D. Stone, Ed. Papers of the Peabody Museum of Archaeology and Ethnology, Harvard University, pp. 87–92.

Puchalski, J. T., R. W. Robinson. 1990. Electrophoretic analysis of isozymes in *Cucurbita* and *Cucumis* and its application for phylogenetic studies. In *Biology and Utilization of the Cucurbitaceae,* D. M. Bates, R. W. Robinson, and C. Jeffrey, Eds. Cornell University Press, Ithaca, NY, pp. 60–76.

Purseglove, J. W. 1968. *Tropical Crops: Dicotyledons,* John Wiley & Sons, New York, 719 pp.

Purseglove, J. W. 1972. *Tropical Crops: Monocotyledons,* John Wiley & Sons, New York, 607 pp.

Purseglove, J. W. 1976. Millets: *Eleusine coracana, Pennisetum americanum.* In *Evolution of Crop Plants,* N. W. Simmonds, Ed. Longman, London, pp. 91–93.

Ramanujam, L. 1976. Chickpea. In *Evolution of Crop Plants,* N. W. Simmonds, Ed. Longman, London, pp. 157–159.

Raskin, I., H. Kende. 1985. Mechanism of aeration in rice. *Science* 228:327–329.

Renfrew, J. M. 1973. *Paleoethnobotany,* Columbia University Press, New York, 248 pp.

Richardson, S. D. 1990. *Forestry and Forests in China: Changing Patterns of Resource Development,* Island Press, Washington D.C., 353 pp.

Rick, C. M. 1976. Tomato. In *Evolution of Crop Plants,* N. W. Simmonds, Ed. Longman, London.

Riesenberg, L. H., G. J. Seiler. 1990. Molecular evidence and the origin and development of the domesticated sunflower. *Econ. Bot.* 44(3 Supplement):79–91.

Rikli, M. 1943–1948. *Das Pflanzenkleid der Mittelmeerländer,* 3 Vols. H. Huber, Bern.

Rindos, D. 1984. *The Origins of Agriculture: An Evolutionary Perspective,* Academic Press, Orlando, FL.

Risi C. J., N. Galwey. 1984. *Chenopodium. Adv. Appl. Biol.* 10:145–216.

Rivera, M. 1991. The prehistory of northern Chile: A synthesis. *J. World Prehistory* 5:1–47.

Roach, F. A. 1985. *Cultivated Fruits of Britain: Their Origin and History,* Basil Blackwell, Oxford, 349 pp.

Robertson, N. F. 1991. The challenge of *Phytophthora infestans. Adv. Plant Pathol.* 7:1–30.

Robinson, R. W., T. W. Whitaker. 1974. *Cucumis.* In *Handbook of Genetics, Vol. 2,* Robert C. King, Ed. Plenum Press, New York, pp. 145–150.

Rodriguez, D. W. 1961. Coffee: A short economic history with special reference to Jamaica. *Jam. Minist. Agric, Commodity Bull.* 2:1–77.

Roe, D. A. 1974. The sharecropper's plague. *Nat. Hist.* 83 (10):52–63.

Rogers, D. J. 1963. Studies of *Manihot esculenta* and related species. *Bull. Torrey Bot. Club* 90:43–54.

Rogers, D. J. 1965. Some botanical and ethnological considerations of *Manihot esculenta. Econ. Bot.* 19:369–3970.

Roosevelt, A. C. 1980. *Parmana: Prehistoric Maize and Manioc Subsistence Along the Amazon and Orinoco,* Academic Press, New York, 320 pp.

Rutger, J. N., D. M. Brandon. 1981. California rice culture. *Sci. Am.* 244(2):42–51.

Ryder, E. J. 1986. Lettuce breeding. In *Breeding Vegetable Crops,* M. J. Bassett, Ed. AVI Publishing Company, Westport, CT, pp. 436–476.

Ryder, E. J., T. W. Whitaker. 1976. Lettuce. In *Evolution of Crop Plants,* N. W. Simmonds, Ed. Longman, London.

Ryder, E. J., T. W. Whitaker. 1980. The lettuce industry in California: A quarter century of change, 1954–1979. *Hortic. Rev.* 2:164–207.

Salaman, R. N. 1949. *The History and Social Influence of the Potato,* Cambridge University Press, New York, 685 pp.

Santhanam, V., J. B. Hutchinson. 1974. Cotton. In *Evolutionary Studies in World Crops: Diversity and Change in the Indian Subcontinent,* J. B. Hutchinson, Ed. Cambridge University Press, New York, pp. 89–100.

Sarkar, K. R., B. K. Mukherjee, D. Gupta, H. K. Jain. 1974. Maize. In *Evolutionary Studies in World Crops: Diversity and Change in the Indian Subcontinent,* J. B. Hutchinson, Ed. Cambridge University Press, New York, pp. 121–127.

Sauer, C. O. 1966. *The Early Spanish Main,* University of California Press, Berkeley, 306 pp.

Sauer, C. O. 1971. *Sixteenth Century North America,* University of California Press, Berkeley, 319 pp.

Sauer, J. D. 1950. The grain amaranths: A survey of their history and classification. *Mo. Bot. Garden Ann.* 37:561–626.

Sauer, J. D. 1965. The Seychelles archipelago and its palms. *Mo. Bot. Garden Bull.* 53:8–15.

Sauer, J. D. 1967b. The grain amaranths and their relatives: A revised taxonomic and geographic survey. *Mo. Bot. Garden Ann.* 54:103–137.

Sauer, J. D. 1976. Grain amaranths. In *Evolution of Crop Plants,* N. W. Simmonds, Ed. Longman, London, pp. 4–7.

Sauer, J. D. 1983. *Cocos nucifera*. In *Costa Rican Natural History,* D. H. Janzen, Ed. University of Chicago Press, IL, pp. 216–219.

Schreiner, E. J. 1969. Tree breeding in United States forestry practice. *Unasylva* 24(2–3):96–108.

Schroeder, C. A. 1967. Avocado introduction in California. In *Proceedings of the International Symposium on Plant Introduction, Escuela Agricola Panamericana, Tegucigalpa, Honduras,* pp. 61–69.

Schroeder, C. A., W. A. Fletcher. 1967. The Chinese gooseberry (*Actinidia chinensis*) in New Zealand. *Econ. Bot.* 21:81–92.

Scott, C. W. 1960. *Pinus radiata. United Nations, Food and Agriculture Organization, Forestry and Forest Products Studies,* Vol. 14.

Scott, D. R. W. 1980. The Pacific Northwest region. In *Regional Silviculture of the United States*, J. W. Barrett, Ed. John Wiley & Sons, New York, pp. 447–494.

Scott, D. H., F. J. Lawrence. 1975. Strawberries. In *Advances in Fruit Breeding,* J. Janick and J. N. Moore, Eds. Purdue University Press, West Lafayette IN, pp. 71–97.

Scott, R. 1961. *Limuria: The Lesser Dependencies of Mauritius,* Oxford University Press, 308 pp.

Seibert, R. 1950. The importance of palms to Latin America: Pejibaye, a notable example. *Ceiba* 1:65–74.

Silbernagel, M. J. 1986. Snap-bean breeding. In *Breeding Vegetable Crops,* M. J. Bassett, Ed. AVI Publishing Company, Westport, CT, pp. 246–283.

Simmonds, N. W. 1964. Studies of the tetraploid potatoes. II. Factors in the evolution of the Tuberosum Group. *J. Linnaean Soc. London, Bot.* 59:43–56.

Simmonds, N. W. 1966. *Bananas*, 2nd ed. Longman, London.

Simmonds, N. W. 1976a. Quinoa and relatives. In *Evolution of Crop Plants,* N. W. Simmonds, Ed. Longman, London, pp. 29–30.

Simmonds, N. W. 1976b. Potatoes. In *Evolution of Crop Plants,* N. W. Simmonds, Ed. Longman, London, pp. 279–283.
Simmonds, N. W. 1976c. Bananas. In *Evolution of Crop Plants,* N. W. Simmonds, Ed. Longman, London, pp. 211–215.
Simmonds, N. W. 1976d. Sugarcanes. In *Evolution of Crop Plants,* N. W. Simmonds, Ed. Longman, London, pp. 104–108.
Simoons, F. J. 1990. *Food in China: A Cultural and Historical Inquiry,* CRC Press, Boca Raton, FL, 570 pp.
Singh, L. R. 1976. Mango. In *Evolution of Crop Plants,* N. W. Simmonds, Ed. Longman, London, pp. 7–9.
Singh, S. P. 1989. Patterns of variation in cultivated common bean. *Econ. Bot.* 32:39–57.
Singh, S. P., P. Gepts, D. G. Debouck. 1991. Races of common bean (*Phaseolus vulgaris. Fabaceae*). *Econ. Bot.* 45:379–396.
Smartt, J. 1990. *Grain Legumes: Evolution and Genetic Resources,* Cambridge University Press, New York, 379 pp.
Smith, B. D. 1987. The economic potential of *Chenopodium berlandieri* in prehistoric eastern North America. *J. Ethnobiol.* 7:29–54.
Smith, B. D. 1989. Origins of agriculture in eastern North America. *Science* 246:1566–1571.
Smith, C. E., Jr. 1966. Archaeologic evidence for selection in avocado. *Econ. Bot.* 20:169–175.
Smith, N. 1983a. Triticale: The birth of a new cereal. *New Sci.* 97:98–99.
Smith, N. 1983b. New genes from wild potatoes. *New Sci.* 98:558–565.
Smith, P. M. 1976. Minor crops. In *Evolution of Crop Plants,* N. W. Simmonds, Ed. Longman, London, pp. 301–324.
Sokolov, R. P. 1981. Blueberry blues. *Nat. Hist.* 89(3):116–120.
Sokolov, R. P. 1989. The peripatetic potato. *Nat. Hist.* 98(3):86–88.
Soost, R. K., J. W. Cameron. 1975. Citrus. In *Advances in Fruit Breeding,* J. Janick and J. N. Moore, Eds. Purdue University Press, West Lafayette, IN, pp. 507–540.
Spain, Ministerio de Agricultura. 1966. *Mapa Forestal de España,* Ministerio General de Montes, Caza, y Pesca Fluvial, Madrid.
Spencer, J. E. 1975. The rise of maize as a major crop plant in the Philippines. *J. Hist. Geogr.* 1:1–16.
Stephens, S. G. 1967a. Evolution under domestication of the New World cottons. *Ciencia e Cultura* 19:118–134.
Stephens, S. G. 1967b. A cotton boll segment from Coxcatlan Cave. In *The Prehistory of the Tehuacan Valley. I. Environment and Subsistence,* D. S. Byers, Ed. University of Texas Press, Austin, pp. 256–260.
Stephens, S. G. 1973. Geographical distribution of cultivated cottons relative to probable centers of domestication in the New World. In *Genes, Enzymes, and Populations,* A. M. Srb, Ed. Plenum, New York, pp. 239–254.
Stephens, S. G. 1975. A reexamination of the cotton remains from Huaca Prieta, north coastal Peru. *Am. Antiq.* 40:496–418.

Stephens, S. G. 1976. The origin of Sea Island cotton. *Agric. Hist.* 50:391–399.

Stone, D. 1984. Pre-Columbian migration of *Theobroma cacao* Linnaeus, and *Manihot esculenta* Crantz, from northern South America into Mesoamerica: a partially hypothetical view. In *Pre-Columbian Plant Migration, Vol. 76.*, D. Stone, Ed. Papers of the Peabody Museum of Archaeology and Ethnology, Harvard University, pp. 67–83.

Storey, W. B. 1965. The ternifolia group of *Macadamia* species. *Pac. Sci.* 19:507–514.

Stover, R. H., N. W. Simmonds. 1987. *Bananas,* 3rd ed. Wiley, New York.

Streets, R. J. 1962. *Exotic Forest Trees in the British Commonwealth,* Clarendon Press, Oxford, 765 pp.

Struever, S., E. D. Vickery. 1973. The beginnings of cultivation in the Midwest-riverine area of the United States. *Am. Anthropol.* 75:1197–1220.

Tanaka, T. 1954. *Species Problem in Citrus,* Japanese Society for Promotion of Science. Jeno, Tokyo, 155 pp.

Thompson, K. F. 1976. Cabbages, kales, etc. In *Evolution of Crop Plants,* N. W. Simmonds, Ed. Longman, London, pp. 49–52.

Tigchelaar, E. C. 1986. Tomato breeding. In *Breeding Vegetable Crops,* M. J. Bassett, Ed. AVI Publications, Westport, CT, pp. 135–171.

Tjomsland, A. 1950. The white potato. *CIBA Symposia* 11:1254–1284.

Totman, C. 1985. The origins of Japan's modern forests: The case of Akita. *Asian Stud. Hawaii* 31:1–87. University of Hawaii Press.

Towle, M. A. 1961. *The Ethnobotany of Pre-Columbian Peru,* Aldine, Chicago, IL, 179 pp.

Tsukada, M. 1982. *Cryptomeria japonica*: Glacial refugia and late-glacial and postglacial migrations. *Ecology* 63:1091–1105.

U, N. 1935. Genome analysis in *Brassica. Jpn. J. Bot.* 7:389–452.

Ugent, D., S. Pozorsky, T. Pozorsky. 1982. Archaeologic potato tuber remains in the Casma valley of Peru. *Econ. Bot.* 36:182–192.

Ugent, D., S. Pozorsky, T. Pozorsky. 1986. Archaeologic manioc (*Manihot*) from coastal Peru. *Econ. Bot.* 40:78–102.

Ugent, D., T. Dillehay, C. Ramirez. 1987. Potato remains from a Late Pleistocene settlement in south central Chile. *Econ. Bot.* 41:17–27.

United Nations, Food and Agriculture Organization. 1964. *World Cocoa Survey,* Rome, 242 pp.

United States, National Research Council. 1975. *Underexploited Tropical Crops With Promising Economic Value,* Board on Science and Technology for International Development, Washington D.C., 186 pp.

United States, National Research Council. 1989. *Lost Crops of the Incas: Little Known Crops of the Andes with Promise for Worldwide Cultivation,* Board on Science and Technology for International Development, Washington D.C., 415 pp.

Upham, S., R. S. MacNeish, W. C. Galinat, C. Stevens. 1987. Evidence concerning the origin of *maiz de ocho. Am. Anthropol.* 89:410–419.

Urquhart, D. H. 1955. *Cocoa,* Longmans, London, 230 pp.
Van der Vossen, H. A. M. 1985. Coffee selection and breeding. In *Coffee: Botany, Biochemistry and Production of Beans and Beverage,* M. N. Clifford and K. C. Wilson, Eds. AVI Publications, Westport, CT, pp. 48–96.
Vishnu-Mittre. 1977. Changing economy in ancient India. In *Origins of Agriculture,* C. A. Reed, Ed. Mouton, The Hague, pp. 569–588.
Walker, L. C. 1980. The Southern pine region. In *Regional Silviculture of the United States,* 2nd ed. J. W. Barrett, John Wiley & Sons, New York, pp. 233–276.
Watkins, R. 1976a. Apple and pear. In *Evolution of Crop Plants,* N. W. Simmonds, Ed. Longman, London, pp. 247–250.
Watkins, R. 1976b. Cherry, plum, peach, apricot, and almond. In *Evolution of Crop Plants,* N. W. Simmonds, Ed. Longman, London, pp. 242–246.
Watson, A. M. 1983. *Agricultural Innovation in the Early Islamic World: The Diffusion of Crops and Foraging Techniques 700–1100,* Cambridge University Press, New York.
Watts, D. 1966. Man's influence on the vegetatin of Barbados, 1627 to 1800. *Hull Univ. Occas. Pap. Geogr.* 4:1–95.
Webster, C. C., W. J. Baulkwill, Eds. 1989. *Rubber,* Longman and Wiley, London, 614 pp.
Weeden, N. F., R. W. Robinson. 1990. Isozyme studies in *Cucurbita.* In *Biology and Utilization of the Cucurbitaceae,* D. M. Bates, R. Robinson, and C. Jeffrey, Eds. Cornell University Press, Ithaca, NY, pp. 51–59.
Weinberger, J. H. 1975. Plums. In *Advances in Fruit Breeding,* J. Janick and J. N. Moore, Eds. Purdue University Press, West Lafayette, IN, pp. 336–347.
Weiss, E. A. 1983. *Oilseed Crops,* Longman, London, 660 pp.
Wellman, F. L. 1972. *Tropical American Plant Disease,* Scarecrow Press, Metuchen, NJ, 989 pp.
Wells, O. O., C. C. Lambeth. 1983. Loblolly pine provenance test in southern Arkansas: 25th year results. *South. J. Appl. For.* 7:71–75.
Wendel, J. F., P. D. Olson, J. McD. Stewart. 1989. Genetic diversity, introgression, and independent domestication of Old World cultivated cottons. *Am. J. Bot.* 76:1795–1806.
Wendel, J. F., R. G. Percy. 1990. Allozyme diversity and introgression in the Galapagos Islands endemic *Gossypium darwinii* and its relationship to continental *G. barbadense. Biochem. Syst. Ecol.* 18:517–528.
West, R. C., J. P. Augelli. 1966. *Middle America,* Prentice-Hall, Englewood Cliffs, NJ, 482 pp.
Weyerhaeuser, G. H. 1981. *Forests for the Future,* Newcomer Society, New York.
Whealy, K. 1989. *Fruit, Berry, and Nut Inventory,* Seed Saver Publications, Decorah, IA, 366 pp.
Whistler, W. A. 1991. Polynesian plant introductions. In *Islands, Plants, and Polynesians: An Introduction to Polynesian Ethnobotany,* P. A. Cox and S. A. Banack, Eds. Dioscorides Press, Portland, OR, pp 41–66.

Whitaker, T. W. 1964. Gourds and people. *Am. Hortic. Mag.* 43:207–213.

Whitaker, T. W. 1969. Salads for everyone — a look at the lettuce plant. *Econ. Bot.* 23:261–264.

Whitaker, T. W. 1983. Cucurbits in Andean prehistory. *Am. Antiq.* 48:576–585.

Whitaker, T. W., W. P. Bemis. 1976. Cucurbits. In *Evolution of Crop Plants,* N. W. Simmonds, Ed. Longman, London, pp. 64–69.

Whitaker, T. W., G. F. Carter. 1954. Oceanic drift of gourds. *Am. J. Bot.* 41:697–700.

Whitaker, T. W., H. C. Cutler. 1967. Pottery and cucurbit species. *Am. Antiq.* 32:225–226.

Whitaker, T. W., R. W. Robinson. 1986. Squash breeding. In *Breeding Vegetable Crops,* M. J. Bassett, Ed. AVI Publishing Company, Westport, CT, pp. 209–242.

White, D. F. 1987. New Zealand forestry: Privatization gets a test. *J. For.* 85(3):41–43.

Wiehe, G. A. 1968. Introduction of barley into the New World. *U.S. Dep. Agric., Agric. Handbook* 338:2–8.

Wilhelm, S. 1974. The garden strawberry: A study of its origin. *Am. Sci.* 62:264–271.

Wilkes, H. G. 1967. *Teosinte: The Closest Relative of Maize,* Bussey Institution of Harvard University, Cambridge, MA, 159 pp.

Wilkes, H. G. 1977. Hybridization of maize and teosinte in Mexico and Guatemala, and the improvement of maize. *Econ. Bot.* 31:254–293.

Wilsie, C. P. 1962. *Crop Adaptation and Distribution,* Freeman, San Francisco, CA, 448 pp.

Wilson, C. M. 1947. *Empire in Green and Gold: The Story of the American Banana Trade,* Henry Holt & Company, New York, 303 pp.

Wilson, H. 1981. Genetic variation among South American populations of tetraploid *Chenopodium* subsect. *Cellulata. Syst. Bot.* 6:380–398.

Wilson, H. 1988. Quinoa biosystematics. *Econ. Bot.* 42:461–464.

Wilson, H. 1990. *Quinua* and relatives (*Chenopodium* sect. *Chenopodium* subsect. *Cellulata*). *Econ. Bot.* 44(3 Supplement):92–110.

Winberry, J. J. 1980. The sorghum syrup industry, 1874–1975. *Agric. Hist.* 54:343–352.

Wing, J. E. 1912. *Alfalfa Farming in America,* Sanders, Chicago, IL, 528 pp.

Worsley, A. T., F. Oldfield. 1988. Palaeoecological study of three lakes in the highlands of Papua New Guinea. II. Vegetational history over the last 1600 years. *J. Ecol.* 76:1–18.

Wrigley, G. 1988. *Coffee,* Longman Scientific and Technical, Burnt Mill, Essex, 639 pp.

Wycherley, P. R. 1976. Rubber. In *Evolution of Crop Plants,* N. W. Simmonds, Ed. Longman, London.

Yarnell, R. A. 1976. Early plant husbandry in eastern North America. In *Essays in Anthropology in Honor of James Bennett Griffin,* C. E. Cleland, Ed. Academic Press, New York, pp. 265–273.

Yen, D. E. 1971. Construction of the hypothesis for distribution of the sweet potato. In *Man Across the Sea: Problems of Pre-Columbian Contacts,* C. L. Riley, J. C. Kelley, C. Pennington, and R. L. Rands, Eds. University of Texas Press, Austin, pp. 328–342.

Yen, D. E. 1974. The sweet potato and Oceania: An essay in ethnobotany. *Bernice P. Bishop Mus. Bull.* 236:1–389.

Yen, D. E. 1976. Sweet potato. In *Evolution of Crop Plants,* N. W. Simmonds, Ed. Longman, London, pp. 42–44.

Yen, D. E. 1991. Polynesian cultigens and cultivars: the question of origins. In *Islands, Plants, and Polynesians: An Introduction to Polynesian Ethnobotany,* P. A. Cox and S. A. Banack, Eds. Dioscorides Press, Portland, OR, pp 67–95.

Yen, D. E., J. M. Wheeler. 1968. Introduction of taro into the Pacific: The indication of chromosome numbers. *Ethnology* 8:259–267.

Zimmerer, K. 1991. The regional biogeography of native potato cultivation in highland Peru. *J. Biogeogr.* 18:165–178.

Zobel, B. 1964. Seed orchards for the production of genetically improved seed. *Silva Genet.* 13:4–11.

Zobel, B. 1971. The genetic improvement of Southern pines. *Sci. Am.* 225(5):95–102.

Zohary, D. 1976. Lentil. In *Evolution of Crop Plants,* N. W. Simmonds, Ed. Longman, London, pp. 163–164.

Zohary, D. 1989. Pulse domestication and cereal domestication: How different are they? *Econ. Bot.* 43:31–34.

Zohary, D., M. Hopf. 1988. *Domestication of Plants in the Old World,* Clarendon Press, Oxford, 249 pp.

Zohary, D., P. Spegel-Roy. 1975. Beginnings of fruit growing in the Old World. *Science* 187:319–327.

INDEX TO GENERA AND SPECIES

Index to Common[a] and Cultivar[b] Names

[a] Lower case.
[b] Upper case.

C

Q

R

S

T